Animal Biodiversity and Ecology

Animal Biodiversity and Ecology

Edited by **Hector Carling**

SYRAWOOD
PUBLISHING HOUSE

New York

Published by Syrawood Publishing House,
750 Third Avenue, 9th Floor,
New York, NY 10017, USA
www.syrawoodpublishinghouse.com

Animal Biodiversity and Ecology
Edited by Hector Carling

International Standard Book Number: 978-1-68286-074-8 (Hardback)

Printed in the United States of America.

Contents

Preface

This book has been an outcome of determined endeavour from a group of educationists in the field. The primary objective was to involve a broad spectrum of professionals from diverse cultural background involved in the field for developing new researches. The book not only targets students but also scholars pursuing higher research for further enhancement of the theoretical and practical applications of the subject.

Animal biodiversity is a vast field of study. This book covers the interdisciplinary aspects of animal biodiversity by integrating concepts from evolutionary ecology, conservation biology, etc. It aims to explain in detail the systematics, morphology, physiology, genetics, etc. related to animal species and their ecosystems. This book will bring forth some innovative concepts in the field and help students, researchers, biologists, zoologists, etc. to better understand this discipline.

It was an honour to edit such a profound book and also a challenging task to compile and examine all the relevant data for accuracy and originality. I wish to acknowledge the efforts of the contributors for submitting such brilliant and diverse chapters in the field and for endlessly working for the completion of the book. Last, but not the least; I thank my family for being a constant source of support in all my research endeavours.

Editor

Living scaphopods from the Valencian coast (E Spain) and description of *Antalis caprottii* n. sp. (Dentaliidae)

A. Martínez–Ortí & L. Cádiz

Martínez–Ortí, A. & Cádiz, L., 2012. Living scaphopods from the Valencian coast (E Spain) and description of *Antalis caprottii* n. sp. (Dentaliidae). *Animal Biodiversity and Conservation*, 35.1: 71–94.

Abstract

Living scaphopods from the Valencian coast (E Spain) and description of Antalis caprottii *n. sp. (Dentaliidae).*— This paper reports on eight scaphopod species found at 128 sampling stations near the coast of Valencia (Spain) during the campaigns of the Water Framework Directive (2000/60/CE) 2005, 2006 and 2008. Samples deposited in several Valencian institutions and private collections are also described. The identified species belong to four families: Dentaliidae (*Antalis dentalis, A. inaequicostata, A. novemcostata, A. vulgaris,* and *A. caprottii* n. sp., a new species described from material found on the coasts of the province of Castellón), Fustiariidae (*Fustiaria rubescens*), Entalinidae (*Entalina tetragona*) and Gadilidae (*Dischides politus*). We describe the characteristics and conchiological variations for each species and give geographic distribution maps on the Valencian coast for each species.

Key words: Scaphopods, Dentalida, Gadilida, *Antalis caprottii*, New species, Mediterranean Sea.

Resumen

Escafópodos de la costa valenciana (E España) y descripción de Antalis caprotti *sp. n. (Dentalidae).*— Se citan y describen en profundidad ocho especies de escafópodos halladas en 128 puntos de muestreo próximos a la costa de la Comunidad Valenciana (España), durante las campañas de la Directiva Marco del Agua (2000/60/ CE) de 2005, 2006 y 2008, en las muestras depositadas en diversas instituciones valencianas y colecciones privadas. Las especies identificadas pertenecen a cuatro familias: Dentaliidae (*Antalis inaequicostata, A. vulgaris, A. dentalis, A. novemcostata* y *A. caprottii* sp. n., especie nueva que se describe a partir de material encontrado en las costas de la provincia de Castellón), Fustiariidae (*Fustiaria rubescens*), Entalinidae (*Entalina tetragona*) y Gadilidae (*Dischides politus*). Todas las especies halladas han sido estudiadas y caracterizadas conchiológicamente y se muestran los mapas de distribución geográfica para cada una de ellas en la costa valenciana.

Palabras clave: Escafópodos, Dentalida, Gadilida, *Antalis caprottii*, Nueva especie, Mar Mediterráneo.

A. Martínez–Ortí, Museu Valencià d'Història Natural, Passeig de la Petxina 15, 46008 Valencia, España (Spain); Dept. de Zoologia, Fac. de Biologia, Univ. de València, Av. Dr. Moliner 50, 46100 Burjassot, Valencia, España (Spain).– L. Cádiz, Dept. de Ciencias Experimentales y Matemáticas. Fac. de Ciencias Experimentales, Univ. Católica de Valencia, c/. Guillem de Castro 94, 46001 Valencia, España (Spain).

Corresponding author: A. Martínez–Ortí. E–mail: amorti@uv.es

Introduction

Many species of scaphopod molluscs have been reported from the Mediterranean Sea since the late 19th century. Bucquoy et al. (1882–89) first cited three species of scaphopods in Roussillon (France): *A. dentalis* (Linnaeus, 1758), *A. novemcostata* (Lamarck, 1818) and *A. vulgaris* (Da Costa, 1778). Perrier (1930) cited *A. novemcostata* and *A. vulgaris* on the French Mediterranean coast, Caprotti (1966a, 1966b, 1966c, 1966d, 1966e, 1966f, 1967, 1979, 2009) cited nine species of the order Dentalida and 16 of the order Gadilida on the Italian coast,, and Ozturk (2011) reported 10 species in the Levantine and Aegean seas. Very recently, Gruppo Malacologico Romagnolo (GMR) (2010) recorded 13 species in the Mediterranean Sea, and Cossignani & Ardovini (2011) recorded 14 species, including a possible fossil species *C. ovulum*.

Scaphopod fauna from the coasts of the Iberian Peninsula and the Balearic islands are poorly known, and only 28 species have been recorded to date. Of these, 12 correspond to the order Dentaliida (families Fustiariidae, Dentaliidae and Gadilinidae) and 16 to the order Gadilida (families Entalinidae, Gadilidae and Pulsellidae) (table 1).

Respect to our study area, there are several historical antecedents. Boscá (1916) cited a scaphopod for the first time here, concretely *A. dentalis* (as *Dentalium dentalis*), and Pardo (1920) cited *Fustiaria rubescens* (as *Dentalium rubescens*) from the Gulf of Valencia. Roselló (1934) cited several species from the Valencia and Alicante coasts: *A. dentalis* (as *Dentalium dentalis*), *A. inaequicosta* (as *D. inaequicostata, D. novemcostatum* v. *inaequicostatum* and *D. fasciatum*), *A. vulgaris* (as *A. vulgare*) from the Valencia coast only, *F. rubescens* (as *Dentalium rubescens*) and *D. politus* (as *Siphonodentalium bifissus*), and *Dentalium elephantinum* from the Pliocene of the Alicante coast. The author reported the presence of *Pulsellum lofotense* but it was later found to correspond to the polychaete *Ditrupa arietina* (Müller, 1776). We also found a specimen of *A. novemcostata* (as *Dentalium novemcostatum*) from the Cantabrian Sea in this collection. Later, Giner (1989) also cited the species *A. inaequicostata* and *F. rubescens* from the coast of Valencia, and Rubio & Rodríguez (1996) cited three unidentified scaphopods from the coast of Dénia (Alicante). And more recently, Plá (2006) cited *A. inaequicostata* and *F. rubescens* from the coast of La Marina Alta (Alicante), Ibiza and Formentera (Balearic Islands), confusing *F. rubescens* with the polychaete *D. arietina*.

Some of these recent species have been recorded as fossils from several Spanish deposits and published in several papers. For example, the species *A. dentalis, A. inaequicostata, A. sexangulum* (Gmelin, 1789), *A. vulgaris, E. tetragona, F. emersoni* (Caprotti, 1979), *F. rubescens* and *P. lofotense* were cited by Vera–Peláez & Lozano (1993) and Jordá et al. (2010) from various Pliocene–Lower Holocene cave deposits located in the province of Málaga. Moreno (1995) cited

A. dentalis, A. novemcostata, A. vulgaris and *Antalis* sp. from deposits ranging from the Lower Palaeolithic until medieval times. In recent months, Cádiz & Martínez–Ortí (2011) reported *A. novemcostata* from the Valencian coast.

This study presents the seven scaphopod species collected between 2005 and 2008 along 454 km of coastline covering 60 municipalities of the Valencian coast. We also include data from samples we studied from various institutions and private collections in Valencia, recording one additional species. The total eight species have been conchiologically characterized, highlighting the discovery and description of a new species, the dentalid *Antalis caprottii* n. sp. The geographic distribution maps for each species in the study area are provided, except for *E. tetragona* due to its imprecise locations.

Material and methods

The study area includes the coast of Valencia from Vinaroz (Castellón), in the north, to Orihuela (Alicante), in the south (fig. 1, table 2). The coastal morphology is very varied and irregular, from coastal lowlands with long beaches to cliffs with varying topography. South of Dénia (Alicante), there are fewer long beaches; small coves predominate and then the beaches become longer again at the southern most end of the province.

Scaphopod samples studied here were obtained in the Water Framework Directive (2000/60/CE) campaigns in July and August in 2005, 2006 and 2008, coordinated by the Universitat Politécnica of Valencia and the 'Universidad Católica de Valencia–San Vicente Mártir'. Topographic features were defined and the water depth was monitored by continuous–recording echo–sounder. Samples were taken from a boat with a bilateral anchor dredge, covering a total of 172 sampling stations, but scaphopod specimens were found in only 63 (fig. 1, table 2: 1–63). The stations were mapped using a global positioning system (GPS) Garmin and coordinates were expressed in UTM units (table 2). These sampling stations vary between 5 and 20 m in depth; they are located on substrates of medium to fine sand, and many of them are dominated by the endemic Mediterranean seagrass *Posidonia oceanica* ((L.) Delile, 1813). All samples obtained from the dredging were fixed in 5% formaldehyde and screened using 0.5 mm mesh sieves. We then proceeded to the screening of the diverse biological groups obtaining the scaphopod assemblages, which were preserved in 70% ethanol.

We also revised 69 samples from several institutions: 7 from the Siro de Fez collection deposited in the 'Museu Valencià d'Història Natural' (MVHN), 16 from the Roselló collection deposited in the 'Museo de Ciencias Naturales' of Valencia (MCNV), 38 from the 'Instituto de Ecologia Litoral' of Alicante (IEL), and 8 from two private collections (table 2: 64–92).

The species were determined using the taxonomic criteria of Bucquoy et al. (1882–89), Hidalgo (1917), Caprotti (1966a, 1966b, 1966c, 1966d,

Table 1. Scaphopods recorded from the coasts of the Iberian peninsula and Balearic Islands: A. Jeffreys (1877, 1882); B. Locard (1898); C. Hidalgo (1917); D. Alzuria (1984, 1985a, 1985b, 1986); E. Steiner (1997); F. Giribet & Peñas (1997); G. Tarruella Ruestes & Fontanet Giner (2001); H. Salas et al. (1985); I. Templado et al. (2006); J. Salas & Gofas (2011).

Tabla 1. Escafópodos citados en las costas de la península Ibérica e Islas Baleares. (Para las abreviaturas, ver arriba.)

	A	B	C	D	E	F	G	H	I	J
Dentaliidae										
A. agilis		X	X		X	X				
A. dentalis	X		X		X					
A. entalis			X		X	X				
A. inaequicostata			X	X	X	X	X	X		
A. panorma		X	X		X					X
A. novemcostata			X		X					X
A. rossati				X	X					
A. vulgaris			X	X	X	X	X		X	X
F capillosum	X				X					
F. camdidum	X				X					
Fustiiaridae										
F. rubescens		X	X		X	X	X	X		X
Gadilinidae										
E. filum	X		X		X					
Entalinidae										
E. tetragona	X				X					
B. ensiculus	X				X					
H. subterfissus		X			X					
Gadilidae										
C. amphora		X			X					
C. artatus		X			X					
C. jeffreysi		X			X					X
C. subfusiformis		X			X					
C. gracilis		X			X					
C. propinquus		X			X					
C. cylindratus		X			X					
C. monterosatoi		X			X					
C. gibbus		X			X					
C. ovulum		X			X					
C. tumidosus					X					
D. politus	X		X		X		X			X
Pulsellidae										
P. lofotense					X					

Table 2. List of species found in the sampling stations together with those from private and public collections: S. Station; D. Depth (in m); MHNV. Museu Valencià d´Història Natural of Valencia; FD. Framework Directive; MCNV. Museo de Ciencias Naturales of Valencia; IEL. Instituto de Ecología Litoral of Alicante.

Tabla 2. Listado de especies encontradas en los muestreos junto a las muestras procedentes de colecciones privadas y museísticas: S. Estación; D. Profundidad (en m); MHNV. Museu Valencià d'Història Natural de Valencia; FD. Directiva Marco; MCNV. Museo de Ciencias Naturales de Valencia; IEL. Instituto de Ecología Litoral de Alicante.

S	Species	Date	UTM coordinates	Locality	D(m)
1	*A. dentalis*	15 VII 2008	31T289532/4487500	Vinaroz	14
2	*A. dentalis*	17 VII 2008	31T293764/4485082	Vinaroz	12
3	*A. dentalis*				
	F. rubescens	17 VII 2008	31T291942/4480865	Vinaroz	17
4	*A. inaequicostata*				
	A. dentalis				
	F. rubescens	15 VII 2008	31T284512/4478992	Benicarló	16
5	*A. inaequicostata*				
	A. dentalis	16 VII 2008	31T289562/4478425	Benicarló	18
6	*A. dentalis*				
	A. novemcostata	09 III 2008	31T282013/4475005	Benicarló	18
7	*A. dentalis*	16 VII 2008	31T286069/4473710	Benicarló	19
8	*A. inaequicostata*				
	F. rubescens	17 VII 2008	31T280011/470467	Peñíscola	20
9	*A. inaequicostata*	17 VII 2008	31T281498/4468312	Peñíscola	18
10	*A. inaequicostata*				
	A. dentalis				
	F. rubescens	18 VII 2008	31T 278236/468664	Peñíscola	13
11	*A. inaequicostata*				
	A. dentalis	18 VII 2008	31T276831/465776	Peñíscola	16
12	*A. inaequicostata*	18 VII 2008	31T278305/464744	Peñíscola	18
13	*A. inaequicostata*				
	A. dentalis	22 VII 2008	31T274966/464094	Peñíscola	17
14	*A. dentalis*	11VIII 2008	31T273274/459447	Alcala de Chivert	16
15	*A. inaequicostata*				
	A. dentalis	22 VII 2008	31T269000/457000	Alcala de Chivert	18
16	*A. dentalis*	12 VIII 2008	31T267000/453500	Torreblanca	12
17	*A. inaequicostata*	12 VIII 2008	31T267539/452554	Torreblanca	16
18	*A. inaequicostata*				
	A. dentalis	12 VIII 2008	31T264212/449040	Torreblanca	14
19	*A. inaequicostata*				
	F. rubescens	27 VIII 2008	31T261939/446420	Cabanes	17
20	*A. dentalis*	14 VII 2008	31T258010/443506	Oropesa	15
21	*A. inaequicostata*				
	F. rubescens	15 VII /2008	31T259306/442677	Oropesa	20
22	*A. inaequicostata*				
	F. rubescens	14 VII 2008	31T256537/440067	Oropesa	13

Table 2. (Cont.)

S	Species	Date	UTM coordinates	Locality	D(m)
23	A. inaequicostata				
	A. dentalis				
	A. novemcostata	27 VIII 2008	31T257121/439613	Benicassim	15
24	A. inaequicostata				
	A. dentalis				
	F. rubescens	27 VIII 2008	31T252102/433367	Castellón	16
25	A. caprottii n. sp.				
	A. vulgaris	28 VIII 2008	31S246666/424870	Burriana	18
26	F. rubescens	29 VIII 2008	30S755572/420716	Burriana	14
27	A. inaequicostata	29 VIII 2008	30S750279/409239	Nules	19
28	A. inaequicostata	26 VIII 2008	30S743528/403001	La Llosa	14
29	A. dentalis	26 VIII 2008	30S741523/399021	Sagunto	17
30	A. inaequicostata				
	F. rubescens	26 VIII 2008	30S743507/398010	Sagunto	19
31	A. inaequicostata				
	A. dentalis	22 VIII 2008	30S740551/395044	Sagunto	12
32	A. dentalis				
	F. rubescens	22 VIII 2008	30S737807/389923	Sagunto	19
33	A. inaequicostata				
	A. dentalis				
	A. novemcostata				
	F. rubescens	25 VIII 2008	30S739414/387687	Sagunto	20
34	F. rubescens	22 VIII 2008	30S736506/388027	Puzol	20
35	A. inaequicostata				
	A. dentalis				
	F. rubescens	25 VIII 2008	30S739060/386877	Puzol	13
36	A. inaequicostata	25 VII 2008	30S733000/379500	Pobla de Farnals	18
37	A. inaequicostata				
	F. rubescens	22 VIII 2008	30S735350/378672	Pobla de Farnals	16
38	A. dentalis				
	F. rubescens	22 VIII 2008	30S732025/376112	Albuixech	17
39	A. inaequicostata				
	A. dentalis				
	F. rubescens	04 IX 2008	30S733762/373908	Tavernes de Valldigna	12
40	A .dentalis	22 VIII 2008	30S730964/365507	Alfafar	18
41	A. dentalis	21 VIII 2008	30S735992/351504	Silla	19
42	F. rubescens	21 VIII 2008	30S738094/346733	Sueca	18
43	A. dentalis	19 VIII 2008	30S741904/343003	Cullera	19
44	F. rubescens	19 VIII 2008	30S744956/333327	Cullera	20

Table 2. (Cont.)

S	Species	Date	UTM coordinates	Locality	D(m)
45	*A. inaequicostata*				
	F. rubescens	18 VIII 2008	30S746057/329417	Gandia	19
46	*F. rubescens*	18 VIII 2008	30S747656/319524	Gandia	20
47	*A. inaequicostata*	01 IX 2008	30S759976/312059	Oliva	20
48	*A. vulgaris*	23 VII 2008	31S244289/308628	Oliva	18
49	*F. rubescens*	24 VII 2008	31S256775/296954	Jávea	19
50	*A. vulgaris*	08 VIII 2008	31S249775/284739	Jávea	21
51	*A. inaequicostata*	07 VIII 2008	30S758614/276051	Altea	17
52	*A. inaequicostata*	06 VIII 2008	30S742151/264403	Villajoyosa	21
53	*A. inaequicostata*	06 VIII 2008	30S737258/262712	Villajoyosa	20
54	*F. rubescens*	05 VIII 2008	30S727139/251339	Campello	15
55	*A. dentalis*	05 VIII 2008	30S728434/250792	Campello	20
56	*A. vulgaris*	31 VII 2008	30S721374/4240206	Santa Pola	20
57	*A. inaequicostata*				
	A. dentalis				
	A. novemcostata	30 VII 2008	30S707504/4220139	Guardamar del Segura	19
58	*A. dentalis*	30 VII 2008	30S710109/4220574	Guardamar del Segura	20
59	*F. rubescens*	28 VII 2008	30S700853/4200015	Orihuela	18
60	*A. dentalis*	29 VII 2008	30S700854/4199974	Orihuela	17
61	*F. rubescens*	28 VII 2008	30S698818/4197311	Orihuela	19
62	*A. inaequicostata*				
	A. dentalis				
	A. novemcostata				
	F. rubescens	29 VII 2008	30S698813/4197300	Orihuela	17
63	*D. politus*	29 VII 2008	31T257121/439613	Benicassim	15
64	*A. inaequicostata*	01 XII 2008	–	Gandia (MVHN–011211AB01)	–
65	*A. vulgaris*	01 XII 2008	–	Orihuela (MVHN–011211AB02)	–
66	*A. inaequicostata*				
	A. vulgaris				
	F. rubescens	01 XII /2009	–	Cullera (MVHN–011211AB03)	–
67	*A. vulgaris*				
	F. rubescens	01 XI 2008	–	Alboraya (MVHN–011211AB04)	–
68	*F. rubescens*	–	–	Valencia (MVHN–011211AB05)	–
69	*A. inaequicostata*	01 I 2011	–	Denia (MVHN–011211AB06)	–
70	*A. inaequicostata*	–	–	Sierra de Irta (MVHN–011211AB07)	–
71	*A. inaequicostata*	–	–	Piles (MVHN–011211AB08)	–
72	*A. inaequicostata*	1999	–	Sagunto (IEL)	–
73	*A. inaequicostata*	1999	–	El Pilar (IEL)	–
74	*A. inaequicostata*	2008	–	Balmar Morro Toix (IEL)	–
75	*A. dentalis*	2008	–	Balmar Morro Toix (IEL)	–
76	*A. inaequicostata*	2000	–	Torrevieja (IEL)	–

Table 2. (Cont.)

S	Species	Date	UTM coordinates	Locality	D(m)
77	*A. dentalis*	1999	–	Prov. Alicante (IEL)	–
78	*F. rubescens*	–	–	Valencia (MVHN–290610AU01)	–
79	*A. inaequicostata*				
	F. rubescens	–	–	Valencia (MVHN–290610AU02)	–
80	*A. vulgaris*	–	–	Valencia (MVHN–290610AU03)	–
81	*A. inaequicostata*				
	A. vulgaris	–	–	Valencia (MVHN–290610AU04)	–
82	*A. inaequicostata*	–	–	Valencia (MVHN–290610AU05)	–
83	*A. vulgaris*	–	–	Valencia (MVHN–290610AU06)	–
84	*A. inaequicostata*	–	–	Valencia (MVHN–290610AU07)	–
85	*A. dentalis*	–	–	Valencia (MCNV–3438)	–
86	*A. inaequicostata*	–	–	Valencia (MCNV–3447)	–
87	*A. vulgaris*	–	–	Valencia (MCNV–3443)	–
88	*D. politus*	–	–	Valencia MCNV–3456	–
89	*E. tetragona*	–	–	Valencia (MCNV–s/n)	–
90	*F. rubescens*	–	–	Valencia (MCNV–3452)	–
91	*A. novemcostata*	–	–	Valencia (MCNV–3446)	–
92	*A. inaequicostata*	–	–	Valencia (MCNV–3445)	–

1966e, 1966f, 1967, 1979, 2009), Alzuria (1984, 1985a, 1985b, 1986), Steiner (1997) and Ozturk (2011), classified using the supraspecific categories of Steiner & Kabat (2001, 2004), and later quantified. We considered only those shells that could be accurately identified, and we show the abundance of each species collected in sampling stations (fig. 2). For each species, measures from up to 20 adult shell specimens were obtained if available: i) total length, ii) length of the central pipe at the apex that some species have, and iii) diameter of both the apex and the base.

In addition, we propose a new procedure to measure the scaphopod mean curvature angle (α) considered as the chord that joins the apex with the midpoint of the base of scaphopod shells, applying the following formulas: $tg\alpha = b/a$ corresponding $\alpha = tag^{-1}(b/a)$, where 'a' is the angle opposite side and 'b' the adjacent side to angle. In addition, we obtained its standard deviation. All these measures were obtained from six adult specimens for each studied species, except for *A. caprottii* n. sp., for which only two type specimens were measured (fig. 3).

To achieve an accurate taxonomical identification we used a SEM HITACHI S–4100 at the Electron Microscopy Service (SCSIE), University of Valencia. Fund type was determined using a granulometric analysis and a Wudden–Wentworth scale (in Holme & McIntyre, 1971) to calculate the Shepard sedimentological triangle (fig. 4). Whole data were processed using the graphics software Grapher.

Results

We identified a total of 492 specimens; 298 from the sampling stations and 194 from the revised collections. All specimens belonged to six species of the order Dentalida (five species from the family Dentaliidae and one from the family Fustiariidae) and two species of the order Gadilida (one from the family Entalinidae and the other from the family Gadilidae). The most abundant species in our sampling stations were *A. inaequicostata* (53.5%) and *A. dentalis* (26.3%) (fig. 2). Granulometric analysis determined that the most common substrate where scaphopods occurred in our study area was fine to medium sand (fig. 4).

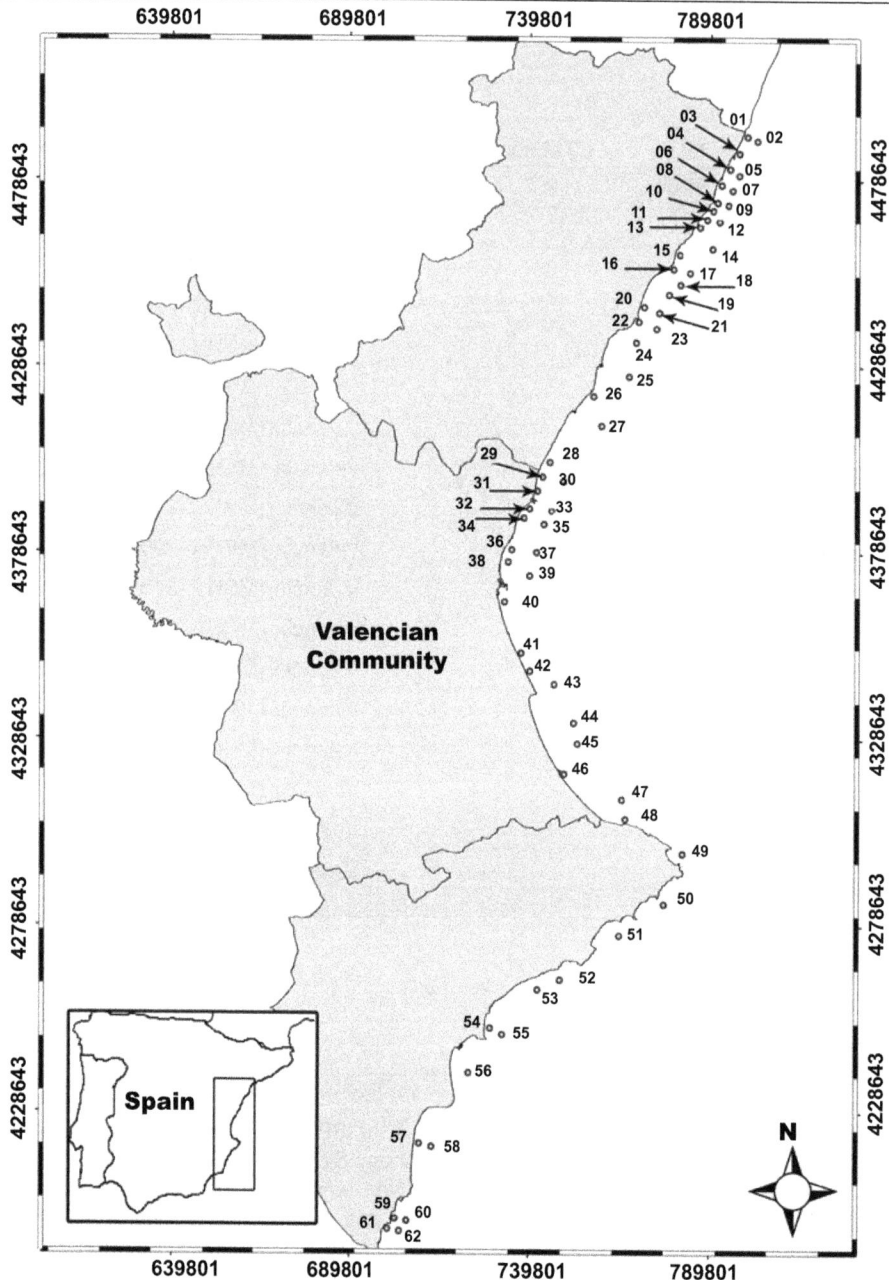

Fig. 1. Sampling stations in the Autonomous Community of Valencia (Spain).

Fig. 1. Estaciones de muestreo en la Comunidad Valenciana (España).

Systematics

Order Dentaliida Starobogatov, 1974
Family Dentaliidae Gray J. E., 1834
Genus *Antalis* Adams H. & Adams A., 1854

Antalis caprottii n. sp. (figs. 5–6).

Type locality
Coast of Burriana (Castellón) (fig. 6, table 2).

Type material
The type series consists of two specimens deposited in the MVHN (Spain). The hototype with the code MVHN–150211UB03A is preserved in alcohol 70% (fig. 5A), and the paratype, a shell, has the code MVHN–50211UB03B (figs. 5B–G).

Etymology
Dedicated to Dr. Erminio Caprotti, an eminent Italian malacologist, specialist in scaphopods.

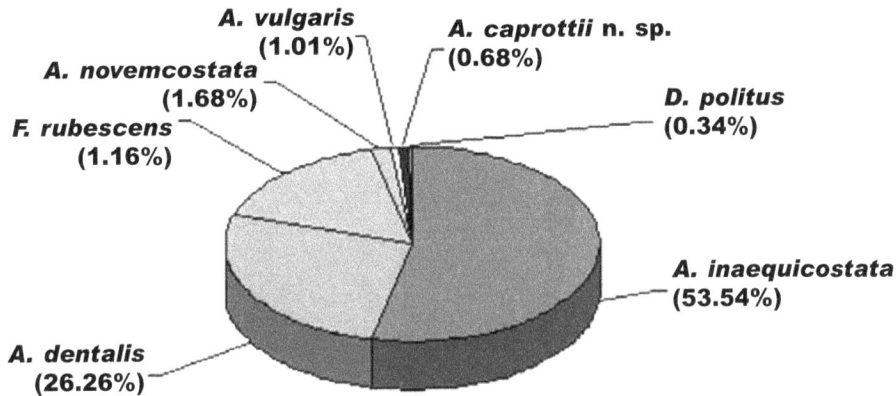

Fig. 2. Pie chart of the abundance of species obtained by dredging in the study area.

Fig. 2. Gráfico de sectores de la abundancia de especies obtenidas mediante dragado en el área de estudio.

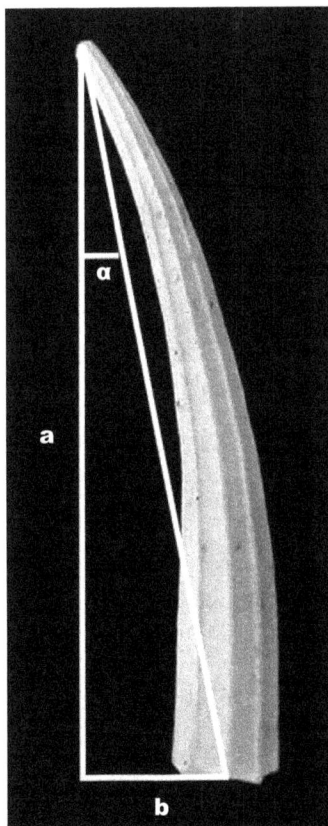

Common name
Valencian scaphopod

Diagnosis
Shell with a central pipe at the apex, from 14 to 16

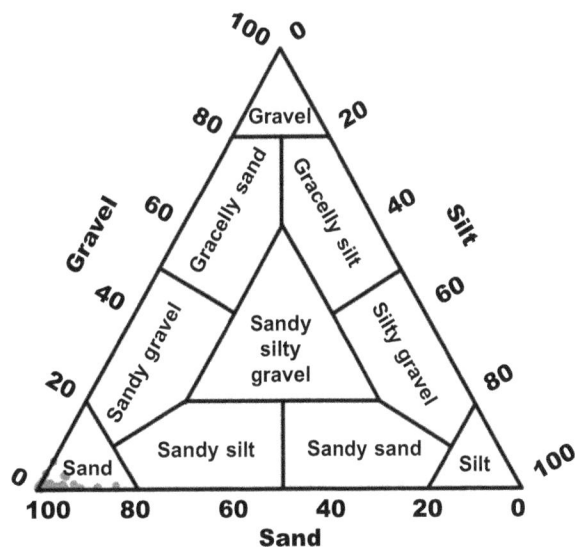

Fig. 3. Method to calculate the mean curvature angle (α) in scaphopods: a. Side opposite angle; b. Side adjacent to angle.

Fig. 3. Método de cálculo del ángulo de curvatura media (α) para escafópodos: a. Cateto opuesto al ángulo; b. Cateto contiguo al ángulo.

Fig. 4. Shepard triangle: sedimentological distribution of grain size (sand, silt and clay).

Fig. 4. Triángulo de Shepard: distribución sedimentológica por tamaño de grano (arenas, limos y arcillas).

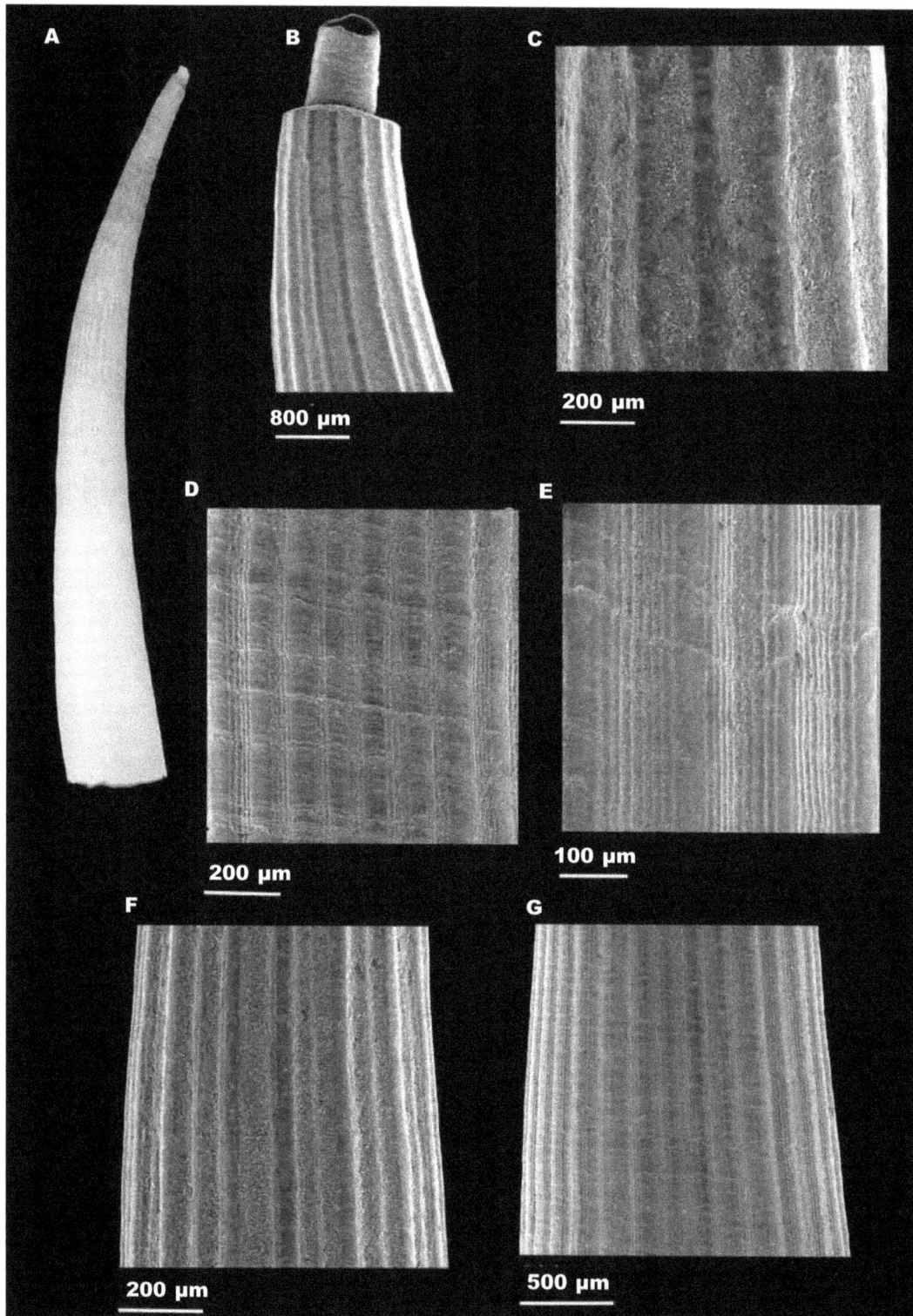

Fig. 5. *Antalis caprottii* n. sp., Burriana (Castellón): A. Holotype (MVHN–150211UB03A). B–G. Paratype (MVHN–150211UB03B): B. Apex with central pipe; C. Detail of the intercostal space in the apical zone; D. Primary ribs; E. Intercostal space detail; F. Middle zone; G. Basal zone.

Fig. 5. Antalis caprottii *sp. n., Burriana (Castellón): A. Holotipo (MVHN–150211UB03A). B–G. Paratipo (MVHN–150211UB03B): B. Ápice con túbulo; C. Detalle del espacio intercostal en la zona apical; D. Costillas primarias; E. Detalle del espacio intercostal; F. Zona media; G. Zona basal.*

primary ribs, increasing to 30 in the basal zone, and eight longitudinal microstriae in the intercostal space.

Description
Opaque, solid shell, moderately curved. Sculpture consisting of 14–16 primary ribs which increase to 30 in the basal zone. Both adults and juveniles present a small short central pipe with a circular section in the apex. In all intercostal spaces there are eight conspicuous longitudinal microstriae. Colour white and orange, more intense at the apex area.

Dimensions
The adult specimen is 22.0 mm in length and 3.0 mm in width at the basal zone. Length of the pipe is 700 µm and the diameter of the apex ranges from 300 to 1,000 µm.

Mean curvature angle
α = 11° 30' 55.99" σ ± 0° 40' 50.61"

Discussion
Because the shell is regularly close to the apex and has a conical foot (Caprotti, 1979) attribution to the family Dentaliidae is clear. It is assigned to the genus *Antalis* because it has a circular section at both the apical and basal ends and lacks an apical cleft (Caprotti, 1979). The specimens of *Antalis caprottii* n. sp. present a combination of features that clearly distinguish them from other known species of the genus. The most similar species are *A. vulgaris* and *A. inaequicostata*. It can be distinguished from these species by its 14 to 16 primary ribs which increase to 30 in the basal zone; *A. vulgaris* has 30 primary ribs which fade towards the base (figs. 13B, 13E, 13H–13I), plus five longitudinal microstriae (Alzuria, 1985a) (fig. 13F) and *A. inaequicostata* has nine primary ribs at the apex (figs. 9A, 9B, 9E increasing to 12 at the base (Steiner, 1997) (figs. 9A, 9H), and around 12 to 20 microstriae(figs. 9C–9D,10D–10E). All three species have a central pipe although in *A. vulgaris* its presence is less frequent (figs. 5A–5B, 9A–9B, 10C, 13C–13D). Besides, the pipe section is circular in *A. caprottii* n. sp. but it is oval in the other two species.

Habitat
Sandy bottoms at depths between 5 and 20 m.

Geographical distribution
Only known from the coast of Burriana (Castellón province).

Antalis dentalis (Linnaeus, 1758) (figs. 7–8)

Material examined
Seventy-eight specimens from 35 sampling stations (table 2).

Description
Species highly variable in size and colour (Caprotti, 1979; Steiner, 1997; GMR, 2010; Ozturk, 2011).

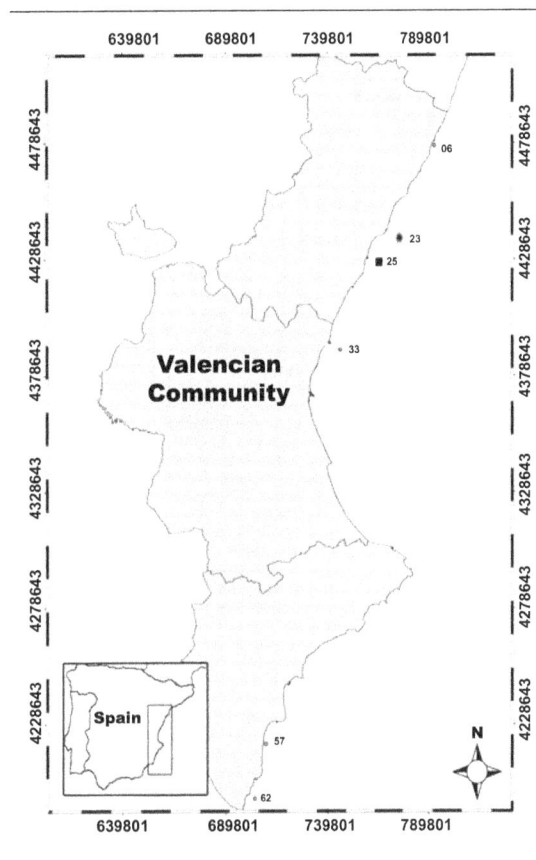

Fig. 6. Geographical distribution of *Antalis caprottii* n. sp. (square), *Antalis novemcostata* (points) and *Dischides politus* (asterisk) on the Valencian coast.

Fig. 6. Distribución geográfica de Antalis caprottii *sp. n. (cuadrado),* Antalis novemcostata *(puntos) y* Dischides politus *(asterisco) en la costa valenciana.*

The shell is slightly curved and fragile, pink, white or beige, and sometimes transparent. It presents a longitudinal sculpture in the apical zone consisting of 9 to 14 primary ribs, which increase to 20 in the basal zone. The intercostal space near the apex shows interrupted longitudinal cords which appear and disappear in the middle zone of the shell.

Dimensions
Length from 10 to 30 mm. Apex diameter from 300 to 500 µm, diameter of the basal zone from 2.0 to 2.5 mm.

Mean curvature angle
α = 9° 38' 19.45" σ ± 1° 44' 10.86"

Remarks
This species is often confused with *A. inaequicostata.*

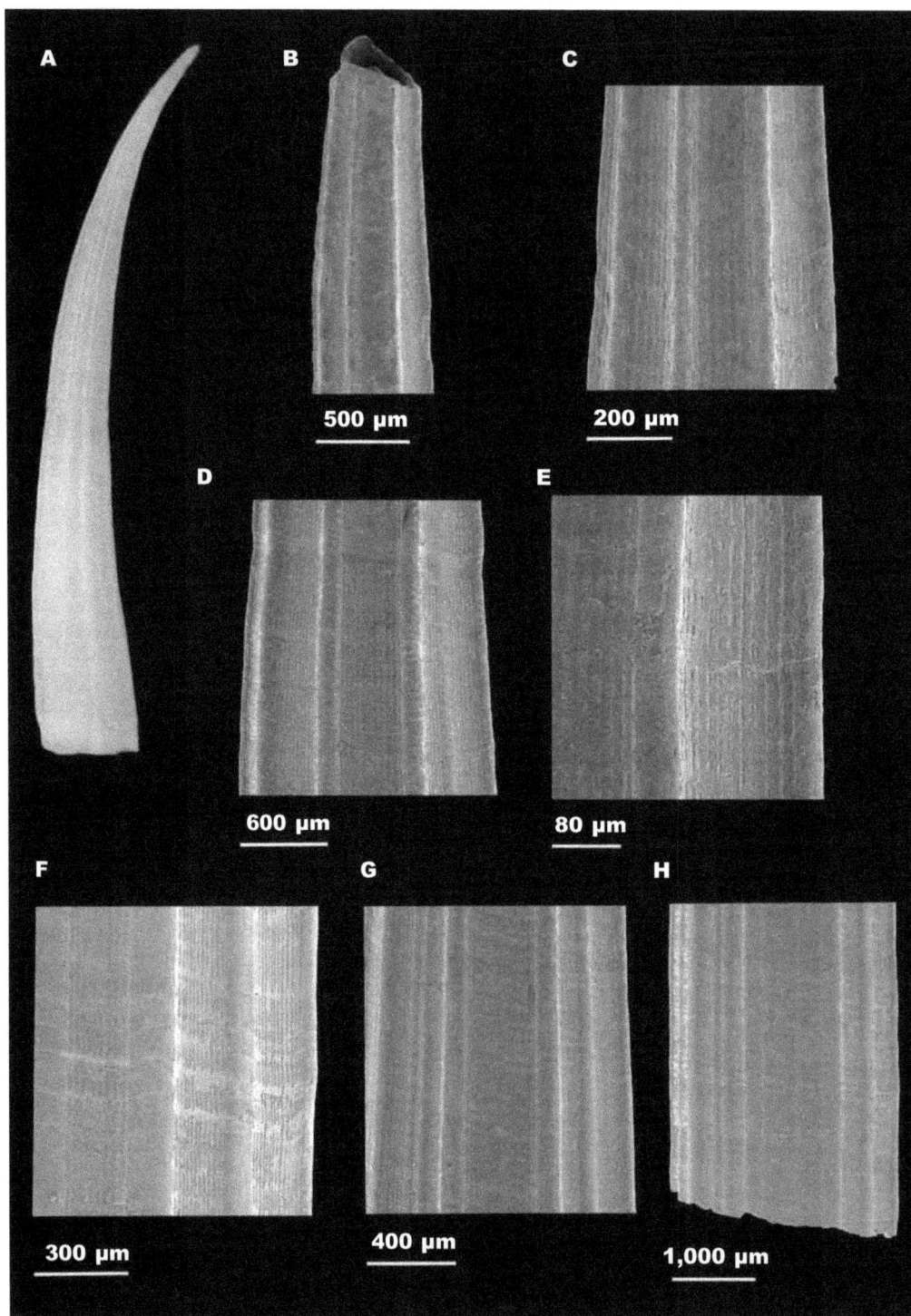

Fig. 7. A–H. *Antalis dentalis:* A. Specimen from Benicarló (Castellón) (MVHN–150211UB02); B–H. Specimen from Tavernes de la Valldigna (Valencia) (MVHN–150211UB01): B. Apex; C. Apex zone; D. Detail of the start of the first secondary rib; E. Intercostal space detail; F. Middle zone; G. Detail of the start of the secondary ribs; H. Basal zone.

Fig. 7. A–H. Antalis dentalis. A. Ejemplar de Benicarló (Castellón) (MVHN–150211UB02); B–H. Ejemplar de Tavernes de la Valldigna (Valencia) (MVHN–150211UB01): B. Ápice; C. Zona apical; D. Detalle del inicio de la primera costilla secundaria; E. Detalle del espacio intercostal; F. Zona media; G. Detalle del inicio de las costillas secundarias; H. Zona basal.

However, *A. dentalis* is smaller and more fragile, having a more curved shell and a greater number of ribs without interruptions (Caprotti, 1979; GMR, 2010). It is also morphologically similar to *A. entails,* a species that in contrast is completely smooth and exclusively distributed in the Atlantic (Caprotti, 1966c, 1979; Ghisotti, 1979; GMR, 2010).

Habitat
Sandy bottoms at depths up to 300 m (Steiner, 1997; Steiner & Kabat, 2004; GMR, 2010).

Geographical distribution
Located in the Mediterranean Sea and Northeast Atlantic (Steiner & Kabat, 2004; CLEMAM, 2012). In the Autonomous Community of Valencia it is common on the coasts of all three provinces (fig. 8).

Antalis inaequicostata (Dautzenberg 1891) (figs. 9–11).

Material examined
159 specimens from 50 sampling stations (table 2).

Description
Solid shell moderately curved, always opaque and white with pinkish tones. It has approximately nine primary ribs at the apex increasing to 12 at the base. The apex has an oval central pipe since juvenile phase (figs. 10A–10C), characteristic of the species. It can be up to 4 mm long in adults(figs. 9A–9B). The intercostal spaces have numerous well–marked longitudinal ridges (figs. 9C–9D). Disruptions in the shell are common (Caprotti, 1979; Steiner, 1997; GMR, 2010; Ozturk, 2011).

Dimensions
Length from 20 to 50 mm. Apical central pipe length is 700–4,000 µm. Diameter of the apex shell is 800–1,500 µm. Basal diameter is 2.5–4.0 mm.

Mean curvature angle
α = 7° 29' 11.18" σ ± 2° 31' 43.55"

Remarks
This species shows a high variability (Vera–Peláez & Lozano, 1983) so even specimens from the same region can show great morphological conchological variations, including differences in shell colour. Some specimens may even be confused with *A. dentalis* or *A. panorma*. However, *A. inaequicostata* is highly characteristic for its abundant primary and secondary ribs (Caprotti, 1979; GMR, 2010).

Habitat
Sandy bottoms at depths of 5 to 120 m (Steiner, 1997; Steiner & Kabat, 2004; GMR, 2010).

Geographical distribution
Located in the Mediterranean Sea and Northeast Atlantic (Steiner & Kabat, 2004; CLEMAM, 2012). In the Valencian Community it is common on the coasts of all three Valencian provinces (fig. 11).

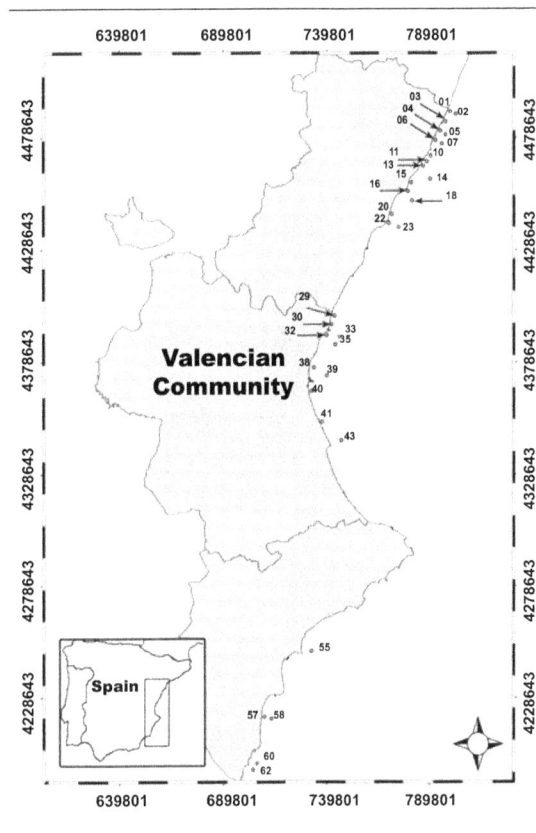

Fig. 8. Geographical distribution of *Antalis dentalis* on the Valencian coast.

Fig. 8. Distribución geográfica de Antalis dentalis *en la costa valenciana.*

Antalis novemcostata (Lamarck, 1818) (figs. 6, 12).

Material examined
Seven specimens from 4 sampling stations (table 2).

Description
Shell slightly curved, small, opaque, usually white or pinkish. Sculpture with nine longitudinal primary ribs at the apex which remain until the end of the base (figs. 12G–12H). The intercostal space is concave, with small, lightly marked, transverse striae plus longitudinal striae (figs. 12D–12E). Sometimes there is a small central pipe at the apex (figs. 12E–12F) (Steiner, 1997).

Dimensions
Shell length between 10 to 20 mm. Apex diameter 500–1,000 µm and diameter of the basal zone 2.0–2.5 mm.

Mean curvature angle
α = 11° 1' 45.35" σ ± 1° 50' 35.34"

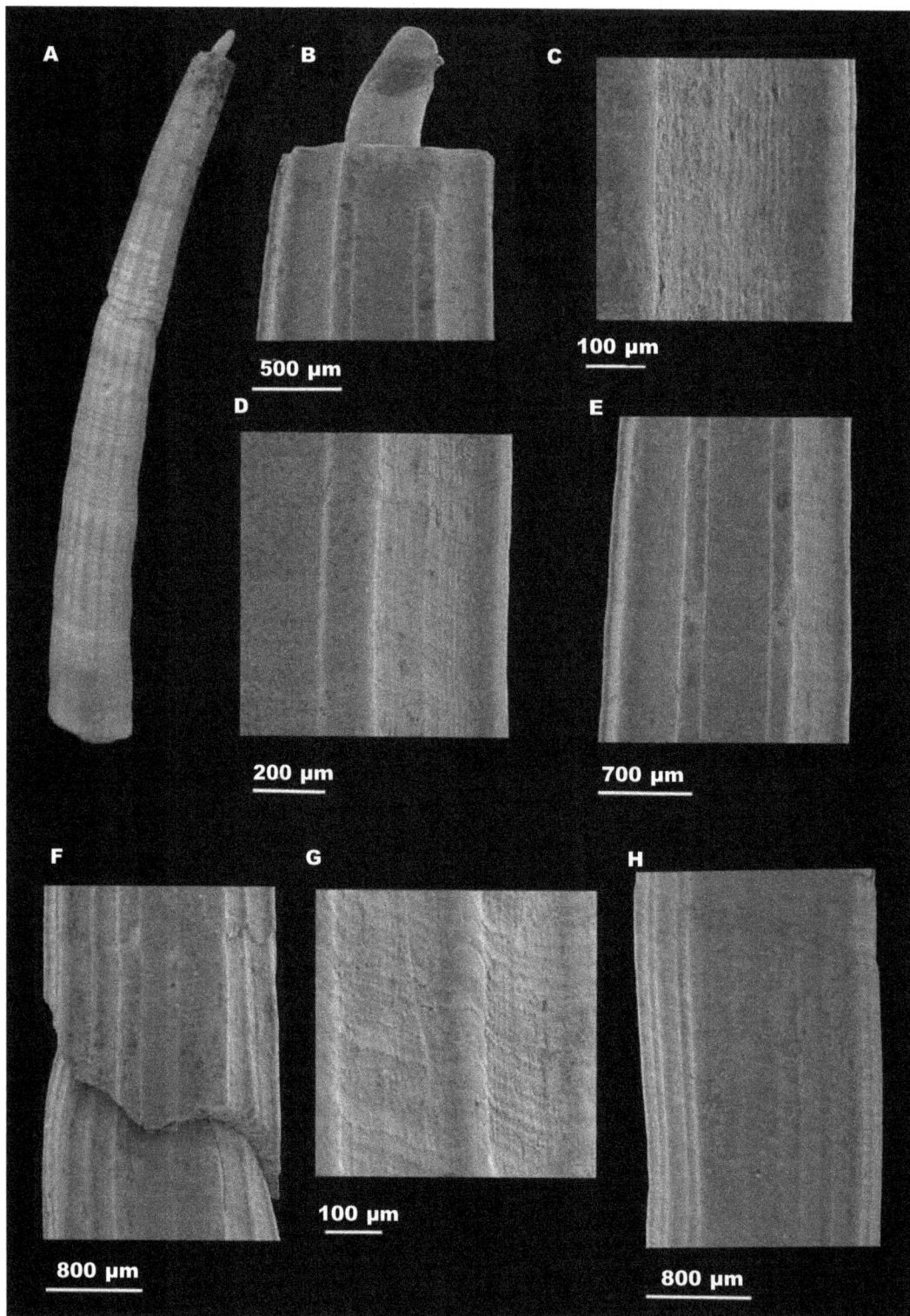

Fig. 9. *Antalis inaequicostata*, Oropesa (Castellón) (MVHN–150211UB08): A. Adult; B. Apex with pipe; C. Intercostal space detail; D. Detail of the beginning of a secondary rib; E. Middle zone; F. Disruption of the shell; G. Details of the growth striations; H. Basal zone.

Fig. 9. Antalis inaequicostata, *Oropesa (Castellón) (MVHN–150211UB08): A. Adulto; B. Ápice con túbulo; C. Detalle del espacio intercostal; D. Inicio de la formación de una costilla secundaria; E. Zona media; F. Interrupción de la concha; G. Detalle de las estrías de crecimiento; H. Zona basal.*

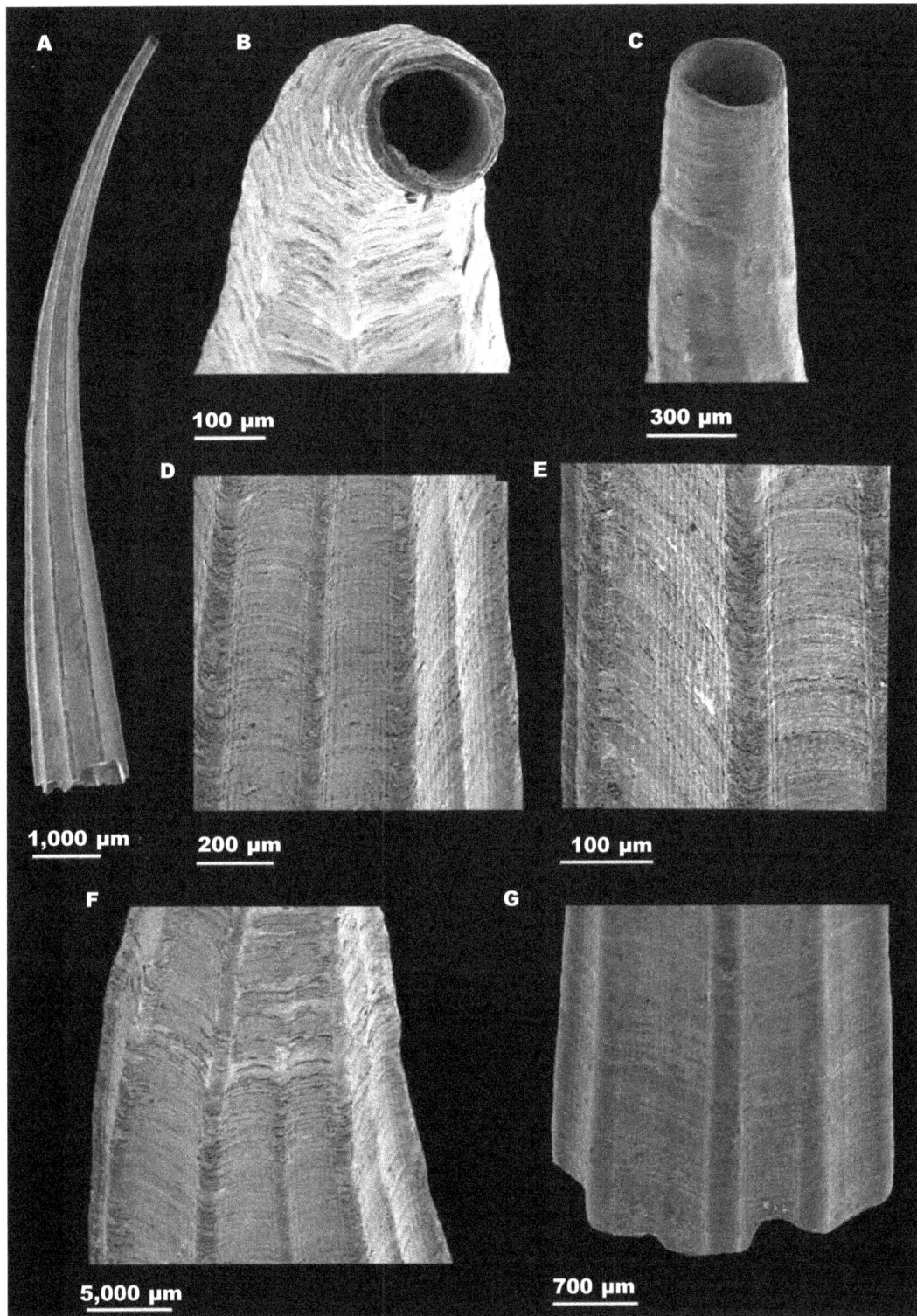

Fig. 10. *Antalis inaequicostata,* Peñíscola (Castellón) (MVHN–211011AC01): A. Juvenile; B, C. Apex with pipe; D. Detail of the beginning of a secondary rib; E. Intercostal space detail; F. Detail of a secondary rib in the middle zone; G. Basal zone.

Fig. 10. Antalis inaequicostata, *Peñíscola (Castellón) (MVHN–211011AC01): A. Juvenil; B, C. Ápice con túbulo; D. Detalle del inicio de una costilla secundaria; E. Detalle del espacio intercostal; F. Detalle de la formación de una costilla secundaria en la zona media; G. Zona basal.*

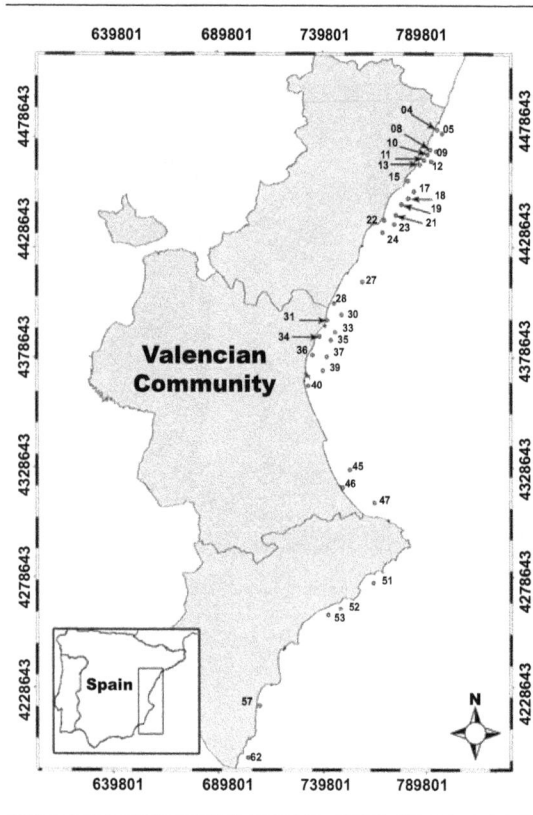

Fig. 11. Geographical distribution of *Antalis inaequicostata* on the Valencian coast.

Fig. 11. Distribución geográfica de Antalis inaequicostata *en la costa valenciana.*

Remarks
According to Caprotti (1979) and Bucquoy et al. (1882–1889) the striae of the intercostal spaces are a criterion to distinguish *A. novemcostata* from *A. inaequicostata*. Both species have a variable number of longitudinal striae (figs. 9C, 10E, 12D, 12F), but only *A. novemcostata* has transversal striae (fig. 12D). Besides, *A. novemcostata* has a more concave intercostal space. Hidalgo (1917) cited this species as recorded at the southeast of the Iberian peninsula, while Perrier (1930) cited it from the French Mediterranean coast. However, Caprotti (1966c, 1979) considered that these records were of *A. inaequicostata*, restricting the distribution of *A. novemcostata* to the Atlantic European waters. Recently, Cádiz & Martínez–Ortí (2011) and Salas & Gofas (2011) reported *A. novemcostata* from the Valencian and Andalucian coasts, respectively, confirming the taxonomical determination by Hidalgo.

Habitat
Sandy funds at depths 7 to 300 m (Steiner, 1997; Steiner & Kabat, 2004).

Geographical distribution
Northeast Atlantic (Steiner & Kabat, 2004; CLEMAM, 2012) and Mediterranean Sea (Perrier, 1930; Cádiz & Martínez–Ortí, 2011; Salas & Gofas, 2011). In the Valencian Community it is present on the coasts of the three Valencian provinces (fig. 6).

Antalis vulgaris (Da Costa, 1778) (figs. 13–14).

Material examined
Two specimens from 12 sampling stations (table 2).

Description
Slightly curved, solid shell, white or yellowish. Shell sculpture consists of numerous primary longitudinal ribs, about 30 in the apical zone, which is almost smooth towards the base (GMR, 2010; Ozturk, 2011). Intercostal space with five longitudinal microstriae (fig. 13F). The apex is thick and may have a central oval pipe (figs. 13C–13D).

Dimensions
Shell length 20–50 mm. Apex diameter 500–1,000 µm; diameter of the basal zone 2–4 mm.

Mean curvature angle
$\alpha = 10° 40' 46.88"$ $\sigma \pm 4° 58' 2.6"$

Remarks
Easily confused with *A. entalis*, a species completely smooth and only having Atlantic distribution, while *A. vulgaris* has many ribs and also lives in the Mediterranean Sea (GMR, 2010). The shell section has four layers as in *A. inaequicostata*, a related species (Alzuria, 1985a).

Habitat
Sandy and muddy bottoms at depths of 5–1,100 m (Steiner, 1997; Steiner & Kabat, 2004; GMR, 2010).

Geographical distribution
Mediterranean Sea and Northeast Atlantic (Steiner & Kabat, 2004; Ozturk, 2011; CLEMAM, 2012). In the Autonomous Community of Valencia it is known on the coasts of both Castellón and Alicante provinces (fig. 14).

Family Fustiariidae Steiner, 1991
Genus *Fustiaria* (Stoliczka, 1868)

Fustiaria rubescens (Deshayes, 1825) (figs. 15–16)

Material examined
Forty-eight specimens from 33 sampling stations (table 2).

Description
Translucent slightly curved shell, completely smooth, without ridges, brown or beige with bright tones (figs. 15A, 15C–15E). The apex has a long

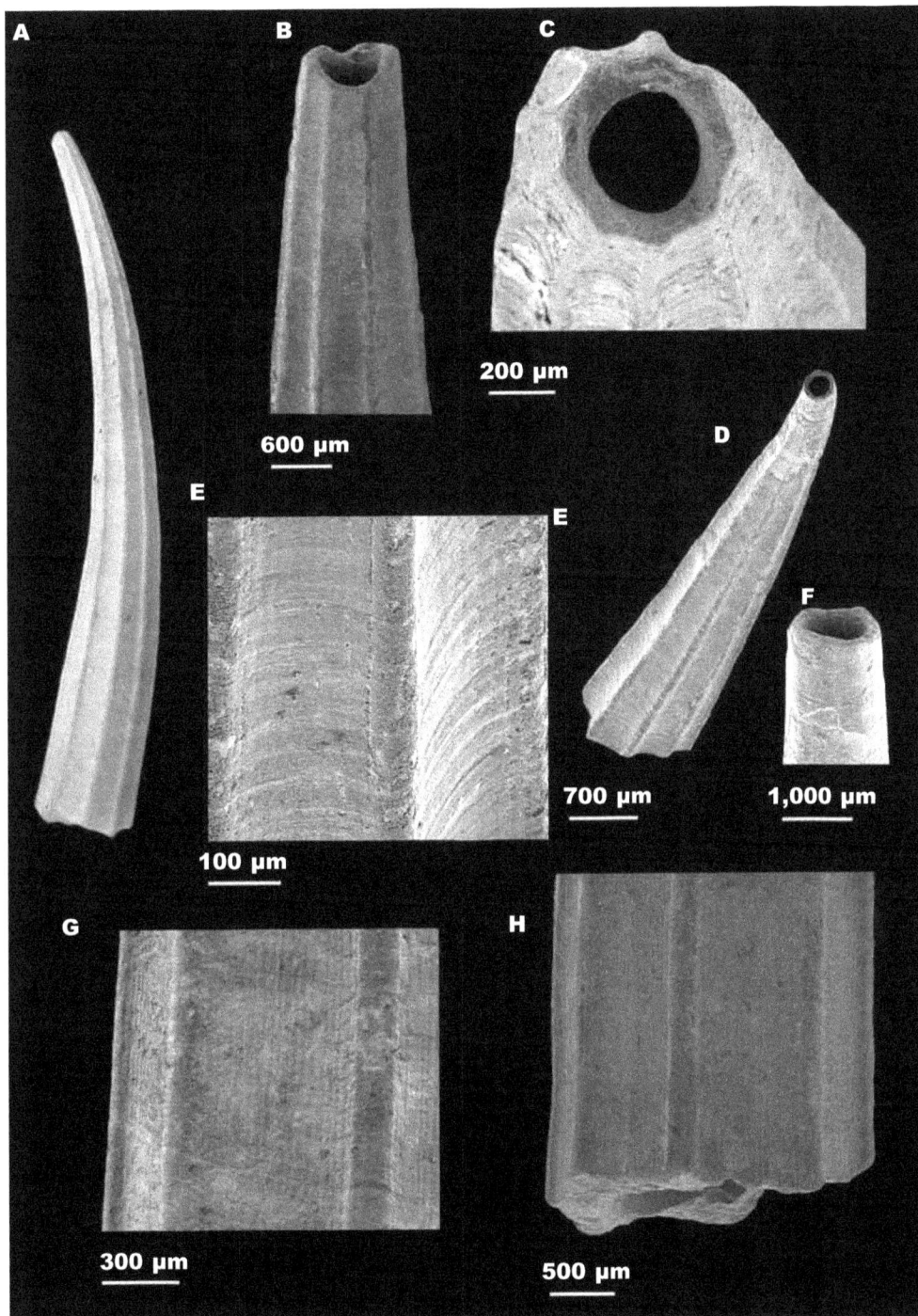

Fig. 12. *Antalis novemcostata:* A. Specimen from Guardamar del Segura (Alicante) (MVHN–150211UB05); B, C. Apex detail of a specimen from Benicassim (Castellón) (MVHN–150211UB04); D. Juvenile from Sagunto (Valencia) (MVHN–211011AC02); E. Intercostal space detail of the juvenile specimen; F. Detail of the apex pipe of the same juvenile; G. Intercostal space detail (Benicassim, Castellón); H. Basal zone (Benicassim, Castellón).

Fig. 12. Antalis novemcostata: *A. Ejemplar de Guardamar del Segura (Alicante) (MVHN–50211UB05); B, C. Detalle del ápice de un ejemplar de Benicassim (Castellón) (MVHN–150211UB04); D. Juvenil de Sagunto (Valencia) (MVHN–211011AC02); E. Detalle del espacio intercostal del ejemplar juvenil; F. Detalle del túbulo apical de dicho ejemplar; G. Detalle del espacio intercostal (Benicassim, Castellón); H. Zona basal (Benicassim, Castellón).*

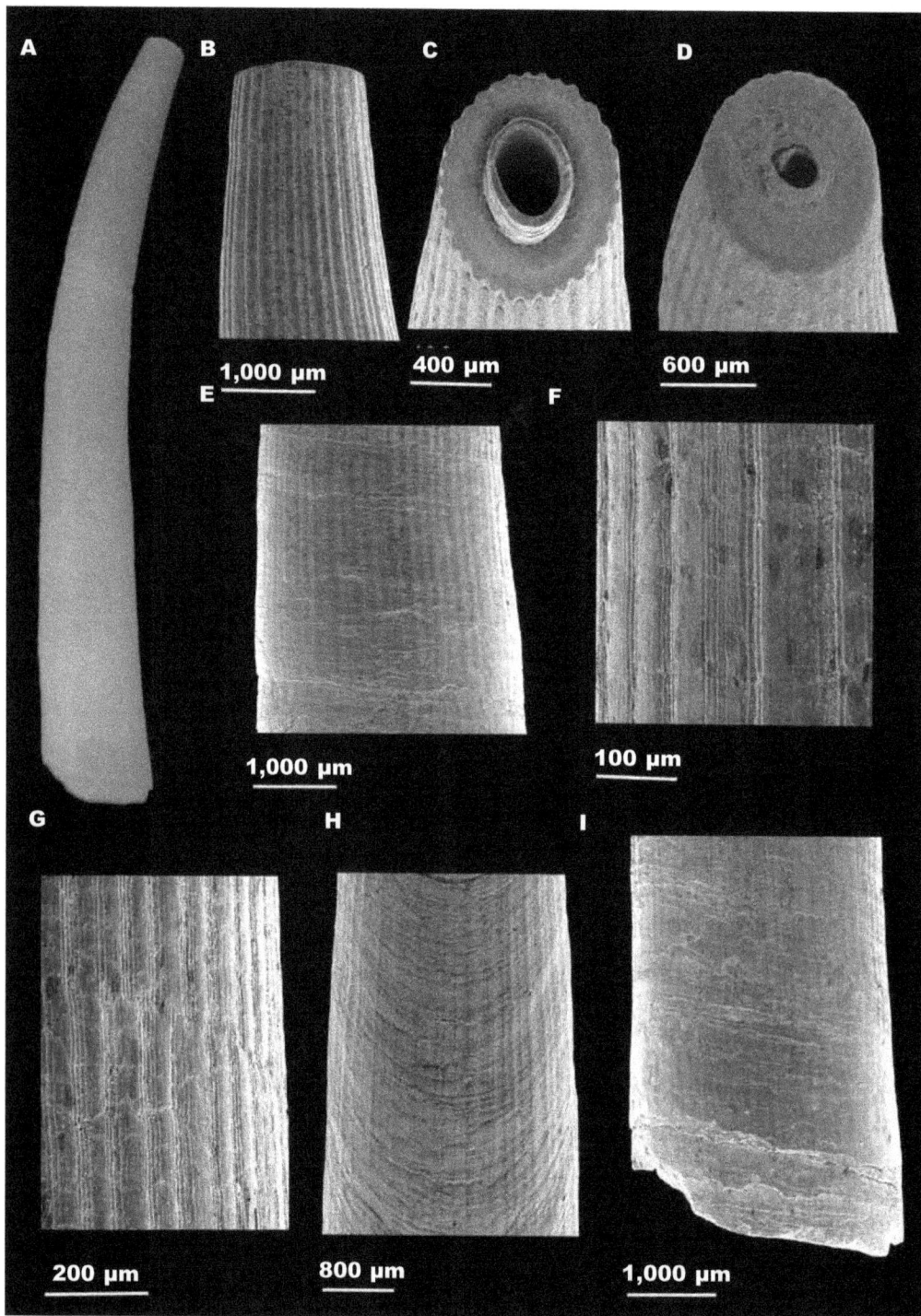

Fig. 13. *Antalis vulgaris:* A. Specimen from Oliva (Alicante) (MVHN–150211UB11). B–I. Specimen from Santa Pola (Alicante) (MVHN–150211UB09): B. Apical zone; C, D. Apex detail; E. Detail of the primary ribs; F. Intercostal space detail; G. Middle zone; I. Basal zone; C, H. Specimen from Jávea (Alicante) (MVHN–211011AC03): C. Apex with pipe; H. Detail of the secondary ribs in the basal zone.

Fig. 13. Antalis vulgaris: *A. Ejemplar de Oliva (Alicante) (MVHN–150211UB11). B–I. Ejemplar de Santa Pola (Alicante) (MVHN–150211UB09): B. Zona apical; C, D. Detalle del ápice; E. Detalle de las costillas primarias; F. Detalle del espacio intercostal; G. Zona media; I. Zona basal; C, H. Ejemplar de Jávea (Alicante) (MVHN–211011AC03): C. Ápice con túbulo; H. Detalle de las costillas secundarias en la zona basal.*

narrow deep cleft, characteristic of this species (fig. 15B) (Caprotti, 1966f, 1979; GMR, 2010; Ozturk, 2011).

Dimensions
Length 10–40 mm. Apex diameter 400–600 μm and diameter at the basal zone 2–3 mm.

Mean curvature angle
$\alpha = 9° 28' 40.95'' \sigma \pm 1° 28' 20.11''$

Habitat
Sandy sediments at depths of 4 to 618 m (Steiner & Kabat, 2004; GMR, 2010).

Geographical distribution
Mediterranean Sea and Northeast Atlantic (Steiner & Kabat, 2004; Ozturk, 2011; CLEMAM, 2012). In the Autonomous Community of Valencian it is common on the coasts of all three provinces

Order Gadilida Starobogatov, 1974
Family Entalinidae Chistikov, 1979
Genus *Entalina* Monterosato, 1872

Entalina tetragona (Brocchi, 1814) (figs. 17A–17E)

Material examined
An unpublished sample found in the Roselló collection, consisting of 12 shells (MCNV–3455b) identified as *Pulsellum quinquangulare* (Forbes, 1844) by Roselló but now considered as a junior synonym of *E. tetragona* (CLEMAM, 2012).

Description
Opaque shell, slightly curved, and white. The apical zone shows a pentagonal section and consists of five primary ribs that reach the end of the basal zone (figs. 17B–17D) (Caprotti, 1979; Steiner, 1997; GMR, 2010; Ozturk, 2011). Intercostal spaces have 28 secondary ribs (fig. 17E), considered as striae by Steiner (1997).

Dimensions
The specimens examined reach a maximum length of 9.2 mm and a maximum width of 1.3 mm.

Mean curvature angle
$\alpha = 19° 33' 20.16'' \sigma \pm 6° 34' 42.49''$

Habitat
It lives in sandy sediments at depths of 26 to 3,500 m (Steiner, 1997; Steiner & Kabat, 2004; GMR, 2010).

Geographical distribution
Mediterranean Sea and Northeast Atlantic (Steiner & Kabat, 2004; Ozturk, 2011; CLEMAM, 2012). This species is present on the Valencian coast since it was collected on the coast of the province of Valencia by Roselló on an undetermined date, but a more exact location remains unknown.

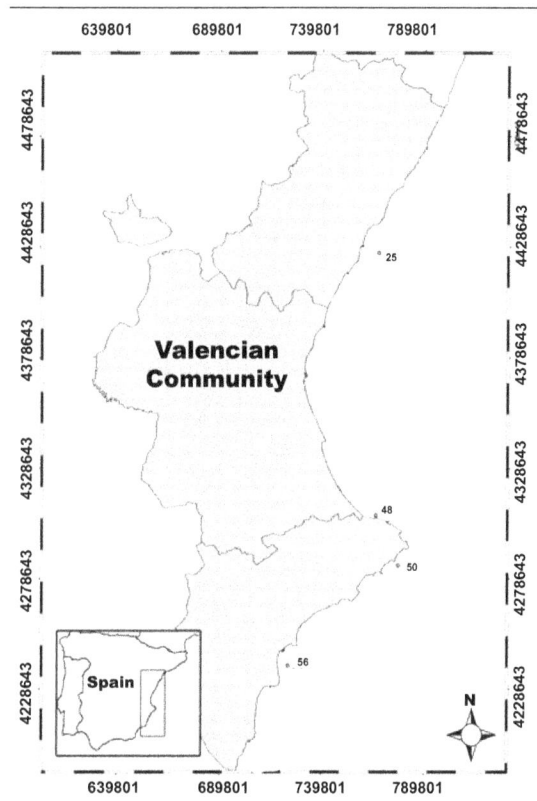

Fig. 14. Geographical distribution of *Antalis vulgaris* on the Valencian coast.

Fig. 14. Distribución geográfica de Antalis vulgaris en la costa valenciana.

Family Gadilidae Stoliczka, 1868
Genus *Dischides* Jeffreys, 1867

Dischides politus (Wood S., 1842) (figs. 6,17F–J)

Material examined
One specimen from one of the sampling stations (table 2). Besides, seven specimens were found in the Roselló collection (MCNV–n° 3456), identified as *Siphonodentalium bifissus* (Jeffreys, 1882) by Roselló (1934), but now considered as a junior synonym of *D. politus* (Steiner, 1997; CLEMAM, 2012).

Description
Translucent shell, slightly curved, subcylindrical, thin and smooth. Characteristics of the species are two short, wide clefts at the apex, dividing the shell into two parts (Caprotti, 1979; Steiner, 1997; GMR, 2010; Ozturk, 2011) (figs. 17F–17I). In adults the shell closes at the basal zone (figs. 17F–17G, 17J), having a similar morphology to the polychaete *D. arietina,* sometimes producing misidentifications.

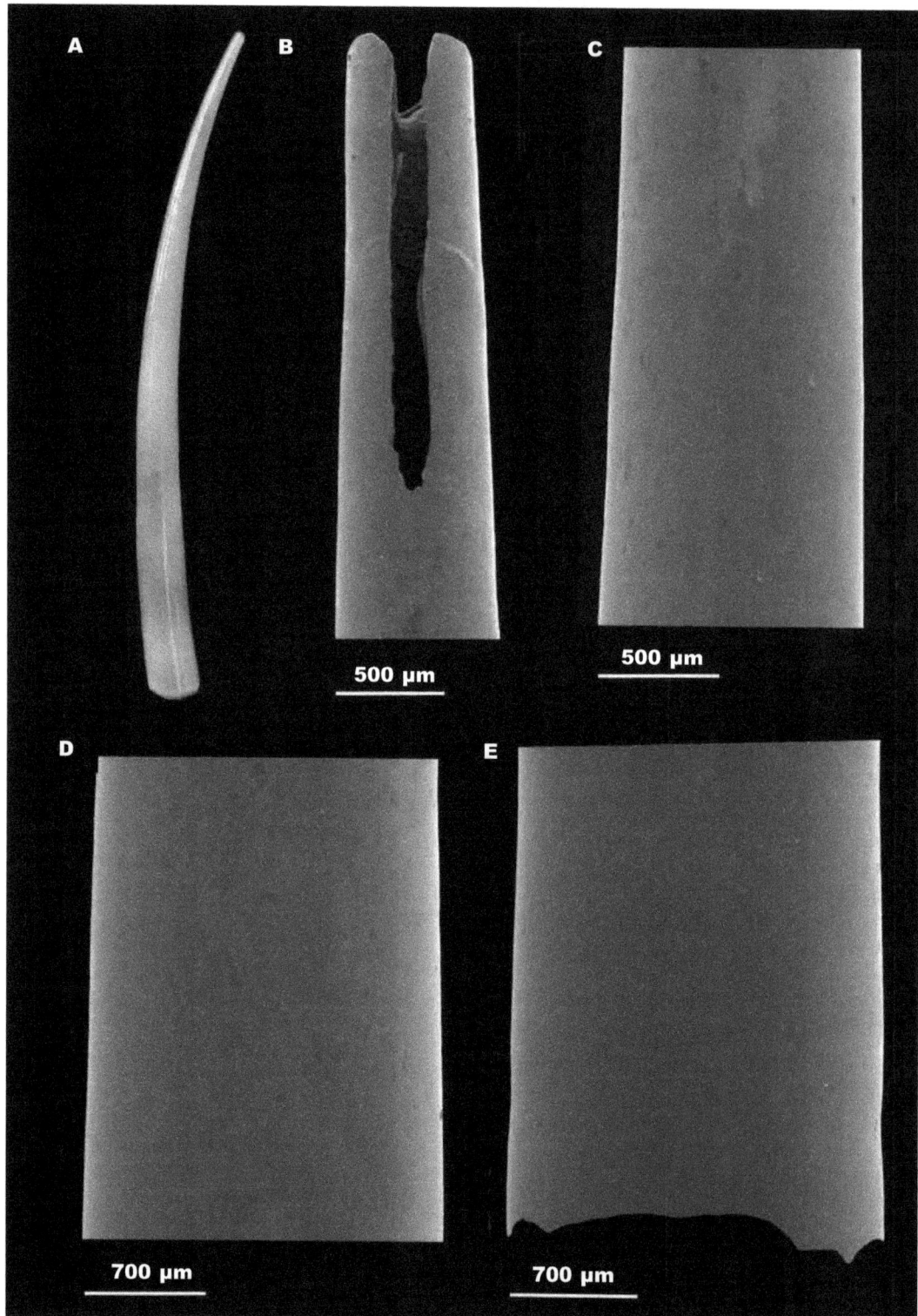

Fig. 15. A–E. *Fustiaria rubescens:* A. Specimen from Sagunto (Valencia) (MVHN–150211UB06). B–E. Specimen from Peníscola (Castellón) (MVHN–150211UB07): B. Apex detail; C. Apical zone; D. Middle zone; E. Basal zone.

Fig. 15. A–E. Fustiaria rubescens: *A. Ejemplar de Sagunto (Valencia) (MVHN–150211UB06). B–E. Ejemplar de Peñíscola (Castellón) (MVHN–150211UB07): B. Detalle del ápice; C. Zona apical; D. Zona media; E. Zona basal.*

Dimensions
Shell length from 5.7 to 8.1 mm and 1 mm in diameter.

Mean curvature angle
$\alpha = 8° 6' 52.84'' \sigma \pm 3° 24' 0.94''$

Habitat
Sandy sediments at depths of 9 to 324 m (Steiner, 1997; Steiner & Kabat, 2004; GMR, 2010).

Geographical distribution
Located in the Mediterranean Sea and Northeast Atlantic (Steiner & Kabat, 2004; Ozturk, 2011; CLEMAM, 2012). On the Valencian coast it has been found at one sampling station in Benicassim (Castellón province) (fig. 6), and there is also a record on the coast of the Valencia province reported by Roselló (1934), but detailed location data are lacking.

Conclusions

This study shows the occurrence of eight scaphopod species on the coasts of the Autonomous Community of Valencia: Antalis caprottii n. sp., A. inaequicostata, A. dentalis, A. novemcostata, A. vulgaris, Dischides politus, Entalina tetragona and Fustiaria rubescens.

The species A. novemcostata and E. tetragona are cited for the first time on these coasts. In addition, according to Hidalgo (1917), Perrier (1930), Cádiz & Martínez–Ortí (2011) and Salas & Gofas (2011) we confirm the occurrence of A. novemcostata for the western Mediterranean Sea. The highest abundance in the Valencian coast was recorded on the coast of Castellón province, representing 79% of the collected specimens. The most abundant collected species are A. inaequicostata (53.5%) and A. dentalis (26.3%). In addition, a new species Antalis caprottii n. sp. from the coast in the province of Castelló is reported.

We propose a new method to measure the scaphopod mean curvature angle (α) that we used for the eight studied species, showing that A. inaequicosta, A. vulgaris, D. politus and F. rubescens present smaller mean curvature angles than A. caprotti n. sp., A. dentalis, A. novemcostata and E. tetragona.

This is a preliminary conclusion, however, and extensive measures of additional specimens are needed to maximize the potential of this tool to delimit and separate species.

The common species in the Mediterranean Sea, such as A. agilis, A. panorma, C. jeffreysi, C. subfusiformis and P. lofotense, were absent in both our sampling stations and revised collections. These species live at deeper substrates, over 100 m, than our sampling stations, which are up to 20 m deep. We expect that at least some of these species will appear in the coasts of the Autonomous Community of Valencia if future studies use suitable tools to obtain samples at deeper points. Such studies

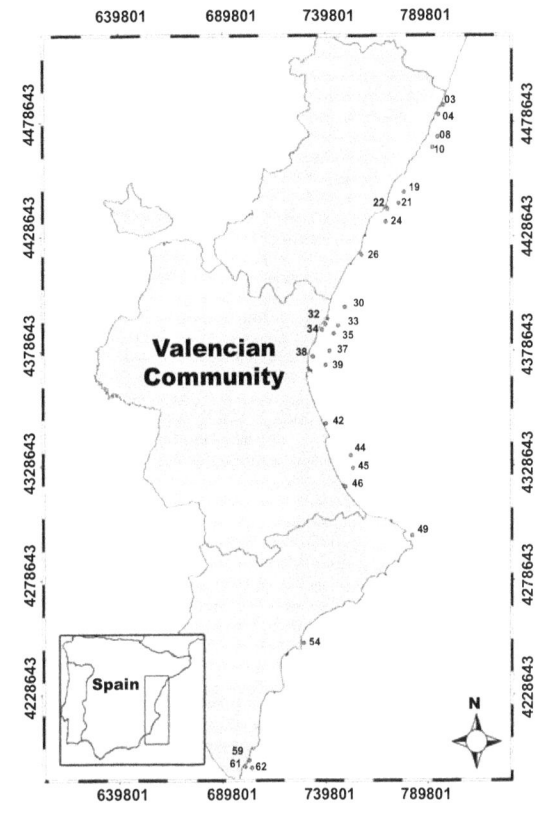

Fig. 16. Geographical distribution of *Fustiaria rubescens* on the Valencian coast.

Fig. 16. Distribución geográfica de Fustiaria rubescens en la costa valenciana.

will increase the knowledge of both Valencian and Mediterranean scaphopods.

Acknowledgements

This work was carried thanks to the support of the Water Framework Directive issued by the 'Dirección General de Calidad del Agua' of the 'Conselleria de Medi Ambient, Aigua, Urbanisme i Habitatge'. We thank Dr. Margarita Belinchón, Director of the Museo de Ciencias Naturales de Valencia, the 'Instituto de Ecología Litoral' at Campello (Alicante) and Vicente Escutia and Antonio López Alabau for the loan of samples. We especially thank Vicente Escutia, mathematician and noted amateur malacologist for his collaboration in the study of mean curvature angle of the studied species. Thanks also to David Osca, from the Museo Nacional de Ciencias Naturales de Madrid, Miguel Royo, Laura Gómez and Borja Mercado from the Universidad

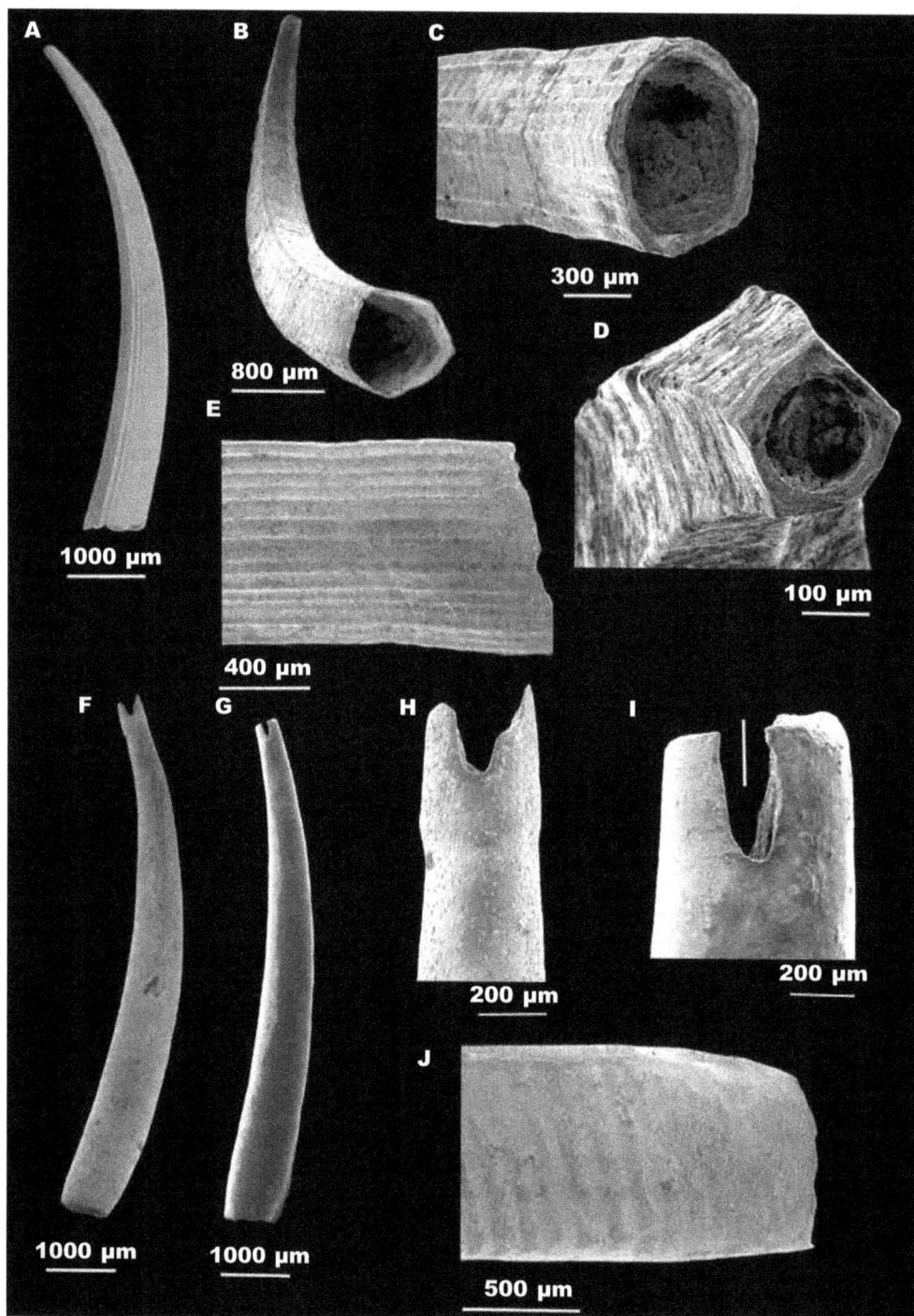

Fig. 17. A–E. *Entalina tetragona,* Valencia (MCNV–3455b): A. Adult; B. View from basal zone; C. Detail of the basal zone; D. Apex detail; E. Detail of the secondary ribs in the basal zone. F–J. *Dischides politus,* Valencia (MCNV–3456); F, G. Adults; H, I. Detail of the apical cleft; J. Detail of the shell narrowness in the basal zone. (Shells were not covered with a gold–paladium layer to avoid alteration as they are important museum specimens.)

Fig. 17. A–E. Entalina tetragona, *Valencia (MCNV–3455b): A. Adulto; B. Vista desde la zona basal; C. Detalle de la zona basal; D. Detalle del ápice; E. Detalle de las costillas secundarias en la zona basal. F–J.* Dischides politus, *Valencia (MCNV–3456); F, G. Adultos; H, I. Detalle de la fisura apical; J. Detalle del estrechamiento de la concha en la zona basal. (Los conchas no han sido cubiertas por una capa de oro–paladio para evitar su alteración por tratarse de importantes muestras museísticas.)*

Católica de Valencia–San Vicente Mártir and Dr. Enrique Peñalver, IGME, for their help during the development of this work. We are also grateful to the Electron Microscopy Service of the S. C. S. I. E. at the University of Valencia for help obtaining the photographs with the Hitachi S–4100.

References

Alzuria, M., 1984. Nota sobre la fracción mineral en *Dentalium mutabile inaequicostatum* (Dautzemberg, 1891) (Mollusca, Scaphopoda). *Publicaciones del Departamento de Zoología de la Universidad de Barcelona*, 10: 23–25.

– 1985a. Ultraestructura de la concha en *Dentalium vulgare* (De Costa 1778) (Mollusca: Scaphopoda). *Iberus*, 5: 11–19.

– 1985b. Ultrastructura de la concha en *Dentalium mutabile inaequicostatum* (Dautzemberg, 1891) (Mollusca; Scaphopoda). *Publicaciones del Departamento de Zoología de la Universidad de Barcelona*, 11: 15–22.

– 1986. *Antalis rossati* (Caprotti, 1966), nuevo escafópodo para la fauna española. *Publicaciones del Departamento de Zoología de la Universidad de Barcelona)*, 12: 37–39.

Boscá, A., 1916. Fauna Valenciana. In: *Geografía General del Reino de Valencia:* 48–50 (A. Martín, Ed.). Casa Editorial de Alberto Martín, Barcelona.

Bucquoy, E., Dautzenberg, P. & Dollfus, G., 1882–1889. Les Mollusques Marins du Rousillon. J. B. Baillière & Fils, Paris.

Cádiz, L. & Martínez–Ortí, A., 2011. The living Scaphopods from the Valencian Community coast (Spain). *6th Congress of the European Malacological Societies*. Vitoria–Gasteiz. p. 59.

Caprotti, E., 1966a. *Dentalium (Antalis) agile* M. Sars, 1872. *Schede Malacologiche Del Mediterraneo*, 86 Aa 06, 2 pp.

– 1966b. *Dentalium (Antalis) dentalis* Linnaeus, 1766. *Schede Malacologiche del Mediterraneo*, 86 Aa 01, 4 pp.

– 1966c. *Dentalium (Antalis) inaequicostatum* Dautzenberg, 1981. *Schede Malacologiche del Mediterraneo*, 86 Aa 03, 3 pp.

– 1966d. *Dentalium (Antalis) panormum* Chenu, 1842. Schede Malacologiche del Mediterraneo, 86 Aa 04, 2 pp.

– 1966e. *Dentalium (Antalis) vulgare* Da Costa, 1778. *Schede Malacologiche del Mediterrâneo* 86 Aa 06, 2 pp.

– 1966f. *Dentalium (Pseudantalis) rubescens* Deshayes, 1825. *Schede Malacologiche del Mediterrâneo* 86 Aa 05, 2 pp.

– 1967. Scafopodi jonici. *Thalassia Salentina*, 2: 134–137.

– 1979. Scafopodi noegenici e recenti del bacino mediterraneo, Iconografia ed epítome. *Bolletino Malacologico*, 15(1–10): 213–288.

– 2009. Osservazioni e aggiornamenti su alcune specie di scafopodi neogenici e quaternari del bacino mediterrâneo. *Bolletino Malacologico*, 45(1):

31–44.

CLEMAM, 2012. Check List of European Marine Mollusca http://www.somali.asso.fr/clemam/biotaxis.php

Cossignani, T. & Ardovini, R., 2011. *Malacologia Mediterranea*. L'informatore Piceno, Ancona.

Giner, I., 1989. Moluscos y comunidades bentónicas de la costa de Alboraya–Albuixech (Golfo de Valencia, Mediterráneo Occidental). Tesis Doctoral, Univ. de València.

Giribet, G. & Peñas, A., 1997. Fauna malacológica del litoral del Garraf (NE de la Península Ibérica). *Iberus*, 15(1): 41–93.

Ghisotti, F.,1979. Chiavi di determinazione degli Scaphopoda del bacino mediterraneo. *Bollettino Malacologico*, 15 (9–10): 289–294,

GMR (Gruppo Malacologico Romagnolo), 2010. La Classe Scaphopoda in Mediterraneo. *Noticiario SIM*, 28(2): 17–22.

Hidalgo, J. G., 1917. Fauna malacologica de España, Portugal y las Baleares. Moluscos testáceos marinos. *Trabajos del Museo Nacional de Ciencias Naturales*, Serie Zoologia, 30: 1–752.

Holme, N. A. & McIntyre, A. D., 1971. *Methods for the Study of the Marine Benthos*. Blackwell Scientific Publications, Oxford.

Jefreys, J. G., 1877. New and peculiar Mollusca of the order Solenoconchia procured by the "Valorous" expedition. *Annals and Magazines of Natural History*, (4)19: 153–158.

– 1882. On the Mollusca procured during the 'Lightning' and 'Porcupine' Expeditions, 1868–70. *Proceedings of the Royal Society of London*, 1882: 656–687.

Jordá, J. F., Aura, J. E., Martin, C. & Avezuela, B., 2010. Restos arqueomalacológicos del Pleistoceno superior Holoceno inferior del Vestíbulo de la cueva de Nerja (Málaga, España). *Munibe*, 31: 78–87.

Locard, A., 1898. Mollusques Testacés. *Expeditions Scientifiques Travailleur et Talisman 1880–1883*, 2: 1–1031.

Moreno, R., 1995. Arqueomalacofaunas de la Península Ibérica: Un ensayo de síntesis. *Complutum*, 6: 353–382.

Ozturk, B., 2011. Scaphopod species (Mollusca) of the Turkish Levantine and Aegean seas. *Turkish Journal of Zoology*, 35(2): 199–211.

Pardo, L., 1920. Las colecciones de animales inferiores, Moluscos y Artrópodos del Instituto General y Técnico de Valencia. *Anales del Instituto General y Técnico de Valencia*, 7: 95–104.

Perrier, R., 1930. *La faune de la France*. Vol. IX. Paris, Librairie Delagrave, 15.

Plá, E., 2006. *Subvida y foto. Guía submarina de La Marina Alta, Ibiza y Formentera*. Plá ed. Alcoy (Alicante).

Roselló, E., 1934. *Catálogo de la colección conquiológica donada a la ciudad de Valencia por D. Eduardo Roselló Bru*. Ed. Diputación de Valencia, Valencia.

Rubio, F. & Rodríguez, C., 1996. Moluscos marinos infralitorales de la Comunidad Valenciana. Parte 1: Cabo San Antonio (Denia). *Libro de Resúmenes*.

XI Congreso Nacional de Malacología, Almería: 110–111.

Salas, C., García Raso, J. E. & López–Ibor, A., 1985. Estudio del macrobentos infralitoral (Mollusca, Crustacea Decapoda y Echinodermata) de la Bahía de Málaga (España). Actas do IV Simposio Iberico de Estudos do Benthos Marinho. Lisboa; 21–25 Maio 1984, 1: 123–146.

Salas, C. & Gofas, S., 2011. Clase Scaphopoda. In: *Moluscos marinos de Andalucía*: 713–716 (S. Gofas, D. Moreno & C. Salas, coords.). Málaga: Servicio de Publicaciones e intercambio Científico, Univ. de Málaga.

Steiner, G., 1997. Scaphopoda from the Spanish coast. *Iberus*, 15(1): 95–111.

Steiner, G. & Kabat, A., 2001. Catalogue of supraspecific taxa of Scaphopoda (Mollusca). *Zoosystema*, 23(3): 433–460.

– 2004. Catalogue of species–group names of Recent and fossil Scaphopoda (Mollusca). *Zoosystema*, 26(4): 549–726.

Tarruella Ruestes, A. & Fontanet Giner, M., 2001. Moluscos marinos del Baix Camp (Tarragona, NE Península Ibérica). *Spira*, 1(1): 1–5.

Templado, J., Calvo, M., Moreno, D., Flores, A., Conde, F., Abad, R., Rubio, J., López–Fé, C. M. & Ortiz, M., 2006. Flora y fauna de la Reserva Marina y Reserva de Pesca de la isla de Alborán. *Museo Nacional de Ciencias Naturales. Consejo Superior de Investigaciones Científicas,* Madrid.

Vera–Peláez, J. & Lozano F. M., 1993. Escafópodos (Mollusca, Scaphopoda) del Plioceno de la Provincia de Málaga, España. *Treballs del Museu de Geologia de Barcelona*, 3: 117–156.

2

Development of urban bird indicators using data from monitoring schemes in two large European cities

S. Herrando, A. Weiserbs, J. Quesada, X. Ferrer & J–Y. Paquet

Herrando, S., Weiserbs, A., Quesada, J., Ferrer X. & Paquet J.–Y., 2012. Development of urban bird indicators using data from monitoring schemes in two large European cities. Animal Biodiversiy and Conservation, 35.1: 141–150.

Abstract

Development of urban bird indicators using data from monitoring schemes in two large European cities.— Bird monitoring projects have provided valuable data for developing biological indicators to evaluate the state of natural and agricultural habitats. However, fewer advances have been made in urban environments. In this study we used bird monitoring data from 2002 to 2012 in two cities with different climates (Brussels and Barcelona), to generate two multi–species urban indicators to evaluate temporal trends on abundance of urban avifauna. To do this we used two different conceptual approaches, one based on a list of widespread species in European cities (WSEC) and another based exclusively on species widespread at city level (WCS) regardless of the birds occurring in other cities. The two indicators gave a similar general pattern, although we found a 3% difference in the mean annual change in both cities, thus suggesting that the values provided by urban indicators may differ depending on the conceptual approach and, hence, by the species list used to generate them. However, both indicators may have their own value and could be treated as complementary indices.

Key words: Urban indicator, Biodiversity, Bird monitoring, Species selection, Barcelona, Brussels.

Resumen

Desarrollo de indicadores de aves urbanas a partir de datos de sistemas de monitoreo en dos grandes ciudades europeas.— Los proyectos de monitoreo de aves han proporcionado datos valiosos para el desarrollo de indicadores biológicos que evalúan el estado de los hábitats naturales y agrícolas; sin embargo, los avances han sido menores en los ambientes urbanos. En este estudio se utilizaron los datos del monitoreo de aves de dos ciudades climáticamente diferentes (Bruselas y Barcelona; período 2002–2010) para generar dos indicadores urbanos multiespecíficos que valorasen las tendencias temporales en la abundancia del conjunto de las aves urbanas. Para hacer esto, utilizamos dos enfoques conceptuales distintos, uno basado en una lista de especies de amplia distribución en las ciudades europeas (WSEC) y otro basado exclusivamente en especies de amplia distribución a nivel de ciudad (WCS), independientemente de las aves de otras ciudades. Los dos indicadores dieron un patrón general similar, aunque un 3% de diferencia entre ellos en cuanto a los valores de cambio promedio anual se encontró en ambas ciudades. Esto sugiere que los valores producidos por los indicadores urbanos pueden diferir dependiendo de la aproximación conceptual y, por tanto, por la lista de especies utilizada para generarlos. Ambos indicadores pueden tener su propio interés y pueden ser tratados como complementarios.

Palabras clave: Indicador urbano, Biodiversidad, Monitoreo de aves, Selección de especies, Barcelona, Bruselas.

Sergi Herrando, Catalan Ornithological Inst., Natural History Museum of Barcelona, Psg. Picasso s/n., E–08003 Barcelona (Spain).– Anne Weiserbs & Jean–Yves Paquet, Aves–Natagora, Rue Nanon 98, 5000 Namur (Belgium).– Javier Quesada, Natural History Museum of Barcelona. Psg. Picasso s/n., E–08003 Barcelona (Spain).– Xavier Ferrer, Dept. of Animal Biology, Univ. of Barcelona, Av. Diagonal 645, E–08028 Barcelona (Spain).

Corresponding author: S. Herrando. E–mail: ornitologia@ornitologia.org

Introduction

Thanks to the many large–scale monitoring schemes, birds currently constitute one of the backbones of biodiversity monitoring in Europe (Schmeller, 2008). Many institutions run volunteer–based bird monitoring projects at national or regional level. Trends of European common birds are updated annually within the framework of the Pan–European Common Bird Monitoring Scheme, which combines the results of these projects to provide trends at a continental scale for 145 common bird species (Voříšek et al., 2008; PECBMS, 2011). Data on trends in bird populations have been increasingly used in recent times to develop indicators of environmental health (Gregory et al., 2005), since experience shows that habitats in which bird numbers are declining tend also to be losing species belonging to other faunal groups (e.g. Robinson & Sutherland, 2002). This has led to the launch of a policy to devise relevant synthetic indicators, and the Farmland Bird Index has even been included in EUROSTAT as one of the continent's sustainability indicators (http://epp.eurostat.ec.europa.eu).

To date, indicators of environmental health for particular habitats have been developed basically for farmland and woodland ecosystems (PECBMS, 2011). Nevertheless, as most human population in Europe live in urban centres, the development of indicators of the biodiversity in cities and towns would also seem to be relevant. These indicators may be an important tool to measure the process of adaptation of biodiversity in this new environment, and also to determine the readiness of design and planning in urban areas to harbour biological diversity (Adams et al., 2006). This is particularly important if we consider that urban habitats grow year after year. Furthermore, given the extent of city environments in Europe and their influence on the quality of life and education of urban dwellers, the development of such indicators may also facilitate the preservation of biodiversity in more natural ecosystems (Savard et al., 2000; Fuller et al., 2009).

Generation of an urban indicator based on bird monitoring data has traditionally been hindered by the definition of the urban ecology of species. European cities and towns provide suitable habitats for many bird species (Kelcey & Rheinwald, 2005; Caula et al., 2010). Most of these species are generalists that can be found in other environments (Clergeau et al., 2006; Devictor et al., 2007) and have only relatively recently colonized and adapted to urban areas (Blair, 1996; Evans et al., 2009; Møller, 2009; Sattler et al., 2010). Thus, they could be described as 'urban adapters'. Also, in a few cases, this process of colonization has led to a shift in a species' populations in urban areas to a degree that their numbers have become higher than in nearby natural areas (Blair, 1996); these species could be referred to as 'urban exploiters'. Using this latter quantitative concept, several attempts have been made to classify species as elements of a multi–species urban indicator (e.g. DEFRA, 2002; Zbinden et al., 2005; SEO/BirdLife,

2010). However, including only 'urban exploiters' means that the list of urban species is very short and mostly contains those species that use buildings for nesting (e.g. House Martin Delichon urbicum, House Sparrow Passer domesticus, Common Swift Apus apus and Feral Pigeon Columba livia). Yet, the largest proportion of urban bird richness comes from greener urban habitats such as parks, avenues with trees, and gardens (Kelcey & Rheinwald, 2005). Indicators of urban biodiversity should therefore probably include not only the 'exploiters' but also, in some way, the 'adapters'. The inclusion or otherwise of the 'urban adapters' in the indicator list is a crucial question, since many of the species inhabiting both urban and other habitats have different behavioural traits that could imply different population dynamics (Adams et al., 2006). Consequently, the development of bird indicators for urban areas is complicated by the choice of an appropriate species set whose numbers show what is happening specifically in urban areas and at the same time, also represent urban bird biodiversity as a whole.

An urban bird indicator may have more than one objective and serve to highlight the health of urban bird populations, changes in populations of special conservation interest, the degree of 'urbanization' of the local avifauna, or the impact of certain environmental pressures. As shown by Gregory et al. (2005) for farmland indicators, common birds could be good candidates for developing bird indicators aimed at evaluating the general state of urban bird populations. In addition, bird species may provide information as a proxy for the state of other taxa in urban gradients (e.g. Blair, 1999; but see Gagné & Fahrig, 2011). This framework could be particularly useful for the study of European urban areas and, in particular, the large cities where breeding bird monitoring projects are currently carried out.

As for the Pan–European Common Bird indicators (Voříšek et al., 2008), in practice, urban indictors could be calculated as aggregated population trends using the geometric mean of annual population indices of a group of species. At this point, it is essential to establish which species set is to be included in the indicator, taking into account that a low number of species in an indicator would make it susceptible to single species fluctuations, and thus it would be less relevant as an indicator of the general state of the environment (Butler et al., 2012). For urban areas, we can use two different conceptual approaches that differ in focus, thereby maximizing the possibilities to compare results between cities at both taxonomic (species that are present in many cities) and ecological (species considered functionally relevant because of their great abundance) levels. In the first approach, the urban indicator could include species that are widespread across many European cities, while in the second, the urban indicator of each city could include only the species that are widespread in a particular city, independently of whether they are present in other cities or not. Nevertheless, both indicators are likely to indicate different things. The first is more about the overall state of common European urban

Fig. 1. Location of the sampling plots: point-counts in Brussels, and line-transects in Barcelona. Grey areas correspond to green spaces; in the case of Brussels, the grey area to the south is the Forest de Soignes, while in Barcelona, Collserola Natural Park lies to the north–west. Sampling plots located in these two natural areas were excluded from the analyses and only plots situated in the built-up areas and urban parks are shown.

Fig. 1. Localización de las áreas de muestreo: estaciones de escucha en Bruselas y transectos lineales en Barcelona. Las áreas grises corresponden a espacios verdes, en el caso de Bruselas, el área gris situada al sur corresponde al bosque de Soignes, mientras que en Barcelona, Parque Natural de Collserola se encuentra al noroeste. Las áreas de muestreo situadas en estas dos áreas naturales fueron excluidas de los análisis y sólo se muestran aquellas que se encuentran en las áreas urbanizadas y los parques urbanos.

birds (in a set of cities), while the latter is more about the state of urban birds in a specific city and refers to environmental conditions in specific cities.

In this study we developed two multi–species indicators as a means of advancing towards the generation of an urban indicator aimed at revealing the response of urban birds to the overall environmental changes occurring in urban habitats. Specifically, we calculated and compared these two indicators (widespread species in European cities and widespread species in each particular city) using bird monitoring data from Brussels (Belgium) and Barcelona (Catalonia, Spain). We also discuss their outcomes in the light of the methodological limitations and applications.

Material and methods

Study areas

Taking into account that the driving forces affecting species dynamics can be very distinct inside and outside cities (Adams et al., 2006), we generated urban bird indicators using data collected exclusively inside cities and rejected data from agricultural and natural areas from outside cities (peri–urban areas). We believe that the cities of Brussels and Barcelona represent an interesting study framework given their distinct biogeographical locations within Europe, the former in the Eurosiberian region and the latter in the Mediterranean.

Brussels

Brussels is located close to the Atlantic coast of Europe (in the centre of Belgium; fig. 1). The city covers 162 km^2 and contains a mosaic of districts whose green spaces cover 53% of the territory (numerous parks, gardens, small woodlands and a large beech forest 'the Forest of Soignes', which represents a tenth of the Brussels' surface area). Parks and gardens are often highly managed, with large lawns, even though the management of an ever–increasing part of public green spaces is beginning to take biodiversity into account. Most of the urban parks and woodland were planted with beech *Fagus sylvatica*, ash *Fraxinus excelsior* or a variety of exotic species at the end of the nineteenth century and so most trees are today very old; active regeneration is under way. The neighbouring areas mainly consist of residential areas, farmland and small towns.

Changes in common bird populations in the Brussels region have been monitored using point–counts (Bibby et al., 2000) since 1992. In practice, 98 point–counts located mostly in green areas throughout the city are sampled twice a year during the breeding season (Weiserbs & Jacob, 2007). Given our aim of focusing on species living in urban habitats, the present analysis did not take into account the 31 points located in the Forest of Soignes. Thus, a total of 67 point–counts was used in this study, each of which was used as a sample unit in subsequent analyses (fig. 1).

Barcelona

Barcelona is located in the western Mediterranean Basin (north–east Spain; fig. 1). It covers 101 km^2 and is dominated by built–up areas, although the Collserola Natural Park in the west of the city is a large natural area. Apart from this site, the network of green areas includes urban parks (mainly small, < 3 ha) scattered among buildings, and private gardens. In total (including Collserola), green spaces cover 36% of the city and its municipal area. Urban parks have a mixture of autochthonous and exotic plants, and many of the city's streets are tree–lined. Trees in public parks and gardens were mainly planted from 1980 onwards. The city of Barcelona itself is at the centre of a highly urbanized metropolitan area covering 636 km^2.

The monitoring of common birds in Barcelona started in 2002. As in Brussels, censuses are conducted twice during the breeding season. The system adopted is the line–transect method (Bibby et al., 2000) and 11 3–km transects are currently conducted, all as part of the Catalan Common Bird Survey (SOCC) that covers the whole of Catalonia (NE Spain). In this study, we did not take into account the two transects located in Collserola Natural Park, nor a transect located in the large urban park of Montjuic for which some degree of spatial overlap occurs. Thus, a total of eight 3–km transects were taken into consideration, each one taken as a sample unit (fig. 1).

Data analysis

We calculated the trends of common species separately for each of the cities using the time–effects model of the TRIM program (Pannekoek & van Strien, 2005). In these analyses at species level, the period taken into account was 2002–2010, the years for which data was available for both monitoring projects. Every species for which the sample size was sufficient was analysed by TRIM (with a minimum presence of 10 point counts in Brussels and four line–transects in Barcelona). However, introduced species (e.g. Red–necked Parakeet *Psittacula krameri* and Monk Parakeet *Myiopsitta monachus*) and feral pigeons (*Columba livia*), whose population dynamics are strongly influenced either by exponential growth at the initial stages of invasion (Crooks, 2005) or by specific management (Sol & Senar, 1995), were not included in the analyses. We also excluded swifts (*Apus apus*, *A. pallidus* and *A. melba*) because sampling bias probably existed (serious mobility and aggregation effects) in the censuses. Given their abundance, swifts could probably be a highly relevant species in an urban context, but a species–specific monitoring scheme would have to be set up if data from these species were to be included in the analyses.

We selected different multi–species urban indicators for each conceptual approach. The first one considered that to advance towards the generation of an urban indicator that would be comparable across European cities, this should minimise the taxonomic variance by containing species that are widespread in European cities (widespread species in European cities, hereafter WSEC). Thus, we used information collected by Kelcey & Rheinwald (2005) in 16 European cities (St. Petersburg, Moscow, Warsaw, Lublin, Sofia, Bratislava, Vienna, Prague, Berlin, Bonn, Hamburg, Brussels, Florence, Rome, Valencia and Lisbon) and assumed that this sample represented the main environmental gradients in European cities. Specifically, we included in this approach all species breeding in at least 14 of these 16 cities, that is, a total of 37 species (table 1). The threshold of 14 instead of the total 16 was chosen to avoid the exclusion of some fairly common species that were not present in the extremes of the ecological gradient represented by this set of cities, mainly in the two cities of the northeast (St. Petersburg and Moscow) or southwest (Valencia and Lisbon). Thus, in this first conceptual approach all species present in a given city on the list could be used to build the multi–species urban indicator, although to be definitively included as part of the indicator in a given city they should be abundant enough to provide reliable information through the monitoring project. The second approach indicates that all widespread urban species in each city should contribute to the index, regardless of how they are distributed in other European cities, thereby maximising urban habitat or ecological coverage in comparisons between cities. In this context, we considered that species present in at least 75% of monitoring plots in a given city during the study period (2002–2010 in our case) could be included as species that are widespread in the

Table 1. Species considered in the widespread species in European cities (WSEC) urban index. This list of bird species was elaborated using the information compiled by Kelcey & Rheinwald (2005) for 16 European cities (St. Petersburg, Moscow, Warsaw, Lublin, Sofia, Bratislava, Vienna, Prague, Berlin, Bonn, Hamburg, Brussels, Florence, Rome, Valencia and Lisbon). Specifically, the list includes 37 species breeding in at least 14 of the 16 cities (see Material and methods). In the cases of House Sparrow *Passer domesticus* and Italian Sparrow *P. italiae*, and Common Starling *Sturnus vulgaris* and Spotless Starling *S. unicolor*, these pairs of species were treated as one because of their very similar ecology and almost non–overlapping distributions.

Tabla 1. Especies consideradas en el indicador urbano de especies de amplia distribución en las ciudades europeas (WSEC). Esta lista de especies de aves fue elaborado utilizando la información recopilada por Kelcey & Rheinwald (2005) para 16 ciudades europeas (San Petersburgo, Moscú, Varsovia, Lublin, Sofía, Bratislava, Viena, Praga, Berlín, Bonn, Hamburgo, Bruselas, Florencia, Roma, Valencia y Lisboa). En concreto, la lista incluye 37 especies que se reproducen en al menos 14 de las 16 ciudades (ver Material y métodos). En el caso de gorrión común Passer domesticus *y el gorrión italiano* P. Italiae *y de los estorninos pintos* Sturnus vulgaris *y negro* S. unicolor, *estos pares de especies fueron tratados como una sola a causa de su ecología muy similar y de que casi no se superponen las distribuciones.*

English name	Scientific name	English name	Scientific name
Mallard	*Anas platyrhynchos*	Swift	*Apus apus*
Blackcap	*Sylvia atricapilla*	Jay	*Garrulus glandarius*
Kestrel	*Falco tinnunculus*	Wryneck	*Jynx torquilla*
Wren	*Troglodytes troglodytes*	Common Starling	*Sturnus vulgaris*
Moorhen	*Gallinula chloropus*	Spotless Starling	*Sturnus unicolor*
Spotted Flycatcher	*Muscicapa striata*	Green Woodpecker	*Picus viridis*
Coot	*Fulica atra*	House Sparrow	*Passer domesticus*
Great Tit	*Parus major*	Italian Sparrow	Passer italiae
Little Ringed Plover	*Charadrius dubius*	Great Spotted Woodpecker	*Dendrocopos major*
Coal Tit	*Periparus ater*	Tree Sparrow	*Passer montanus*
Wood Pigeon	*Columba palumbus*	Swallow	*Hirundo rustica*
Blue Tit	*Cyanistes caeruleus*	Chaffinch	*Fringilla coelebs*
Collared Dove	*Streptopelia decaocto*	House Martin	*Delichon urbica*
Long tailed Tit	*Aegithalos caudatus*	Goldfinch	*Carduelis carduelis*
Turtle Dove	*Streptopelia turtur*	Pied Wagtail	*Motacilla alba*
Nuthatch	*Sitta europaea*	Greenfinch	*Carduelis chloris*
Cuckoo	*Cuculus canorus*	Robin	*Erithacus rubecula*
Red–backed Shrike	*Lanius collurio*	Serin	*Serinus serinus*
Tawny Owl	*Strix aluco*	Blackbird	*Turdus merula*
Magpie	*Pica pica*		

habitats of the city (widespread species in each city, hereafter WSC). This quantitative criterion selected the commoner species; scarcer species, while being potentially interesting urban indicators, are more difficult to monitor properly. For each of these two candidates (WSEC and WSC), we assessed two multi–species urban indicators for Brussels and Barcelona using the procedure developed by Gregory et al. (2005). In this approach, for a particular set of species a multi–species index for a given year can be obtained as the geometrical mean of the species population index obtained by TRIM, while standard errors can be obtained by a Taylor linearization of the nonlinear geometric mean (Gregory et al., 2005). The statistical significance of the changes shown by the indicators was evaluated using 95% confidence intervals (95% CI); if the 95% CI of a given annual value did not include the reference initial value of the temporal

Table 2. Species with large enough sample size to be considered in the analyses of population trends in each city. The species that fitted the criteria to be considered as widespread in European cities (WSEC) or widespread in each city (WSC) and that have been use to build these indicators are marked (see Materials and methods). According to the TRIM results (see Materials and methods), mean annual change (%) and significant decreases and increases over the period 2002–2010 are also marked: moderate decline (↓), moderate increase (↑), stable (–) and uncertain (?). These four trend categories follow the classification reported in Pannekoek & Van Strien (2005), in which 'moderate decrease' and 'moderate increase' correspond to significant trends and 'stable' and 'uncertain' correspond to non–significant trends; species considered 'stable' were those for which their mean annual changes are clearly less than 5% per year, whereas 'uncertain' includes species whose mean annual changes are clearly not less than 5%.

Tabla 2. Especies con tamaño de muestra suficientemente grande como para ser consideradas en el análisis de las tendencias demográficas en cada ciudad. Las especies que se ajustaron a los criterios para ser consideradas como especies de amplia distribución en las ciudades europeas (WSEC) o de amplia distribución en cada ciudad (WSC) están marcadas (ver Material y métodos). De acuerdo con los resultados TRIM (ver Material y métodos), la variación promedio anual (%) y las disminuciones y los incrementos significativos durante el período 2002–2010 también están marcados: disminución moderada (↓), incremento moderado (↑), estable (–) e incierto (?). Estas cuatro categorías de tendencia siguen la clasificación mostrada en Pannekoek & Van Strien (2005), en las cuales 'disminución moderada' e 'Incremento moderado' corresponden a tendencias significativas y 'estable' e 'Incierto' corresponden a no significativas, siendo consideradas 'estable' aquellas especies para las cuales su tasa promedio de cambio es con certeza menos del 5% anual, mientras que las que tienen la categoría de 'incierto' hacen referencia a aquellas en las que su tasa promedio de cambio anual no es seguro que sea menor del 5%.

English name	Scientific name	Brussels		Barcelona	
		Trend	Indicator	Trend	Indicator
Stock Dove	*Columba oenas*	–10%,↓	WSC		
Wood Pigeon	*Columba palumbus*	0%,–	WSC,WSEC		
Collared Dove	*Streptopelia decaocto*	–8%,↓	WSC,WSEC	+9%,?	WSC,WSEC
Green Woodpecker	*Picus viridis*	–5%,?	WSC,WSEC		
Great Spotted Woodpecker	*Dendrocopos major*	–2%,?	WSC,WSEC		
Swallow	*Hirundo rustica*			–4%,?	WSC,WSEC
Pied Wagtail	*Motacilla alba*	+7%,?	WSEC	–4%,?	WSC,WSEC
Dunnock	*Prunella modularis*	+4%,?	WSC		
Robin	*Erithacus rubecula*	–2%,?	WSC,WSEC	+7%,?	WSC,WSEC
Song Thrush	*Turdus philomelos*	–2%,?	WSC		
Blackbird	*Turdus merula*	–2%,↓	WSC,WSEC	+1%,?	WSC,WSEC
Garden Warbler	*Sylvia borin*	–11%,?			
Blackcap	*Sylvia atricapilla*	–3%,↓	WSC,WSEC		
Sardinian Warbler	*Sylvia melanocephala*			+5%,?	WSC
Willow Warbler	*Phylloscopus trochilus*	–5%,?			
Chiffchaff	*Phylloscopus collybita*	–6%,↓	WSC		
Wren	*Troglodytes troglodytes*	–2%,↓	WSC,WSEC		
Spotted Flycatcher	*Muscicapa striata*			+6%,?	WSEC
Great Tit	*Parus major*	+2%,–	WSC,WSEC	+19%,?	WSC,WSEC
Blue Tit	*Cyanistes caeruleus*	–2%,–	WSC,WSEC	+30%,↑	WSC,WSEC
Marsh Tit	*Poecile palustris*	–4%,?			
Long–tailed Tit	*Aegithalos caudatus*	+3%,?	WSEC	–15%,?	WSEC
Nuthatch	*Sitta europaea*	–1%,?	WSEC		

Table 2. (Cont.)

English name	Scientific name	Brussels		Barcelona	
		Trend	Indicator	Trend	Indicator
Short–toed Treecreeper	*Certhia brachydactyla*	−5%,↓	WSC		
Magpie	*Pica pica*	−3%,↓	WSC,WSEC	+10%,?	WSC,WSEC
Jay	*Garrulus glandarius*	+2%,?	WSC,WSEC		
Jackdaw	*Corvus monedula*	+11%,↑		+2%,?	
Carrion Crow	*Corvus corone*	+4%,↑	WSC		
Starling	*Sturnus vulgaris*	−8%,↓	WSC,WSEC	+8%,↑	WSC,WSEC
House Sparrow	*Passer domesticus*	+8%,↑	WSEC	−5%,↓	WSC,WSEC
Chaffinch	*Fringilla coelebs*	+8%,↑	WSEC		
Goldcrest	*Regulus regulus*	−8%,?			
Goldfinch	*Carduelis carduelis*			−7%,?	WSC,WSEC
Greenfinch	*Carduelis chloris*	+15%,↑	WSEC	−3%,?	WSC,WSEC
Serin	*Serinus serinus*			+5%,?	WSC,WSEC

series, then these two values were considered to be significantly different (see Pannekoek & Van Strien, 2005 for the same approach at species level). Finally, we assessed a magnitude of yearly average change in the indicators (WSC and WSEC) by calculating the parameter (slope) in the simple regression model between the yearly value of the indicator (dependent variable) and time (predictor).

Results

During the study–period a total of 84 native breeding species were recorded in Brussels and 76 in Barcelona. Only for some of these species (30 in Brussels and 17 in Barcelona), was sample size considered sufficient (see Material and methods) to run TRIM over the period 2002–2010 (table 2). Species trends in Brussels showed that in the period 2002–2010, seven species (28%) decreased significantly, two (8%) were stable, and five (20%) increased significantly, whereas in Barcelona, where most species' trends were non–significant, only one species (6%) decreased and two (12%) increased significantly (table 2).

We compared the two approaches to develop urban indicators (WCS and WSEC), which varied according to the species included in each case (table 2). The two indicators gave similar temporal patterns for each of the cities (fig. 2). Overall, the change was non–significant over the study period in both cities, although there was a slight increase in Barcelona (5% annual increase for WSC and 2% for WSEC), while the indicators for Brussels showed a slight decrease or remained stable (3% annual decrease for WSC and 0% for WSEC) over the study period (fig. 2).

Discussion

The development of a reliable, urban multi–species indicator based on bird monitoring data is not a simple task. Starting with data gathering, urban habitats are often under–represented in large–scale monitoring schemes since they are less interesting for ornithologists than more natural areas (*e.g.* Saris et al., 2004; McCaffrey, 2005; but see also Ferrer et al., 2006). This is partially compensated for by the efforts of some local councils, as in Barcelona and Brussels. Nevertheless, monitoring schemes specifically designed for cities have to cope with relatively low sample sizes compared to whole regions or countries, and this often limits the number of species in the data set to just a few dozen (see table 2 for the studied cities). This small set of species could grow if the survey efforts (either in common bird censuses or in species specific schemes) and/or the number of species adapted to such artificial environment increases over time. Hence, in a few years' time the number of available species to generate an urban indicator may also increase, and so it would be useful to establish procedures that describe when and how such species should be included in the indicators, and what the consequences will be in relation to the results of former indices.

Within this context, the selection of a group of bird species to provide better information on changes in urban biodiversity is also hampered by the definition of the urban ecology of the species, above all if we consider that an important component of urban variability depends on the avifauna in surrounding habitats (Sattler et al., 2010). Even within Europe, the number of urban adapters varies from one city

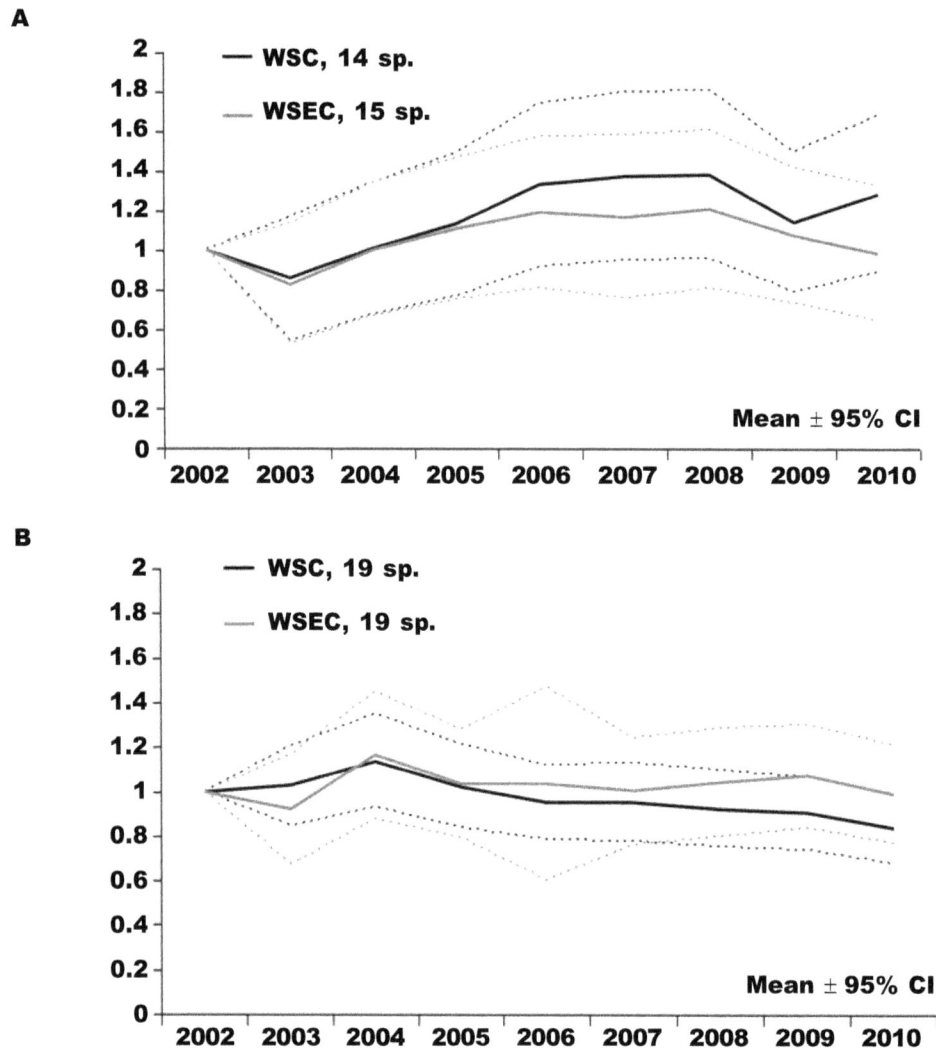

Fig. 2. Changes revealed by the two different candidates for an urban indicator (WSC and WSEC) during the study period in Barcelona (A) and Brussels (B). (For abbreviations see material and methods.)

Fig. 2. Cambios mostrados por los dos distintos candidatos a indicador urbano (WSC y WSEC) durante el periodo de estudio en Barcelona (A) y Bruselas (B). (Para las abreviaturas ver Material y métodos.)

to another (Kelcey & Rheinwald, 2005) and gradually increases as additional species invade and adapt to urban areas (Rutz, 2008; Evans et al., 2009). Therefore, in this study we focused fundamentally on the urban character of the study sites rather than that of the bird species, thereby rejecting non–urban sites and focusing on urban sites, mainly consisting of built–up areas and city parks. This approach is different from that of other Pan–European indicators such as the Farmland Bird Index (Voříšek et al., 2008) that uses species lists whatever the habitat that the monitoring data is collected in.

In this study we used data from monitoring projects carried out in Barcelona and Brussels to derive two urban multi–species indicators that could potentially

be applied in other European cities. We took into account the fact that inter–city comparisons could be maximized either at species level (using lists of bird species that are as similar as possible to minimize taxonomic variation) or at ecological level (regardless of the number of species shared among cities and trying to maximize the information provided by birds on the state of their habitats in each city). In the first approach, we used a set of species that are widespread in the 16 European cities cited in Kelcey & Rheinwald (2005). However, it could be argued that these cities are not totally representative of the overall European urban avifauna since 50% of them are located in central Europe, and there are, for example, few western, southern and northern European cities. Thus, although we considered that this was probably among the best sources of information, this potential weakness should be taken into account in future studies. The second approach did not present such limitations because it was city–specific, but both approaches had a subjective threshold for a given species to be included in the indicator (present in at least 75% of monitoring plots in a particular city, or species breeding in at least 14 of the 16 cities), and hence these criteria would also deserve further investigation.

Although several important issues on conservation rely on the trends of a particular species (*e.g.* threatened species), multi–species indicators better capture ecosystem complexity than indicators based on one or a few species (Buckland et al., 2005; Gregory et al., 2005; Butler et al., 2012). In our case, the analysis by species gave relatively little information and statistically significant trends were only obtained for a small number of species, especially in Barcelona. This could be caused, in part, by the short time framework, as illustrated by the fact that in Brussels an analysis including the 10 previous years of sampling provided more significant results at species level (Weiserbs, 2010). Nevertheless, trends in the multi–species indicators generated in this study seem to be more robust than the individual species trends. Overall, the values shown by the indicators did not change significantly over the period 2002–2010 in either of the two cities, although there was a slight non–significant increase in Barcelona and the indicators for Brussels showed a slight non–significant decrease or remained stable (fig. 2). Regardless of the city, the pattern revealed by the two indicators (WSC and WSEC) was relatively similar. Nevertheless, the detected 3% difference in the overall trend could be considered relevant and reveals the importance of the species–selection procedure and the criteria used. The WSEC indicator shows performance of European species that are widespread in urban environments at a continental scale, whereas the WSC focuses on the species of a particular city. Thus, the two types of indicators presented in this study give different messages. We consider that both indicators have their value and should be treated as complementary indicators rather than competing indicators. Nevertheless, these indicators do not shed light on their respective accuracies with respect to what they are expected to indicate, and more studies are needed to analyse the relation between these patterns and other independent sources of information about the state of the environment (*i.e.* revealing relationships between indicators and environmental predictors relevant for population dynamics).

Further studies are also obviously needed if we are to define a set of the most suitable species for creating a multi–species urban indicator, and collaboration between European cities will be crucial if this is to be to achieved. Indeed, this may eventually result in the generation of biodiversity indicators not only for specific cities, but also for all urban areas in a country or, even, in a whole continent.

Acknowledgements

The data analysed in this study was obtained by volunteer ornithologists, without whom bird monitoring in Barcelona and Brussels would not exist. The Common Bird Monitoring Scheme carried out in Brussels is coordinated by Aves–Natagora within the Monitoring Programme of the State of the Environment run by Bruxelles Environnement–IBGE. The Common Bird Monitoring in Barcelona is coordinated by the Catalan Ornithological Institute and run by Barcelona City Council and the University of Barcelona; this programme is integrated into the Catalan Common Bird Survey (SOCC), which is run by the Catalan Government. We would like to thank all these institutions for their continued support. We would like to thank Petr Voříšek and three anonymous reviewers for their interesting comments on earlier drafts of this manuscript. JQ and SH received partial funding from the BIOCAT–BB CGL2009–08798 and BIONOVEL CGL2011–29539 projects from the Spanish Science and Innovation Secretariat. XF received partial funding from Barcelona Zoo and advice from Daniel Sol.

References

Adams, C. E., Linsdsey, K. J. & Ash, S. J., 2006. *Urban wildlife management.* Taylor & Francis, Boca Raton.

Bibby, C. J., Burgess, N. D., Hill, D. A. & Mustoe, S. H., 2000. *Bird Census Techniques.* Elsevier, London.

Blair, R. B., 1996. Land use and avian species diversity along an urban gradient. *Ecol. Appl.,* 6: 506–519.

– 1999. Birds and butterflies along an urban gradient: surrogate taxa for assessing biodiversity? *Ecol. Appl.,* 9: 164–170.

Buckland, S. T., Magurran, A. E., Green, R. E. & Fewster, R. M., 2005. Monitoring change in biodiversity through composite indices. *Phil. Trans. R. Soc. B.,* 360: 243–254.

Butler, S. J., Freckleton, R. P., Renwick, A. R. & Norris, K., 2012. An objective, niche–based approach to indicator species selection. *Methods in Ecology and Evolution,* 3(2): 317–326.

Caula, S. A., Sirami, C., Marty, P. & Martin, J.–L., 2010. Value of an urban habitat for the native Mediterranean avifauna. *Urban Ecosystems,* 13: 73–89.

Clergeau, P., Croci, S., Jokimäki, J., Kaisanlahti–Jokimäki, M.–L. & Dinetti, M., 2006. Avifauna homogenisation by urbanisation: Analysis at different European latitudes. *Biological Conservation,* 127: 336–344.

Crooks, J. A., 2005. Lag times and exotic species: the ecology and management of biological invasions in slow–motion. *Ecoscience,* 12: 316–329.

DEFRA, 2002. *Working with the Grain of Nature.* DEFRA Publications, London.

Devictor, V., Julliard, R., Couvet, D., Lee, A. & Jiguet, F., 2007. Functional homogenization effect of urbanization on bird communities. *Conservation Biology,* 21(3): 741–751.

Evans, K. L., Gaston, K. J., Frantz, A. C., Simeoni, M., Sharp, S. P., McGowan, A., Dawson, D. A., Walasz, K., Partecke, J., Burke, T. & Hatchwell, B. J., 2009. Independent colonization of multiple urban centres by a formerly forest specialist bird species *Proc. R. Soc. B,* 276: 2403–2410.

Ferrer, X., Carrascal, L., Gordo, O. & Pino, J., 2006. Bias in avian sampling effort due to human preferences: an analysis with Catalonian birds (1900–2002). *Ardeola,* 53(2): 213–227.

Fuller, R. A., Tratalos, J. & Gaston, K. J., 2009, How many birds are there in a city of half a million people? *Diversity and Distributions,* 15: 328–337.

Gagné, S. A., & Fahrig, L., 2011. Do birds and beetles show similar responses to urbanization? *Ecol. Appl.,* 21, 2297–2312.

Gregory, R. D., van Strien, A., Voříšek, P., Gmelig Meyling, A., Noble, D., Foppen, R. & Gibbons, D., 2005. Developing indicators for European birds. *Phil. Trans. R. Soc. B,* 360: 269–288.

Kelcey, J. & Rheinwald, G., 2005. *Birds in European Cities.* Ginster Verlag, St. Katharinen.

McCaffrey, R. E., 2005. Using Citizen Science in Urban Bird Studies. *Urban Habitats,* 3(1): 70–86.

Møller, A. P., 2009. Successful city dwellers: a comparative study of the ecological characteristics of urban birds in the Western Palearctic. *Oecologia,* 159: 849–858.

Pannekoek, J. & Van Strien, A., 2005. *TRIM 3 Manual (Trends & Indices for Monitoring data).* Statistics Netherlands, Voorburg.

PECBMS, 2011. *Population Trends of Common European Breeding Birds 2011.* CSO, Prague.

Robinson, R. A. & Sutherland, W. J., 2002. Post–war changes in arable farming and biodiversity in Great Britain. *Journal of Applied Ecology,* 39: 157–176.

Rutz, C., 2008. The establishment of an urban bird population. *J. Anim. Ecol.,* 77: 1008–1019.

Saris, F., Hustings, F., Hagemeijer, W., Van Dijk, A., Sierdsema, H. & Verstrael, T., 2004. The Dutch breeding bird monitoring scheme: Evaluation, new objectives and its merits for conservation. *Bird Census News,* 13: 113–121.

Sattler, T., Borcard, D., Arlettaz, R., Bontadina, F., Legendre, P., Obrist, M. K. & Moretti, M., 2010. Spider, bee, and bird communities in cities are shaped by environmental control and high stochasticity. *Ecology,* 91(11): 3343–3353.

Savard, J.–P. L., Clergeau, P. & Mennechez, G., 2000. Biodiversity concepts and urban ecosystems. *Landscape and Urban Planning,* 48: 131–142.

Schmeller, D. S., 2008. European species and habitat monitoring: where are we now? *Biodiversity and Conservation,* 17: 3321–3326.

SEO/BirdLife, 2010. *Estado de conservación de las aves en España en 2010.* SEO/BirdLife, Madrid (in Spanish).

Sol, D. & Senar, J. C., 1995. Urban pigeon populations: stability, home range, and the effect of removing individuals. *Can. J. Zool.,* 73: 1154–1160.

Voříšek, P., Klvanová, A., Wotton, S. & Gregory, R. D. (Eds.), 2008. *A best practice guide for wild bird monitoring schemes.* CSO/RSPB, Trebon.

Weiserbs, A. & Jacob, J. P., 2007. Analyse des résultats 1992–2005 de la surveillance des oiseaux nicheurs «communs» dans la Région de Bruxelles–Capitale. *Aves,* 44: 65–78 (in French).

Weiserbs, A., 2010. *Oiseaux communs de Bruxelles – Cartographie des tendances. Oiseaux de Bruxelles n°2.* Aves, Liège (in French).

Zbinden, N., Schmid, H., Kéry, M. & Keller, V., 2005. Swiss Bird Index SBI–Kombinierte Indices für die Bestandsentwicklung von Artengruppen regelmässig brütender Vogelarten der Schweiz 1990–2004. *Der Ornithologische Beobachter,* 102: 283–291 (in German).

3

Effects of migratory status and habitat on the prevalence and intensity of infection by haemoparasites in passerines in eastern Spain

J. Rivera, E. Barba, A. Mestre, J. Rueda, M. Sasa, P. Vera & J. S. Monrós

Rivera, J., Barba, E., Mestre, A., Rueda, J., Sasa, M., Vera, P. & Monrós, J. S., 2013. Effects of migratory status and habitat on the prevalence and intensity of infection by haemoparasites in passerines in eastern Spain. *Animal Biodiversity and Conservation*, 36.1: 113–121.

Abstract

Effects of migratory status and habitat on the prevalence and intensity of infection by haemoparasites in passerines in eastern Spain.— The Iberian peninsula is a suitable place to study the effects of migratory condition on the prevalence of blood parasites in avian communities as resident, local populations cohabit with migratory species and with abundant vector populations. In this study we examined the incidence of avian blood parasites in three localities in the Mediterranean region (east Spain), in relation to the migratory status of the species. We analyzed 333 blood smears from 11 avian species, and obtained an overall prevalence of 9.6%. The prevalence of parasites varied among the different species studied, although intensity of infection did not. Our results are discussed in terms of population dynamics and abundance of Diptera vectors able to transmit blood parasites to other birds.

Key words: Blood parasites, *Trypanosoma* ssp., *Haemoproteus* spp., Passeriformes, Diptera vectors.

Resumen

Efectos del estatus migratorio y del tipo de hábitat sobre la prevalencia y la intensidad de la infección por hemoparásitos en paseriformes en el este de España.— La península ibérica es un sitio idóneo para estudiar los efectos de la condición migratoria en la prevalencia de hemoparásitos en comunidades de aves, dado que convergen poblaciones residentes locales con especies migratorias y abundantes poblaciones de vectores. En este trabajo examinamos la incidencia de hemoparásitos presentes en aves de tres localidades de la región mediterránea (este de España), con respecto del estatus migratorio. Examinamos 333 frotis sanguíneos de 11 especies, y encontramos una prevalencia global del 9,6%. A diferencia de la intensidad de la infección, la prevalencia de parásitos mostró variación entre las distintas especies estudiadas. Nuestros resultados se interpretan en relación con la dinámica de poblaciones y la abundancia de dípteros vectores capaces de transmitir los hemoparásitos a otras aves.

Palabras clave: Hemoparásitos, *Trypanosoma* spp., *Haemoproteus* spp., Paseriformes, Dípteros vectores.

Jennifer Rivera, Mahmood Sasa, Inst. 'Clodomiro Picado', Univ. de Costa Rica, San José, Costa Rica.– Emilio Barba, Pablo Vera & Juan S. Monrós, Inst. 'Cavanilles' of Biodiversidad y Biología Evolutiva, Univ. de Valencia, AC 22085, E–46071 Valencia, España (Spain).– Alexandre Mestre & Juan Rueda, Dept. de Microbiología y Ecología, Univ. de Valencia, c/ Dr. Moliner 50, E–46100 Burjassot, España (Spain).

Corresponding author: J. S. Monrós. E–mail: monros@uv.es

Introduction

Avian hematozoa parasites (Protista) are a heterogeneous group of organisms widely distributed worldwide (Peirce, 1981; Valkiūnas, 2005). Atkinson & Van Riper (1991) noted that haemoparasites have been recorded in almost 70% of the avian species examined, although prevalence estimates may depend on the method used in their detection (Fallon et al., 2005).

Parasites from the genus *Haemoproteus* are among the most common avian haematozoa (two thirds of the described blood parasite morphospecies, Valkiūnas, 2005). Parasites of the genus *Leucocytozoon, Plasmodium* (Bennett et al., 1993) and some *Trypanosoma* species (Kučera, 1982) are also common in avian species. These haemoparasites exert selective pressure on their hosts (Hamilton & Zuck, 1982), negatively affecting the efficiency of metabolism (Chen et al., 2001), survival, breeding success, and physical aptitude (Marzal et al., 2008; Stjernman et al., 2008; Ruiz de Castañeda et al., 2009; Martínez de la Puente et al., 2010), and body growth (Soler et al., 2003).

The incidence of haemoparasites in avian communities varies geographically (Sol et al., 2000). This variation has been linked to habitat characteristics, species composition in the community, vector–host specificity and ecological requirements of the vectors (Deviche et al., 2005). Prevalence and intensity of parasitic infections on birds may also depend on the migratory status of the host species. The probability of being infected would thus be higher in migratory species than in sedentary species, as they are exposed to more than one parasitic fauna during their life cycle (Figuerola & Green, 2000). Migration may also limit the transmission of parasites to new host species, due to the vector–host specificity (Hellgren et al., 2008).

Habitat features affect the incidence of infections in birds (Martínez–Abraín et al., 2004) due to differences in vector abundance and behavior (Bennett et al., 1982). The incidence of parasitemia can thus be expected to be lower in semi–arid regions (Little & Earlé, 1995) than in humid regions with aquatic environments (Moyer et al., 2002).

A latitudinal gradient related to climatic conditions and their effect on vectors could be involved in the prevalence of blood parasites in birds (Bensch & Åkesson, 2003). Several studies carried out in the north and center of Europe, where seasonal climatic changes are severe, found that the prevalence of haemoparasites was relatively high (*i.e.* Kučera, 1981; Valkiūnas et al., 2003; Shurulinkov & Golemansky, 2003), but other studies found a higher prevalence in the south (Marzal el al., 2011). Thus, in southern Europe no clear pattern in the prevalence of blood parasites has been observed (*i.e.* Merino et al., 1997; Valera et al., 2003). Depending on latitude, the Iberian Peninsula shows peculiar climatic characteristics that make it suitable to host high numbers of migratory birds during the winter (Tellería, 1988). Mild peninsular winters thus provide a great variety of resources both for short distance migrants and resident bird populations (Senar & Borras, 2004).

In Spain, a number of significant studies have been carried out to describe and understand the patterns of haemoparasite infections in birds (Merino et al., 1997; Tomás et al., 2007). Studies in the center of the country have shown that the higher the vector abundance, the higher the haemoparasitic prevalence (Merino & Potti, 1995). Preliminary surveys on the Mediterranean coast showed that haemoparasites are almost absent in the Passeriformes species (*Parus major, Periparus ater, Lophophanes cristatus*; E. Barba, non–published data). Similarly, blood parasites were absent in nigh–jars, *Caprimulgus ruficollis* (Forero et al., 1997) and in storks *Ciconia ciconia* (Jovani et al., 2002) in Doñana National Park, a patchy region with wetland and Mediterranean forests. Absence of blood parasites was also noted in Kentish plover and gulls breeding on the Mediterranean coast of Spain (Figuerola et al., 1996; Martínez–Abraín et al., 2002). However, a high prevalence of infection by blood parasites has been found in both migratory and resident species in the south of the Iberian peninsula (Marzal et al., 2008, 2011; López et al., 2011). Most parasitic infections are transmitted by Diptera (Ceratopogonidae, Culicidae and Simuliidae) and the abundance of these vectors depends on the local climate and water conditions in each season (Valkiūnas et al., 2003).

The present study aimed to determine the haemoparasitic infection prevalence and intensity in Passeriformes in three localities in Eastern Spain, to analyze the differences in prevalence and intensity of infection between resident and wintering species, and to relate these measures with the presence of vectors that are potential transmitters of the parasitic infections.

Material and methods

Birds included in this study were trapped between September and December 2008 in three localities close to the Mediterranean coast in eastern Spain (fig. 1). The first locality was the Marjal Pego–Oliva Natural Park (Pego–Oliva; 38° 52' N, 0° 3' W), on the border between the provinces of Valencia and Alicante. The birds were trapped in a wetland with large reed bed areas with mixed patches of cattails and sedge, next to rice fields. The second study site was an orange grove (*Citrus sinensis*) in Sagunto, province of Valencia (Sagunto; 39° 42' N, 0° 15' W), 4 km from the coast. The third locality was L'Albufera Natural Park (Albufera; 39° 19' N, 0° 21' W), a wetland in the south of the Valencia city, dominated by rice fields and some patches of marshland natural vegetation.

In the three study areas, birds were trapped using mist–nets, operating weekly as part of the constant effort ringing programs. In all three sites, 60 m of mist–nets were set at dawn and were operated for 4 hours, following Belda et al. (2007). Each bird was banded with an individual metal ring. Each species was catalogued as resident (species present throughout the year) or wintering (migratory species that winter but do not breed in the study area).

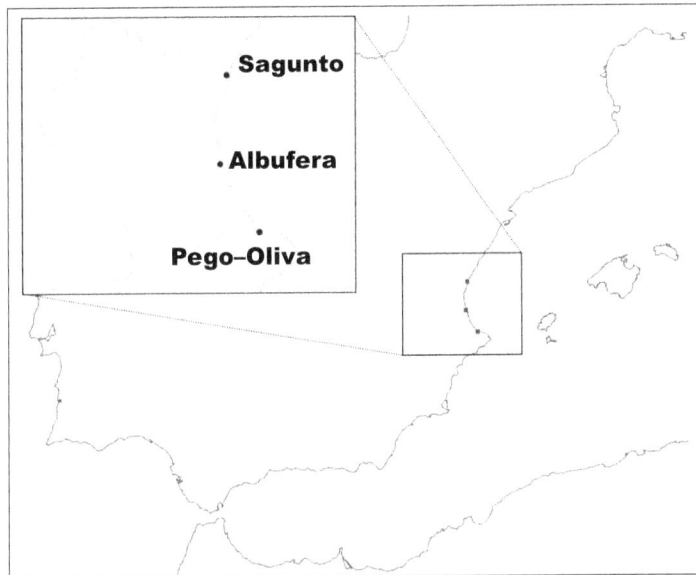

Fig. 1. Location of the study localities in the Iberian peninsula.

Fig. 1. Situación geográfica de las localidades en la península ibérica.

We extracted a drop of blood from the brachial vein of each trapped bird. The drop was placed on a glass slide and dried air. In the laboratory, samples were fixed with absolute methanol and dyed with Giemsa for 45 minutes, following the protocol of Merino et al. (1997). We randomly chose one half of the slide and quantified at 400x the presence of extracellular parasites (*Trypanosoma* spp.) or intracellular parasites (*Leucocytozoon* spp.) along the longitudinal axis. The number of haemotozoa observed in 100 optic fields was recorded. Infection intensity by intracellular parasites (*Haemoproteus* spp. o *Plasmodium* spp.) was obtained as the number of parasites per 2,000 erythrocytes, following Merino & Potti (1995). All the slides were revised by J. R. Parasite identification was based in the morphological characteristics after Valkiūnas (2005).

Complementarily to the bird sampling, water samples were collected in the three study areas to determine the composition of the potential vector community (related to haemoparasite transmission) in different water bodies (such as irrigation ponds, natural springs, and channels). We sampled 52 water bodies, 39 in natural habitats and 13 in artificial ponds. The sampling was performed using a hand net with a square frame of 25 cm per side and a net with pore diameter of 250 µm Each sub–sample was concentrated on a 30 x 40 cm plastic plate. Sampling was concluded when no new taxa were found in the sub–sample. The whole sample (as the set of sub–samples) was stored in a plastic 1 l bottle in 70° alcohol. In the laboratory, samples were washed with water in a 250 µm pore diameter sieve to remove the silt. Species were identified following Tachet et al. (2000) and Rueda & López (2003) using a Motic Digital Microscope DM 143 stereoscopic microscope and a Bresser TrinoLab 40–1,600x microscope.

The prevalence and infection intensity were analyzed at two levels: migratory status and habitat (locality). We analyzed the differences between groups using mixed generalized linear models (GLMMs) fitted by Laplace approximation. Two analyses were made for both *Trypanosoma* spp. and *Haemoproteus* spp. using locality as a fixed factor in the first analysis and migratory status in the second, and individual and species as random factor in both analyses. The individual–level represents a per–observation error term, which captures over–dispersion (Elston et al., 2001; Atkins et al., 2013). We were unable to analyse the two effects together because of zero inflation in the results; and one for *Trypanosoma* sp. only with data obtained in Chiffchaffs (*Phylloscopus collybita*) as it was the only species that was trapped in the three localities (using locality as fixed factor) (Brew & Maddy, 1995). We used binomial models with a logit link function and Poisson models with a logarithmic link function. All tests were performed using the *lme4* package v.0.999375–42 (Bates et al., 2012) for R version 2.14.1 (R Development Core Team, 2009).

Results

A total of 333 birds were trapped, belonging to 11 species and five families. In all three localities, both migratory and sedentary birds were trapped and

Table 1. Infection status for the individuals of the 11 species included in this study: ITry. Infected by *Trypanosoma* spp.; IHae. Infected by *Haemoproteus* spp. Migratory status and locality of capture are given for each species.

Tabla 1. Situación de los individuos de las 11 especies incluidas en el presente estudio respecto de la infección: ITry. Infectado por Trypanosoma *spp.; IHae. Infectado por* Haemoproteus *spp. Para cada especie se indican el estatus migratorio y la localidad de captura.*

| | Number of birds | | | Migratory | |
| | Sampled | Infected by | | status | Locality |
		ITry	IHae		
Sylviidae					
Acrocephalus melanopogon	30	3(10)	0(0)	Resident	Pego–Oliva
Cettia cetti	14	0(0)	0(0)	Resident	Pego–Oliva
Phylloscopus collybita	91	5(5.5)	0(0)	Wintering	Pego–Oliva, Sagunto, Albufera
Sylvia atricapilla	30	1(3.3)	21(70)	Wintering	Sagunto
Sylvia melanocephala	21	1(4.8)	0(0)	Resident	Sagunto
Emberizidae					
Emberiza schoeniclus	45	0(0)	0(0)	Wintering	Albufera
Passeridae					
Passer domesticus	11	0(0)	0(0)	Resident	Albufera
Passer montanus	13	0(0)	0(0)	Resident	Albufera
Turdidae					
Turdus merula	20	0(0)	0(0)	Resident	Sagunto
Erithacus rubecula	30	0(0)	2(6.7)	Wintering	Sagunto
Fringillidae					
Fringilla coelebs	28	0(0)	0(0)	Wintering	Albufera
Total	333	10(3)	23(6.9)		

sampled (table 1). Only 32 birds were infected, with a global prevalence of 9.6% (table 1). The most common parasite was *Haemoproteus* spp., which was identified in 23 birds (6.9%). *Trypanosoma* spp. was detected in 10 birds (3.0%). Only one individual of Blackcap *Sylvia atricapilla* showed both parasites (0.3%). No individual showed infection by *Plasmodium* spp.

The prevalence of infected birds differed between species (χ^2 = 216.5, p < 0.001, df = 10, table 1). Haemoparasites were not detected in six species (Cettia's Warbler *Cettia cetti*, Reed Bunting *Emberiza schoeniclus*, Chaffinch *Fringilla coelebs*, House Sparrow *Passer domesticus*, Tree Sparrow *Passer montanus* and Blackbird *Turdus merula;* table 1). Prevalence did not correlate with the number of samples collected for each species (Spearman's rho; *Trypanosoma* spp.: r = 0.549, p = 0.080; *Haemoproteus* spp.: r = –0.299, p = 0.371). The species that showed the highest abundance of parasites and the highest proportion of infected individuals (n = 22) was the Blackcap, a species that winters in this area, mainly in shrubs and croplands rather than wetlands.

Trypanosoma spp. infections were detected in Sardinian Warbler *Sylvia melanocephala* (n = 1), Chiffchaff (n = 5), Blackcap (n = 1) and Moustached Warbler *Acrocephalus melanopogon* (n = 3), although in an overall analysis no significant differences were found in the infection prevalence between species (χ^2 = 12.20; p = 0.27; df = 10). Taking Pego–Oliva as reference locality level to calculate the estimators of locality effects, we did not find any statistical differences between localities (Wald χ^2 < 0.001; p > 0.999; df = 2), or between migratory status (Wald χ^2 < 0.001; p = 0.988; df = 1) (see the lower coefficient values of each level compared with the higher SE values showed in table 2). In a partial analysis with data collected on Chiffchaffs (the only species found in the three sampling localities), we did not find differences in the prevalence of *Trypanosoma* spp. between localities (Wald χ^2 < 0.001; p > 0.999; df = 2).

Table 2. Results of the *GLMMs* used to analyze the effects of locality and migratory status on the prevalence of *Trypanosoma* spp. and *Haemoproteus* spp.

Tabla 2. Resultados de los modelos lineales generalizados mixtos utilizados para analizar los efectos de la localidad y el estatus migratorio en la prevalencia de Trypanosoma *spp. y* Haemoproteus *spp.*

		Effect	Estimate	SE	Z	p
Trypanosoma spp.	Locality	Intercept	−14.28	28.75	−0.497	0.619
		Sagunto	−0.14	37.05	0.004	0.997
		Albufera	−17.53	$6.3 \cdot 10^6$	< 0.001	> 0.999
		Pego–Oliva	0.00	−	−	−
	Migratory status	Intercept	−15.14	24.97	−0.606	0.544
		Resident	0.60	39.30	0.015	0.988
		Wintering	0.00	−	−	−
Haemoproteus spp.	Locality	Intercept	−21.80	$6.7 \cdot 10^3$	−0.003	0.997
		Sagunto	17.79	$6.7 \cdot 10^3$	0.003	0.998
		Albufera	−0.000001	$8.7 \cdot 10^3$	< 0.001	> 0.999
		Pego–Oliva	0.00	−	−	−
	Migratory status	Intercept	−3.99	1.60	−2.497	0.012
		Resident	−17.40	$4.2 \cdot 10^3$	−0.004	0.997
		Wintering	0.00	−	−	−

Haemoproteus spp. infections were found only in Blackcaps ($n = 21$) and European Robin *Erithacus rubecula* ($n = 2$), and the prevalence differed between all the species ($\chi^2 = 205.99$; $p < 0.001$; df = 10). A *posteriori* test showed that the differences were due to Blackcaps ($\chi^2 = 18.30$; $p = 0.05$; df = 10). Again taking Pego–Oliva as reference locality level to calculate the estimators of locality effects, we did not find statistical differences between localities (Wald $\chi^2 < 0.001$; $p > 0.999$; df = 2) or between migratory status (Wald $\chi^2 < 0.001$; $p = 0.997$; df = 1) (table 2).

Taking Pego–Oliva as reference locality level to calculate the estimators of locality effects, we did not find statistical differences in the intensity of *Trypanosoma* parasitism between localities (Wald $\chi^2 < 0.001$; $p > 0.999$; df = 2), or between migratory status (Wald $\chi^2 = 0.006$; $p = 0.936$; df = 1) (see coefficient values of each level compared with the higher SE values in table 3). Neither did we find differences in the infection intensity between localities when we considered only the data collected on Chiffchaffs (Wald $\chi^2 = 0.169$; $p = 0.919$; df = 2). In the case of *Haemoproteus* infection, intensity results were similar, showing no differences between localities (Wald $\chi^2 < 0.001$; $p > 0.999$; df = 2), or migratory status (Wald $\chi^2 < 0.001$; $p > 0.999$; df = 1) (table 3).

Table 4 shows the composition of the community of Diptera species potentially acting as a vector for haemoparasite transmission in the three sampling areas. The Pego–Oliva area had the highest richness (13 species and eight genera). In Sagunto, only two species (from two different genera) were detected. Unfortunately, our sampling strategy did not allow comparison of species abundance between species or localities.

Discussion

The intensity and prevalence of infection caused by *Trypanosoma* spp. did not differ between species, migratory status, or locality, as shown previously in studies carried out in the center of Spain (Merino et al., 1997) and north of Europe (Hauptmanova et al., 2006). These results may be attributed to several factors: i) low frequency of individuals infected with this parasite (*i.e.* only five individuals of 91 Chiffchaffs sampled showed infection by *Trypanosoma* spp.); ii) problems with the methodology used for the detection could increase the number of false negatives; Apanius (1991) showed that *Trypanosomas* are not commonly found in peripheral blood, but are abundant in the bone marrow of the infected bird; and iii) as sampling was done in autumn, birds may have had low intensity of infection as they successfully passed the peak period of parasitic infection and thus present residual infection rates (Pérez–Tris & Bensch, 2005; Arizaga et al., 2009).

Haemoproteus spp. was the most prevalent infection, with values similar to those reported in other

Table 3. Results of the GLMMs used to analyze the effects of locality and migratory status on the infection intensity of *Trypanosoma* spp. and *Haemoproteus* spp.: PML. Parasite mean load.

Tabla 3. Resultados de los modelos lineales generalizados mixtos utilizados para analizar los efectos de la localidad y el estatus migratorio en la intensidad de la infección causada por Trypanosoma *spp. y* Haemoproteus *spp.: PML. Carga parasitaria media.*

		Effect	Estimate	SE	Z	p	PML
Trypanosoma spp.	Locality	Intercept	−8.18	3.22	−2.54	0.011	−
		Sagunto	−0.006	4.16	−0.001	0.999	0.04
		Albufera	−18.03	$4.5 \cdot 10^4$	< 0.001	> 0.999	0.00
		Pego–Oliva	0.00	−	−	−	0.05
	Migratory status	Intercept	−9.06	2.93	−3.09	0.002	−
		Resident	0.37	4.62	0.080	0.936	0.04
		Wintering	0.00	−	−	−	0.02
Haemoproteus spp.	Locality	Intercept	−22.76	$1.0 \cdot 10^4$	−0.002	0.998	−
		Sagunto	15.89	$1.0 \cdot 10^4$	0.002	0.999	0.90
		Albufera	−0.0003	$1.3 \cdot 10^4$	< 0.001	> 0.999	0.00
		Pego–Oliva	0.00	−	−	−	0.00
	Migratory status	Intercept	−8.36	3.91	−2.14	0.032	−
		Resident	−17.81	$4.6 \cdot 10^4$	< 0.001	> 0.999	0.00
		Wintering	0.00	−	−	−	0.52

studies, with prevalence around 40% during the autumn (Merino et al., 2000; Pérez–Tris & Bensch, 2005; Arizaga et al., 2009). *Haemoproteus* infection is transmitted to birds by *Culicoides* (Diptera: Ceratopogonidae) (Garvin et al., 2006). The life cycle of this parasite develops rapidly, with asexual reproduction stages in the host (Merino et al., 2004), increasing the probability of infection transmission. Our data suggest that prevalence differed between species, although intensity of infection did not differ between species. Differences could also be attributed to the presence of a species with high infection values (Blackcap), although the importance of other variables such as age, sex, immune state, and season could not be tested due to the low sample size.

Migratory status had a significant effect on the prevalence of *Haemoproteus*. Several studies show that *Haemoproteus* infections in migratory birds are common due to the wide distribution range of the parasite (Waldenström et al., 2002; Pérez–Tris & Bensch 2005). Waldenström et al. (2002) also found evidence of blood parasites as a cost of migration in birds, which may have a considerable impact on the evolution of migration.

The highest prevalence of haemoparasites was recorded in Sagunto, although it was the locality with the lowest richness of vectors. This can be explained by the fact that the vector community is not rich but shows high abundance for some species. It is of note that some authors found that the incidence of haemoparasites is correlated with local abundance of vectors (Merilä et al., 1995; Sol et al., 2000). Therefore, if dipteral vectors have a wide distribution and small habitat restrictions, their local distribution and abundance could increase the presence of haemoparasites in different bird populations. According to this hypothesis, we would also expect a high parasitemia in the resident species. However, our results do not show this parasitemia, so we think that in this case, abundance of vectors does not explain the prevalence of haemoparasites.

We think that these results are due to an effect of the host community and the migratory status of the hosts. Our results show that during the winter, Sagunto hosts several species with high blood parasite prevalence, particularly Blackcaps, a wintering species that was only trapped and sampled in Sagunto. In addition, some studies show that the prevalence of haemoparasites is related to macrohabitat characteristics. For example, Tella et al. (1999) noted that species of birds of prey nesting in forests showed a high prevalence of blood parasites. We did not find a high level of parasitemia in the other two localities, possibly due to the different habitat characteristics, as both were wetlands.

Our results highlight the importance of considering migratory status as a possible factor influencing the prevalence of haemoparasites in bird communities.

Table 4. Composition of the community of Diptera vectors in the three study areas: S. Sagunto; P–O. Pego–Oliva; A. Albufera.

Tabla 4. Composición de la comunidad de dípteros vectores en las tres zonas del estudio: S. Sagunto; P–O. Pego–Oliva; A. Albufera.

	S	P–O	A
Anopheles spp.		+	
Culex pipiens	+	+	+
Culex modestus		+	+
Culex theileri			+
Culicoides spp.		+	
Culiseta subochrea		+	
Culiseta longiareolata	+	+	+
Dasyhelea spp.		+	
Forcipomyia spp.		+	
Ochlerotatus caspius		+	
Ochlerotatus detritus		+	
Simulium reptans		+	
Simulium ruficorne		+	
Simulium velutinum		+	
Total species	2	13	4

Acknowledgements

We sincerely thank Santiago Merino and Josué Martínez de la Puente for their helpful collaboration in identifying the haemoparasites, and Rubén Piculo and José Luis Greño for their assistance in the fieldwork. We are also grateful to Aarón Gómez and Fabián Bonilla for providing the sampling equipment and for their comments and revision of the manuscript. Thanks too to Alfonso Marzal, Jordi Figuerola and an anonymous reviewer for valuable comments on this manuscript. J. R. received a grant from the Fundació General de la Universitat de València (Jóvenes Investigadores de Países en Vías de Desarrollo) for this study.

References

Apanius, V., 1991. Avian trypanosomes as models of hemoflagellate evolution. *Parasitology Today*, 7: 87–90.

Arizaga, J., Barba, E. & Hernández, M. A., 2009. Do Haemosporidians affect fuel deposition rate and fuel load in migratory Blackcaps? *Ardeola*, 56: 41–47.

Atkins, D. C., Baldwin, S. A., Zheng, C., Gallop, R. J. & Neighbors, C., 2013. A tutorial on count regression and zero–altered count models for longitudinal substance use data. *Psychology of Addictive Behaviors*. 27: 166–177.

Atkinson, C. T. & Van Riper III, C., 1991. Pathogenicity and epizootiology of avian haematozoa: Plasmodium, Leucocytozoon, and Haemoproteus. In: *Bird–parasite interactions: ecology, evolution and behavior:* 19–48 (J. E. Loye & M. Zuk, Eds.). Oxford Univ. Press, Oxford.

Bates, D., Maechler, M. & Bolker, B., 2012. lme4: linear mixed–effects using S4 classes (Computer software manual). Available from http://lme4.r–forge.r–project.org/ (R package version 0.999375–42).

Belda, E., Monrós, J. S. & Barba, E., 2007. Resident and transient dynamics, site fidelity and survival in wintering blackcaps *Sylvia atricapilla*: evidence from capture–recapture analyses. *Ibis*, 149: 396–404.

Bennett, G. F., Bishop, M. A. & Piece, M. A., 1993. Checklist of the avian species of Plasmodium Marchiafava & Celli, 1885 (Apicomplexa) and their distribution by avian family and Wallacean life zones Systematic. *Parasitology*, 26: 171–179.

Bennett, G. F., Thommes, F., Blancou, J. & Artois, M., 1982. Blood parasites of some Birds from The Lorraine región, France. *Journal of Wildlife Diseases*, 18: 81–88.

Bensch, S. & Åkesson, S., 2003. Temporal and spatial variation of hematozoans in Scandinavian willow warblers. *Journal of Parasitology*, 89: 388–391.

Brew, J. S. & Maddy, D., 1995. Generalized linear modelling. In: *Statistical modelling of quaternary science data:* 125–160 (D. Maddy & J. S. Brew, Eds.). Quaternary Research Association.

Chen, M., Shi, L. & Sullivan, D. Jr., 2001. Haemoproteus and Schitosoma synthesize heme polymers similar to Plasmodium hemozoin and b–hematin. *Molecular Biochemistry and Parasitology*, 113: 1–8.

Deviche, P., McGraw, K. & Greiner, E. C., 2005. Interspecific differences in Hematozoan infection in sonoran desert Aimophila Sparrows. *Journal of Wildlife Disease*, 41: 532–541.

Elston, D. A., Moss, R., Boulinier, T., Arrowsmith, C. & Lambin, X., 2001. Analysis of aggregation, a worked example: numbers of ticks on red grouse chicks. *Parasitology*, 122: 563–569.

Fallon, S., Bermingham, M. E. & Ricklefs, R. E., 2005. Host Specialization and Geographic Localization of Avian Malaria Parasites: A Regional Analysisin the Lesser Antilles. *The American Naturalist*, 165: 466–480.

Figuerola, J. & Green, A. J., 2000. Haematozoan parasites and migratory behaviour in waterfowl. *Evolutionary Ecology*, 14: 143–153.

Figuerola, J., Velarde, R., Bertolero, A. & Cerdá, F., 1996. Abwesenheit von haematozoa bei einer brutpopulation des seeregenpfeifers *Charadrius alexandrinus* in Nordspanien. *Journal fur Ornithologie*, 137: 523–525.

Forero, M., Tella, J. L. & Gajon, A., 1997. Absence of blood parasites in the red–necked nightjar. *Journal of Field Ornithology,* 68: 575–579.

Garvin, M., Szell, C. C. & Moore, F. R., 2006. Blood parasites of Neartic–Neotropical migrant passerines birds during spring Trans–Gulf migration: impact on host body condition. *Journal of Parasitology,* 92: 990–996.

González–Solís, J. S. & Abella, J. C., 1997. Negative record of haematozoa parasites Cory's Shearwater *Calonectris diomedea. Ornis Fennica,* 74: 153–155.

Hamilton, W. D. & Zuk, M., 1982. Heritable true Fitness and bright birds: A role for parasites? *Science,* 218: 384–387.

Hauptmanova, K., Benedikt, V. & Literák, I., 2006. Blood Parasites in Passerine Birds in Slovakian East Carpathians. *Acta Protozoologica,* 45: 105–109.

Hellgren, O., Bensch, S. & Malmqvist, B., 2008. Bird hosts, blood parasites and their vectors –associations uncovered by molecular analyses of blackfly blood meals. *Molecular Ecology,* 17: 1605–1613.

Jovani, R., Tella, J. L., Blanco, G. & Bertellotti, M., 2002. Absence of haematozoa on colonial white storks *Ciconia ciconia* throughout their distribution range in Spain. *Ornis Fennica,* 79: 41–44.

Kučera, J., 1981. Blood parasites of birds in Central Europe 3. Plasmodium and Haemoproteus. *Folia parasitol. (Praha),* 28: 303–312

– 1982. Blood parasites of birds in Central Europe. 4. Trypanosoma, Atoxoplasma, microfilariae and other rare haematozoa. *Folia Parasit. (Praha),* 29: 107–113.

Little, R. M. & Earlé, R. A., 1995. Sandgrouse (Pterocleidae) and sociable weavers Philetarius socius lack avian haematozoa in semi–arid regions of South Africa. *Journal of Arid Environments,* 30: 367–370.

López, G., Soriguer, R. & Figuerola, J., 2011. Is bill colouration in wild male Blackbirds (*Turdus merula*) related to biochemistry parameters and parasitism? *Journal of Ornithology,* 152: 965–973.

Martínez–Abraín, A., Esparza, B. & Oro, D., 2004. Lack of blood parasites in bird species: Does absence of blood parasites vector explain it all? *Ardeola,* 51: 225–232.

Martínez–Abraín, A., Merino, S., Oro, D. & Esperanza, B., 2002. Prevalence of blood parasites in two western–Mediterranean local populations of the yellow–legged Gull *Larus cachinnans michahellis. Ornis Fennica,* 79: 34–40.

Martínez–de la Puente, J., Merino, S., Tomás, G., Moreno, J., Morales, J., Lobato, E., García–Fraile, S. & Belda, E. J., 2010. The blood parasite *Haemoproteus* reduces survival in a wild bird: a medication experiment. *Biology Letters,* 6(5): 663–665.

Marzal, A., Bensch, S., Reviriego, M., Balbontín, J. & de Lope, F., 2008. Effects of malaria double infection in birds: one plus one is not two. *Journal of Evolutionary Biology,* 21: 979–87.

Marzal, A., Ricklefs, R. E., Valkiūnas, G., Albayrak, T., Arriero, E., Bonneaud, C., Czirják, G. A., Ewen, J., Hellgren, O., Hořáková, D., Iezhova, T. A., Jensen, H., Križanauskienė, A., Lima, M. R., De Lope, F., Magnussen, E., Martin, L. B., Møller, A. P., Pali-

nauskas, V., Pap, P. L., Pérez–Tris, J., Sehgal, R. N. M., Soler, M., Szöllősi, E., Westerdahl, H., Zetindjiev, P. & Bensch, S., 2011. Diversity, Loss, and Gain of Malaria Parasites in a Globally Invasive Bird. *PLoS One,* 6(7): e21905

Merilä, J., Björklund, M. & Bennet, G. F., 1995. Geographical and individual variation in the Greenfinch Carduelis chloris. *Canadian Journal of Zoology,* 73: 1798–1804.

Merino, S., Moreno, J., Sanz, J. J. & Arriero, E., 2000. Are avian blood parasites pathogenic in the wild? A medication experiment in blue tits (*Parus caeruleus*). *Proc. R. Soc. Lond. Ser. B,* 267: 2507–2510.

Merino, S. & Potti, J., 1995. High prevalence of hematozoa in nestlings of a passerine species, the pied flycatcher (*Ficedula hypoleuca*). *Auk,* 112: 1041–1043.

Merino, S., Potti, J. & Fargallo, J. A., 1997. Bloods Parasites of birds fron central Spain. *Journal of Wildlife Diseases,* 33: 638–641.

Merino, S., Tomás, G., Moreno, J., Sanz, J. J., Arriero, E. & Folgueira, C., 2004. Changes in Haemoproteus sex ratios: fertility insurance or differential sex lifespan? *Proc. R. Soc. Lond. B,* 271: 1605–1609.

Moyer, B. R., Drown, D. M. & Clayton, D. H., 2002. Low humidity reduces ectoparasite pressure: implications for host life history evolution. *Oikos,* 97: 223–228.

Peirce, M. A., 1981. Distribution and host–parasite check–list of the haematozoa of birds in western Europe. *J. Nat. Hist.,* 15: 419–458.

Pérez–Tris, J. & Bensch, S., 2005. Dispersal increases local transmission of avian malarial parasites. *Ecology Letters,* 8: 838–845.

R Development Core Team, 2009. *R: A language and environment for statistical computing* (Version 2.9.2). R Foundation for Statistical Computing, Vienna.

Rueda, J. & López, C., 2003. Valoración de la calidad biológica de los ríos. Claves de identificación para la enseñanza secundaria. *Didáctica de las ciencias experimentales y sociales,* 17: 107–123.

Ruiz–de Castañeda, R., Morales, J., Moreno, J., Lobato, E., Merino, S., Martínez de la Puente, J. & Tomás, G., 2009. Costs and benefits of early reproduction: *Haemoproteus* prevalence and reproductive success of infected male pied flycatchers in a montane habitat in Central Spain. *Ardeola,* 56: 271–280.

Senar, J. C. & Borras, A., 2004. Sobrevivir al invierno: Estrategias de las aves invernantes en La Península Ibérica. *Ardeola,* 51: 133–168.

Shurulinkov, P. & Golemansky, V., 2003 Plasmodium and Leucocytozoon (Sporozoa: Haemosporida) of Wild Birds in Bulgaria. *Acta Protozool.,* 42: 205–214.

Sol, D., Jovani, R. & Torres, J., 2000. Geographical variation in blood parasites in feral pigeons: the role of vectors. *Ecography,* 23: 307–314.

Soler, J. J., Neve, L., Pérez–Contreras, T., Soler, M. & Sorci, G., 2003. Trade–off between immunocompetence and growth in magpies: an experimental study. *Proc. R. Soc. Lond. B,* 270: 241–248.

Stjernman, M., Råberg, L. & Nilsson, J.–Å., 2008. Maximum host survival at intermediate parasite infection intensities. *PLoS One,* 3(6): e2463.

Tachet, H., Richoux, P., Bournaud, M. & Usseglio-Polatera, P., 2000. *Invertébrés d'eau Douce. Systématique, Biologie, Écologie.* CNRS Editions, Paris.

Tellería, J. L., 1988. *Invernada de aves en la Península Ibérica.* Sociedad Española de Ornitología, Madrid.

Tella, J. L., Blanco, G., Forero, M. G., Gajón, A., Donazár, J. A. & Hidalgo, F., 1999. Habitat, world geographic range, and embryonic development of hosts explain the prevalence of avian hematozoa at small spatial and phylogenetic scales. *PNAS,* 96: 1785–1789.

Tomás, G., Merino, S., Moreno, J., Morales, J. & Martinez–De la Puente, J., 2007. Impact of blood parasites on immunoglobulin level and parental effort: a medication field experiment on a wild passerine. *Functional Ecology,* 21: 125–133.

Valera, F., Carrillo, C. M., Barbosa, A. & Moreno, E., 2003. Low prevalence of haematozoa in Trumpeter finches *Bucaneteus githagineus* from south–eastern Spain: additional support for a restricted distribution of blood parasites in arid lands. *Journal of Arid Environments,* 55: 209–213.

Valkiūnas, G., 2005. *Avian malarial parasites and other haemosporidia.* CRC, Boca Raton, Florida.

Valkiūnas, G., Iezhova, T. A. & Shapoval, A. P., 2003. High prevalence of blood parasites in hawfinch *Coccothraustes coccothraustes. Journal of Natural History,* 37: 2647–2652

Waldenström, J., Bensch, S., Kiboi, S., Hasselquist, D. & Ottosson, U., 2002. Cross–species infection of blood parasites between resident and migratory songbirds in Africa. *Molecular Ecology,* 11: 1545–1554.

4

Distribution patterns of invasive Monk parakeets (*Myiopsitta monachus*) in an urban habitat

R. Rodríguez–Pastor, J. C. Senar, A. Ortega, J. Faus, F. Uribe & T. Montalvo

Rodríguez–Pastor, R., Senar, J. C., Ortega, A., Faus, J., Uribe, F. & Montalvo, T., 2012. Distribution patterns of invasive Monk parakeets (*Myiopsitta monachus*) in an urban habitat. *Animal Biodiversity and Conservation*, 35.1: 107–117.

Abstract

*Distribution patterns of invasive Monk parakeets (*Myiopsitta monachus*) in an urban habitat.*— Several invasive species have been shown to have a marked preference for urban habitats. The study of the variables responsible for the distribution of these species within urban habitats should allow to predict which environmental variables are indicative of preferred habitat, and to design landscape characteristics that make these areas less conducive to these species. The Monk parakeet *Myiopsitta monachus* is an invasive species in many American and European countries, and cities are one of its most usual habitats in invaded areas. The aim of this paper was to identify the main factors that determine distribution of the Monk parakeet in Barcelona, one of the cities in the world with the highest parakeet density. We defined our model based on eight preselected variables using a generalized linear model (GLZ) and evaluated the strength of support for each model using the AIC–based multi–model inference approach. We used parakeet density as a dependent variable, and an analysis restricted to occupied neighbourhoods provided a model with two key variables to explain the distribution of the species. Monk parakeets were more abundant in neighbourhoods with a high density of trees and a high percentage of people over 65 years. This is interpreted by the fact that parakeets use trees as food sources and support for the nests, and that older people often feed the species. Data support the 'human–activity' hypothesis to explain how invasive species can successfully establish in a non–native habitat, and stress how limiting food resources, especially food supplied by humans, may be the easiest way to exert some control on Monk parakeet populations.

Key words: Biological invasions, Urban habitat, *Myiopsitta monachus*, Density of trees, The 'human–activity' hypothesis, Older people.

Resumen

*Patrones de distribución de la cotorra argentina (*Myiopsitta monachus*) en un hábitat urbano.*— Varias especies invasoras han demostrado tener una marcada preferencia por los hábitats urbanos. El estudio de las variables responsables de la distribución de estas especies dentro de hábitats urbanos debe permitir predecir cuáles son las variables ambientales indicativas de su hábitat preferido, y diseñar las características del paisaje que hacen a estas áreas ser menos favorables para estas especies. La cotorra argentina *Myiopsitta monachus* es una especie invasora en muchos países de América y de Europa, siendo las ciudades uno de los hábitats más comunes para esta especie en las áreas invadidas. El propósito de este estudio fue identificar los factores principales que determinan la distribución de la cotorra argentina en Barcelona, una de las ciudades en el mundo con una de las densidades más grande de cotorras. Hemos definido nuestro modelo basado en ocho variables preseleccionadas mediante un modelo lineal generalizado (GLZ) y evaluamos el poder de cada modelo a través de un análisis de inferencia multimodelo basado en el valor AIC. Utilizamos la densidad de cotorra argentina como variable dependiente y restringimos el análisis a aquellos barrios de la ciudad ocupados por la especie, obteniendo un modelo con dos variables clave que explicaban la distribución de la especie. Las cotorras argentinas eran más abundantes en aquellos barrios con alta densidad de árboles y con un alto porcentaje de personas mayores de 65 años. Esto se interpreta por el hecho de que las cotorras utilizan los árboles como fuente de alimento y como lugar de nidificación, y porque las personas mayores a menudo alimentan a la especie. Los datos apoyan la hipótesis de 'la actividad humana' para explicar cómo las especies invasoras pueden exitosamente establecerse en un hábitat no nativo, y subraya cómo la limitación de las fuentes de alimento, especialmente la comida suministrada por los seres humanos, puede ser la forma más sencilla de ejercer cierto control sobre las poblaciones de cotorra argentina.

Palabras clave: Invasiones biológicas, Hábitat urbano, Myiopsitta monachus, Densidad de árboles, Hipótesis de 'la actividad humana', Gente mayor.

Ruth Rodríguez–Pastor, Juan Carlos Senar, Alba Ortega, Jordi Faus & Francesc Uribe, Associate Research Unit, CSIC, Museu de Ciències Naturals de Barcelona, Psg. Picasso s/n., E–08003 Barcelona, Espanya (Spain).– Tomas Montalvo, Servei de Vigilància i Control de Plagues Urbanes, Agència de Salut Pública de Barcelona, Av. Príncep d'Astúries 63, 3r 2a, E–08012 Barcelona, Espanya (Spain).

Corresponding author: R. Rodríguez–Pastor. E–mail: ruth.r.pastor@gmail.com

Introduction

The spread of exotic species is a major threat to native biodiversity and ecosystem functioning (Diamond, 1989; Temple, 1992; Wilcove et al., 1998; Kolar & Lodge, 2001; Stockwell et al., 2003). The main threats of invasive species can be summarized as: alteration of ecosystem processes (Raizada et al., 2008); decrease of native species abundance and richness through competition, predation, hybridization and indirect effects (Blackburn et al., 2004; Gaertner et al., 2009); changes in community structure (Hejda et al., 2009); and alteration of genetic diversity (Ellstrand & Schierenbeck, 2000). When invasive birds establish in a new habitat, they become part of the local biotic community, and this inevitably has an environmental and economic impact: transmission of avian diseases, or crop damages (Temple, 1992). Identifying the mechanisms that enable these species to establish viable populations in their new areas is one of the most important tools for their future management (Kolar & Lodge, 2001). Elucidating distribution patterns and the key variables that explain these patterns is an advisable approach (Franklin, 2009).

A number of hypotheses have been proposed to explain how invasive species can successfully establish in a non–native habitat. The 'climate–matching' hypothesis assumes that invasive species have a higher probability of successful establishment when they are introduced into regions with a climate similar to that in their native area (Williamson, 1996; Leprieur et al., 2008). In contrast with this, the 'human activity' hypothesis argues that it is human activity that favours this success by disturbing natural landscapes and increasing the number of individuals released and the frequency of introduction (Taylor & Irwin, 2004). These two hypothesis are the best supported by empirical data (Blackburn et al., 2009), and Temple (1992) showed that both mechanisms are conducive to successful reproduction. Some other hypothesis have additionally been proposed, as for instance the 'enemy–release' hypothesis, which states that the abundance or impact of some invasive species is related to the scarcity of natural enemies in the introduced range compared with the native range (Keane & Crawley, 2002; Torchin et al., 2002, 2003; Mitchell & Power, 2003).

The Monk parakeet is a common invasive bird species. Due to parrot trade, escapes from captivity, and the accidental or deliberate releases by people that had them as a pet, this species has spread its distribution range from South America (native habitat) to North America and Western Europe (Hyman & Pruett–Jones, 1995; Van Bael & Pruett–Jones, 1996; Cassey et al., 2004). This parakeet is characterized by building many nests within a communal nest structure of tightly inter–twined twigs and sticks (Spreyer & Bucher, 1998; Domènech & Senar, 2006; Burger & Gochfeld, 2009). Each individual nest is called a chamber and is used for breeding and roosting (Domènech & Senar, 2006). Nest structures are often clustered, forming colonies in a group of trees (Burger & Gochfeld, 2009). The nests are used throughout the whole year, and are repaired and reused year after year (Navarro et al., 1992; Eberhard, 1996, 1998).

This species is considered an agricultural pest in its native range (Bucher et al., 1990). In several of the invaded areas, the species has been able to establish increasingly large breeding populations (Bull, 1973; Shields, 1974; Summerour, 1990; Clavell et al., 1991; Schwab & Gwynn III, 1992; Caruso & Scelsi, 1994; Hyman & Pruett–Jones, 1995; Van Bael & Pruett–Jones, 1996). Through building its nests it causes considerable damage to ornamental trees and power lines, as well as to buildings and other structures. Furthermore, it creates problems with noise pollution, falling nests, and damage to agriculture (Bucher & Bedano, 1976; Bucher & Martin, 1983; Bucher, 1984, 1992; Temple, 1992; Conroy & Senar, 2009).

Although recent data on Monk parakeets show that number of frost days is a limiting factor for their establishment (Strubbe & Matthysen, 2009), the species has been able to adapt to places with a cold climate, such as Chicago, northern New Jersey, and Brooklyn (Pruett–Jones & Tarvin, 1998). It has also successfully colonised temperate areas, such as in Europe. Human activity therefore also appears to be highly responsible for the distribution of the species (Muñoz & Real, 2006; Strubbe & Matthysen, 2009). Their success as an invasive species provides the opportunity to examine adaptation not only to new habitats and environments, but also specifically to crowded cities (Burger & Gochfeld, 2009).

To date, however, studies on the patterns and key variables responsible for the distribution and successful spread of the Monk parakeet have been based on large geographic areas (Muñoz & Real, 2006; Strubbe & Matthysen, 2009). Although the role of cities is very important in the spread of this species (Pruett–Jones & Tarvin, 1998; Burger & Gochfeld, 2009), available data about the variables responsible for the success of the species within cities is scarce. Barcelona city holds one of the largest populations of Monk parakeets in Western Europe (Domènech, 1997; Domènech et al., 2003). The bird was detected for the first time in 1975 in the Ciutadella Park (Batllori & Nos, 1985) and since then, the species has dramatically increased its populations and has expanded its range throughout the entire city (Clavell et al., 1991; Domènech, 1997; Sol et al., 1997). In 2001, there were 313 nests in Barcelona (Conroy & Senar, 2009), but by 2010 the number had increased to 650 nests, with 1,876 chambers (J. C. Senar & T. Montalvo, pers. obs.). The aim of this paper was to determine the factors that affect the distribution of the Monk parakeet in Barcelona city so as to obtain information about the habitat preferences of the species within the urban habitat.

Material and methods

Study area

The study was conducted in Barcelona city, in north–eastern Spain. Barcelona has an area of 102.16 km^2,

72.34% of which is built up. It is divided into ten districts, which allows decentralized local administration. Each district consists of several 'barrios' (neighbourhoods). Barcelona has 73 neighbourhoods, each with its own personality, historical tradition, and homogeneous urban structure, making them clearly recognizable. Neighbourhoods are hence our sample unit.

Data on species distribution

A census of Monk parakeet was carried out from October 2009 until March 2010. A group of university biology students in the final year of their graduate programme helped us to take a census of the population of Monk parakeets in the whole city. We visited every street and all green areas within each neighbourhood. We marked the exact position of each nest found, obtaining the coordinates for each one. We also recorded the number of chambers each nest contained, and the tree species or the substrate on which it was built. The nest coordinates were later transferred to the GIS system MiraMon, creating a layer of points representing the nests. This layer was combined with a layer of the neighbourhoods, reverting data to the neighbourhood level. The resulting map allowed us to know the number and characteristics of nests within each neighbourhood.

We used the density of chambers (number of chambers per square kilometre) per neighbourhood to estimate population density of Monk parakeets at each sampling unit, following Sol et al. (1997) and Domènech et al. (2003).

Descriptor variables

Demographic, socioeconomic and urban development data were provided by the Statistics Department at the Barcelona City Council (http://www.bcn.cat/estadistica/angles/index.htm). We used variables created by the Department of Basic Information and Cartography to determine the main characteristics of building volumes and spatial distribution. From all these data, we selected those variables that would most likely affect the distribution of the Monk parakeet and that could better describe the characteristics of the neighbourhoods. These variables included: human population density (PD), the percentage of people over 65 years (PD > 65), the percentage of buildings constructed before 1901 (B < 1901), and the percentage of streets and roads (SR). The main reasons for choosing these variables are explained below. Population density (PD) gives us an idea about how many people inhabit in each neighbourhood. This could be a variable of interest since it has been seen to have a strong influence on feral pigeon density (Senar et al., 2009). The percentage of people over 65 years (PD > 65) was considered because this population group seems to be more likely to feed birds in urban habitats (Montalvo & Senar, pers. obs.). The last two variables were related to urban characteristics: the percentage of buildings constructed before 1901 (B < 1901) shows that old buildings were constructed more closely together with each other, leaving insufficient space between them for urban parks

and trees; and the percentage of streets and roads (SR) indicates that neighbourhoods with wider streets could have many more green areas and trees where Monk parakeets could nest and feed. Finally, we should point out that as our study was based on a small–scale distribution of the Monk parakeets, it was pointless to choose climatic, topographic or lithologic variables as other studies have done (Muñoz & Real, 2006).

The Department of Parks and Gardens in Barcelona provided additional variables describing the vegetation in each neighbourhood: percentage of grass area (grass area per neighbourhood; GA), percentage of shrub area (shrub area per neighbourhood: SA), density of trees (number of trees per square kilometre; DT), and percentage of forest area (forest area per neighbourhood; FA). Barcelona has 295 species of planted trees, but only a few are used by parakeets for feeding or nesting (Carrillo–Ortiz, 2009). So as to determine whether the species of trees had any effect on Monk parakeet distribution we classified the trees into three groups depending on the type of fruit that they produced: trees with fleshy fruit (e.g. Celtis sp., Ficus sp., Prunus sp., Robinia pseudoacacia), trees with dry fruit (e.g. Populus sp., Tilia sp., Tipuana tipu, Ulmus sp.), and trees with conus or strobilus (e.g. Pinus sp., Cupressus sp., Cedrus sp.). We also included the density of Phoenix sp. and the density of Platanus sp. as independent variables. Phoenix sp. is one of the most commonly used trees by parakeets to build nests in Barcelona (Sol et al., 1997). Platanus sp. is the most abundant tree in the city and is increasingly being used to build nests (Carrillo & Senar, pers. obs.). Densities were measured as number of trees per neighbourhood in km^2.

Statistical analyses

Because the dependent variables and several of the independent variables did not follow a normal distribution, all the variables were rank transformed (Conover, 1981). We tested the correlation between the different variables to detect any possible colinearity problem. All the correlations were < 0.50, and in most cases < 0.20–0.30. We can therefore be confident that colinearity does not apply to our data. Next, data were introduced in a generalized linear model (GLZ) adjusted to a recent approach based on the Akaike Information Criterion (AIC) (Burnham & Anderson, 2002; Lukacs et al., 2007). The dependent variable (the density of Monk parakeet measured as density of chambers per neighbourhood) and all the explanatory variables were introduced in the model. The eight independent variables were: human population density (PD), percentage of people over 65 years (PD > 65), percentage of buildings built before 1901 (B < 1901), percentage of road system (SR), percentage of grass area (GA), percentage of shrub area (SA), density of trees (DT), and percentage of forest area (FA). Why some neighbourhoods did not have parakeets could be either because the area was not suitable for the species or just that parakeets had not yet colonized it. We thus preferred to restrict our sample size to those neighbourhoods that had already been occupied (i.e. number of chambers > 1). This left us with 51 neighbourhoods (N = 51).

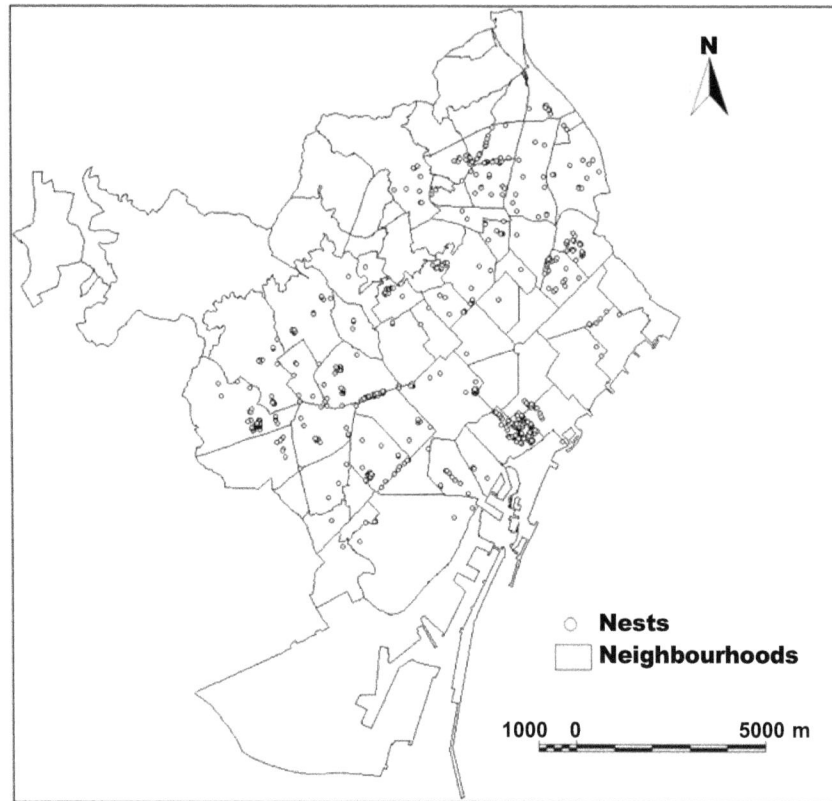

Fig. 1. Distribution map of Monk parakeet nests in the neighbourhoods of Barcelona city. The dots depict the exact localization of the nests.

Fig. 1. Mapa de distribución de los nidos de la cotorra argentina en los barrios de la ciudad de Barcelona. Los puntos representan la localización exacta de los nidos.

As our survey showed that most nests of Monk parakeets were built in palm trees, we decided to additionally test whether the typology of tree species in an area influenced their abundance. We made a multiple regression approach using the density of chambers as the dependent variable, and the density of *Phoenix*, the density of *Platanus*, the density of trees with conus, the density of trees with fleshy fruit, and the density of trees with dry fruit as independent variables. All statistical analyses were carried out using Statistica 10.0 (StatSoft, 2001).

Results

Monk parakeets appeared throughout whole city of Barcelona (fig. 1). The total number of detected nests was 650 and the number of chambers was 1876. The 'Sant Pere, Santa Caterina i la Ribera' neighbourhood had the highest concentration of nests (n = 120), most of which were located within the Ciudadela Park. The average density of nests per neighbourhood in 2010 was 8.27 ± 14.69 (n = 73) and that of chambers was 23.51 ± 38.80 (n = 73) (table 1). Variation between

neighbourhoods was high. Five neighbourhoods had a total density of chambers of above 80, most had a density of less than 20 chambers, and almost one third (22 neighbourhoods) had no chambers (figs. 2, 3A).

The size of the nests, computed as average number of chambers per nest and neighbourhood, was also highly variable. In almost half of the neighbourhoods, nests had an average size of one to four chambers, whereas only seven neighbourhoods had smaller nests (average size < 2 chambers), and nine neighbourhoods had the larger nests (average size > 4 chambers) (fig. 3B).

The AIC–based multi–model inference showed that the best model (with the smallest AIC value) regarding the other models consisted of two variables: density of trees (DT) and percentage of people over 65 years (PD > 65) (table 2, fig. 4). The next best model included one further variable: density of human population (PD) (table 2).

A multiple regression analysis relating the preference of Monk parakeet for any type of tree did not show any significant result. Despite the abundance of palm trees and *Platanus* sp. in the city of Barcelona, neither of them affected the density of parakeets (table 3).

Table 1. Descriptive statistics on the density of nests and the density of chambers of the Monk parakeet in Barcelona, both in all neighbourhoods (n = 73) and restricted to neighbourhoods with at least one chamber (n = 51). Density is measured as number of nests or chambers per square kilometre.

Tabla 1. Estadísticos descriptivos de la densidad de nidos y la densidad de cámaras de la cotorra argentina en Barcelona, tanto en todos los barrios (n = 73) como restringido sólo a los barrios con al menos una cámara (n = 51). La densidad es medida como número de nidos y de cámaras por kilómetro cuadrado.

	Valid N	Mean	Median	Minimum	Maximum	SE
Density of nests	73	8.27	3.64	0.00	109.09	1.72
	51	11.84	7.78	0.77	109.09	16.36
Density of chambers	73	23.51	10.00	0.00	233.64	4.54
	51	33.65	16.67	0.77	233.64	42.66

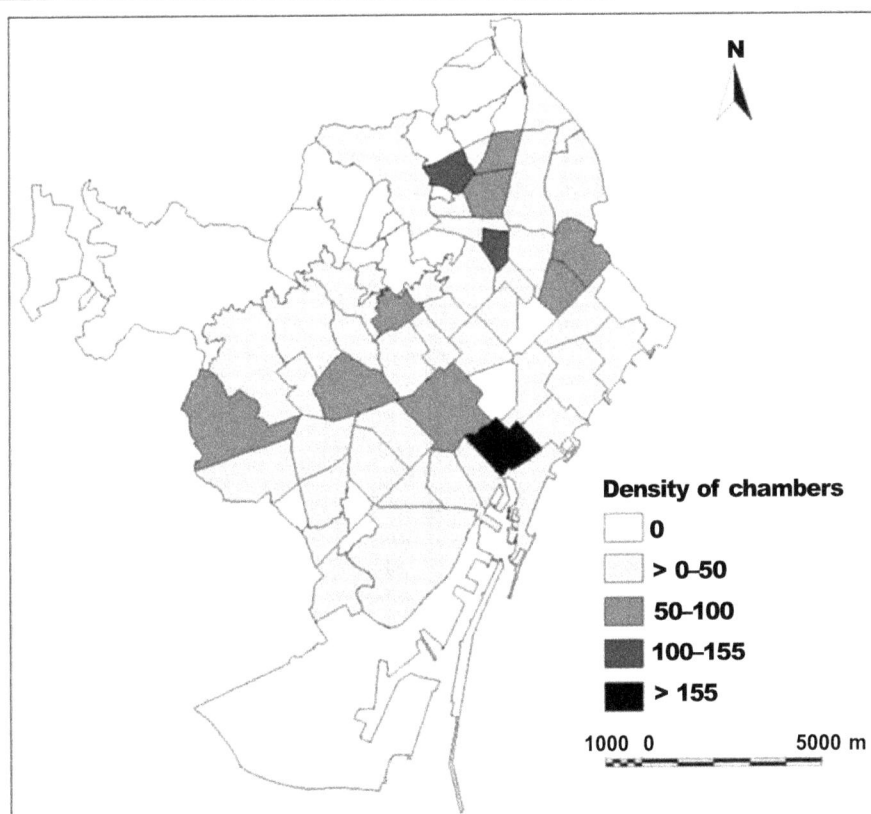

Fig. 2. Distribution map of Myiopsitta monachus in Barcelona city. The presence of Monk parakeet measured as the density of chambers (number of chambers per square kilometre) in the different neighbourhoods, where the species was recorded, is categorized in five classes according to a grey scale.

Fig. 2. Mapa de distribución de Myiopsitta monachus en la ciudad de Barcelona. La presencia de la cotorra argentina medida como la densidad de cámaras (número de cámaras por kilómetro cuadrado) en los diferentes barrios, donde la especie fue registrada, se clasificó en cinco categorías siguiendo una escala de grises.

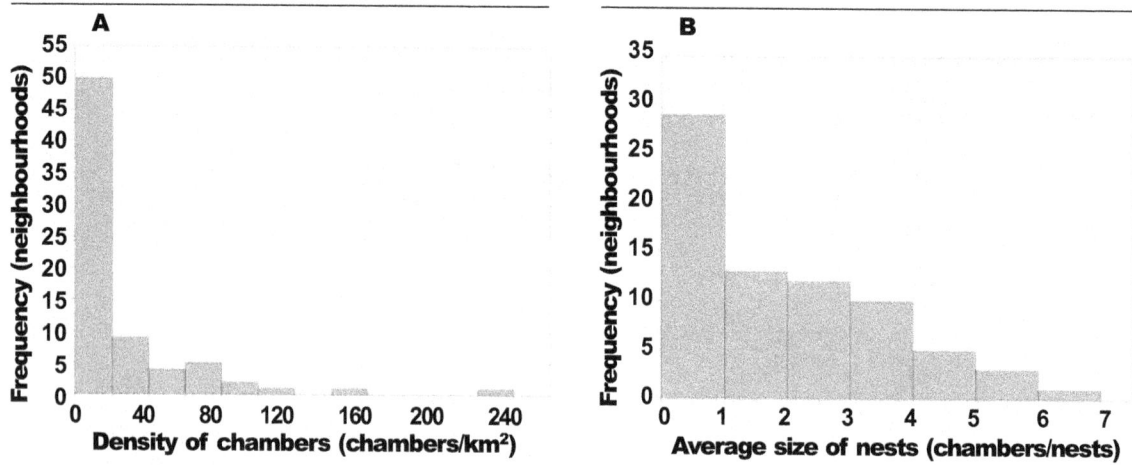

Fig. 3. Frequency distribution of: (A) the density of chambers, and (B) the average size of nests measured as average number of chambers in each neighbourhood in Barcelona (N = 73; 2010 census).

Fig. 3. Distribución de frecuencias de: (A) la densidad de cámaras y (B) el tamaño medio de nidos medido como el número medio de cámaras en cada barrio de Barcelona (N = 73; censo de 2010).

Discussion

Information about the environment in areas of origin of invader species may be important to model their distribution in large, invaded geographic areas (Blackburn et al., 2009). However, it has been recently suggested that the non–native distribution of invading species cannot be predicted from the characteristics of the native niche (Petitpierre et al., 2012). Hence, it becomes highly relevant to obtain

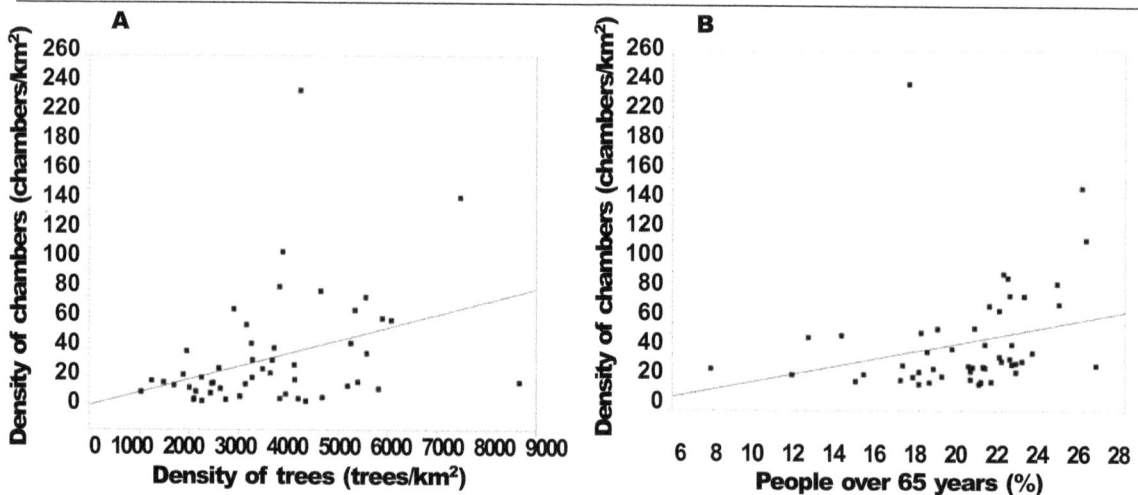

Fig. 4. Relationships between the density of Monk parakeet chambers with: (A) tree density, and (B) the percentage of people over 65 years old. Neighbourhoods with no Monk parakeet nests (# chambers = 0) are not depicted in the graphs. The partial correlations for variables in the multiple regression best model (table 2) were tree density: r = 0.32, p = 0.02 and percentage of people over 65 years: r = 0.36, p = 0.01.

Fig. 4. Correlaciones entre la densidad de cámaras de la cotorra argentina con: (A) la densidad de árboles y (B) el porcentaje de personas mayores de 65 años. Los barrios sin nidos de cotorras (nº cámaras = 0) no se representan en los gráficos. Las correlaciones parciales para las variables incluidas en el mejor modelo (tabla 2) de regresión múltiple fueron densidad de árboles: r = 0,32, p = 0,02 y porcentaje de personas mayores de 65 años: r = 0,36, p =0,01.

Table 2. Results of the multimodel inference using AIC to predict which variables in Barcelona city showed the best approximating model that explained the density of Monk parakeets (measured as number of nest chambers): PD. Human population density; PD > 65. Percentage of people over 65 years; B < 1901. Percentage of buildings constructed before 1901; SR. Percentage of streets and roads; GA. Percentage of grass area; SA. Percentage of shrub area; DT. Density of trees; FA. Percentage of forest area.

Tabla 2. Resultados del análisis de inferencia multimodelo basado en el AIC para predecir qué variables en la ciudad de Barcelona definen el mejor modelo que explica la densidad de cotorra argentina (medida como número de cámaras de nido): PD. Densidad de población humana; PD > 65. Porcentaje de personas mayores de 65 años; B < 1901. Porcentaje de edificios construidos antes de 1901; SR. Porcentaje de calles y carreteras; GA. Porcentaje de zonas con césped; SA. Porcentaje zonas arbustivas; DT. Densidad de árboles; FA. Porcentaje de zona forestal.

Step	Models								DF	AIC
1	PD > 65	DT							2	410.24
2	PD	PD > 65	DT						3	410.95
3	PD > 65	FA	DT						3	411.97
4	PD > 65	SR	DT						3	412.03
5	PD > 65	GA	DT						3	412.15
6	PD > 65	B < 1901	DT						3	412.16
7	PD > 65	SA	DT						3	412.21
8	PD	PD > 65	B < 1901	DT					4	412.51
9	PD	PD > 65	SA	DT					4	412.88
10	PD	PD > 65	GA	DT					4	412.91
11	PD	PD > 65	FA	DT					4	412.94
12	PD	PD > 65	SR	DT					4	412.94
13	PD > 65								1	413.49
14	PD > 65	B < 1901	SR	DT					4	413.80
15	PD > 65	B < 1901	FA	DT					4	413.85
40	DT								1	415.42
149	PD	PD > 65	B < 1901	SR	GA	SA	FA	DT	8	420.16
171	B < 1901								1	421.18
181	SR								1	421.89
182	GA								1	422.14
187	PD								1	422.38
197	SA								1	422.97
198	FA								1	422.98

detailed information about the factors responsible for the distribution of these species in invaded areas. In the case of species that are highly linked to urban habitats, like Monk parakeets (Domènech et al., 2003) or House crows *Corvus splendens* (Soh et al., 2002) it becomes imperative to predict their preferences and distribution within cities.We found that the model that best explained the abundance of Monk parakeets in Barcelona city showed two relevant variables: density of trees and the percentage of people over 65 years. The fact that trees are used as a main food source

and also to support and build nests (Carrillo–Ortiz, 2009) explains why the density of trees is the main variable to account for Monk parakeet abundance in Barcelona city. Interestingly, our results showed that parakeets had no preferences for different species of trees when selecting preferred 'habitat'. This finding supports a recent study on breeding ring–necked parakeets (*Psittacula krameri*) in Brussels, Belgium, that showed that parakeets preferred to forage in city parks and gardens due to the huge variety of exotic and ornamental plants (Strubbe & Matthysen, 2011;

Table 3. Multiple regression using the different trees grouped by type of fruit that they produce as independent variables. *Phoenix* sp. and *Platanus* sp. were separated from the others because of their abundance.

Tabla 3. Regresión múltiple utilizando los diferentes árboles agrupados por el tipo de fruto que producen como variables independientes. Phoenix sp. y Platanus sp. se separaron del resto debido a su abundancia.

	Beta	Partial correlation	t(45)	p–level
Density of *Phoenix*	0.13	0.12	0.80	0.43
Density of *Platanus*	0.09	0.09	0.57	0.57
Density of trees with conus	−0.10	−0.09	−0.60	0.55
Density of trees with fleshy fruit	0.17	0.15	0.99	0.33
Density of trees with dry fruit	0.22	0.20	1.38	0.17

Clergeau & Vergnes, 2012). Moreover, our finding shows that Monk parakeets in Barcelona currently use other tree species for nesting in cases where palm trees are scarce, which was not the case in the earlier stages of colonization by the species (Sol et al., 1997). In Singapore, Soh et al. (2002) found similar results with another urban invasive bird species, the House crow with regard to the yellow flame trees (*Peltophorum pterocarpum*).

As the Monk parakeet is a colonial species, the stages of colonization can be a consequence of historical and contagious effects (*i.e.*, the occurrence of one nest may prompt the colonization of that area by other individuals of the population). Nevertheless, the average dispersal distance of the species in Barcelona city is 1,114 ± 190 m (Carrillo–Ortiz, 2009), which provides ample opportunities for the colonization of new areas within the city. The presence of a good number of nests with only one chamber supports that this colonization is taking place. Hence, we think that our data reflect habitat selection in spite of historical and contagious effects.

Our data support the 'human–activity' hypothesis to explain how invasive species can successfully establish in a non–native habitat. We found that the percentage of people over 65 years old was an important variable in the best fit model, and the density of human population was strong in the second model (AIC < 2). Human presence thus clearly favours the invasion by Monk parakeets, not only because human activity favours escapes and releases of individuals, but also because of the foraging advantage that individuals enjoy because of (older) people supplying food. Monk parakeets have changed their foraging behaviour in recent years, adapting to novel conditions and using novel resources of food offered by humans (Lefebvre et al., 2004; Sol, 2007). A recent report showed that these new food resources in Barcelona are mainly cereals and bread, and they account for up to 40% of parakeet food sources (Carrillo–Ortiz, 2009). The population group most likely to feed birds is usually retired people (Montalvo & Senar, pers.

obs.). Similarly, Pithon (1998) observed that parakeet densities in the UK correlated highly with densities of detached and semi–detached houses inhabited by retired people.

Taken together, our results support the view that the preference and success of invasive species in cities is related to the abundance of food available and nesting sites in these anthropogenic habitats, so that invasive species can easily increase both breeding success and parental survival (Chamberlain et al., 2009). As food sources play a major role in favouring invasive species in general and Monk parakeets in particular, we propose that limiting food resources provided by the population would be a reasonable way to control these species in urban habitats. However, further studies with new variables are needed to understand better and determine why Monk parakeets prefer these areas in Barcelona city.

Acknowledgements

This work was supported by the Barcelona Public Health Agency (ASPB), and by research project CGL2009–10652, Ministry of Science and Innovation, Spanish Research Council. We thank Issac Aparicio and Eugènia Valero (Department of Basic Information and Cartography, Barcelona City Council) and Juan Miguel Pérez Díaz (Department of Parks and Gardens, Barcelona City Council), for providing specific data layers. We also thank Lluïsa Arroyo, Daniel Riba, Margarida Barceló, Ruben Parrilla and all the biology students engaged in the project for their field support.

References

Batllori, X. & Nos, R., 1985. Presencia de la Cotorrita gris (*Myiopsitta monachus*) y de la Cotorrita de collar (*Psittacula krameri*) en el área metropolitana de Barcelona. *Miscel·lània Zoològica,* 9: 407–411.
Blackburn, T. M., Cassey, P., Duncan, R. P., Evans,

K. L. & Gaston, K. J., 2004. Avian extinction and mammalian introductions on oceanic islands. *Science*, 305: 1955–1958.

Blackburn, T. M., Lockwood, J. L. & Cassey, P., 2009. *Avian invasions. The Ecology and Evolution of Exotic Birds*. Oxford Univ. Press, Oxford.

Bucher, E.H., 1984. Las aves como plaga en la Argentina, 3–17. Centro de Zoología Aplicada, Universidad Nacional de Córdoba, Córdoba, Argentina.

– 1992. Neotropical parrots as agricultural pests. In: *New world parrots in crisis: solutions from conservation biology*: 201–219 (S. R. Beissinger & N. F. R. Snyder, Eds.). Smithsonian Intitution Press, Washington D.C.

Bucher, E. H. & Bedano, P. E., 1976. Bird damage problems in Argentina. In: *International studies on Sparrows*: 3–16 (Working group on granivorous birds–intecol, Eds.). PWN– Polish Scientific Publishers, Warsawa, Poland.

Bucher, E. H., & Martin, L. F., 1983. Los nidos de cotorras (*Myiopsitta monachus*) como causa de problemas en líneas de transmisión eléctrica. *Vida Silvestre Neotropical, 50–51.*

Bucher, E. H., Martin, L. F., Martella, M. B. & Navarro, J. L., 1990. Social behaviour and population dynamics of the Monk Parakeet. *Proc. Int. Ornithol. Congr.*, 20: 681–689.

Bull, J., 1973. Exotic birds in the New York City area. *Wilson Bulletin, 85:* 501–505.

Burger, J. & Gochfeld, M., 2009. Exotic monk parakeets (*Myiopsitta monachus*) in New Jersey: nest site selection, rebuilding following removal, and their urban wildlife appeal. *Urban Ecosystems, 12:* 185–196.

Burnham, K. & Anderson, D. R., 2002. *Model selection and multimodel inference: A practical information-theoretic approach*. 2nd Ed., Springer-Verlag, New York, NY, USA.

Carrillo–Ortiz, J., 2009. Dinámica de poblaciones de la cotorra de pecho gris (*Myiopsitta monachus*) en la ciudad de Barcelona. Ph. D. Thesis, Univ. of Barcelona.

Caruso, S. & Scelsi, F., 1994. Nesting of feral Monk Parakeets, *Myiopsitta monachus*, in the town of Catania, Sicily. *Rivista Italiana di Ornitologia, 63:* 213–215.

Cassey, P., Blackburn, T. M., Russell, G. J., Jones, K. E. & Lockwood, J. L., 2004. Influences on the transport and establishment of exotic bird species: an analysis of the parrots (Psittaciformes) of the world. *Global Change Biology,* 10: 417–426.

Chamberlain, D. E., Cannon, A. R., Toms, M. P., Leech, D. I., Hatchwell, B. J. & Gaston, K. J., 2009. Avian productivity in urban landscapes: a review and meta–analysis. *Ibis,* 151: 1–18.

Clavell, J., Martorell, E., Santos, D. M. & Sol, D., 1991. Distribució de la cotorreta de pit gris *Myiopsitta monachus* a Catalunya. *Butlletí Grup Català d'Anellament,* 8: 15–18.

Clergeau, P. & Vergnes, A. 2012. Bird feeders may sustain feral Rose–ringed parakeets *Psittacula krameri* in temperate Europe. *Wildlife Biology,* 17: 248–252.

Conover, W. J. & Iman, R. L., 1981. Rank transformation as a bridge between parametric and nonparametric statistics. *The American Statistician,* 35: 124–129.

Conroy, M. J. & Senar, J. C., 2009. Integration of demographic analyses and decision modelling in support of management of invasive monk parakeets, and urban and agricultural pest. *Environmental and Ecological Statistics,* 3: 491–510.

Diamond, J. M., 1989. Overview of recent extinctions. In: *Conservation for the twenty–first century:* 37–41 (D. Western & M. C. Pearl, Eds.). Oxford Univ. Press, Oxford.

Domènech, J., 1997. Cotorra gris de la argentina (*Myiopsitta monachus*). In: *Animales de nuestras ciudades:* 253 (A. Omedes, J. C. Senar & F. Uribe, Eds.). Planeta, Barcelona.

Domènech, J., Carrillo–Ortiz, J. & Senar, J. C., 2003. Population size of the Monk Parakeet *Myiopsitta monachus* in Catalonia. *Revista Catalana d'Ornitologia,* 20: 1–9.

Domènech, J. & Senar, J. C., 2006. Comportamiento junto al nido de la cotorra argentina. In: *Fauna en acción. Guía para observar comportamiento animal en España:* 205–208 (M. Soler, J. Martín, L. Tocino, J. Carranza, A. Cordero, J. Moreno, J. C. Senar, M. Valdivia & F. Bolívar, Eds.). Lynx Edicions, Bellaterra, Barcelona.

Eberhard, J. R., 1996. Nest adoption by Monk Parakeets. *Wilson Bulletin,* 108: 374–377.

– 1998. Breeding biology of the monk parakeet. *Wilson Bulletin,* 110: 463–473.

Ellstrand, N. C. & Schierenbeck, K. A., 2000. Hybridization as a stimulus for the evolution of invasiveness in plants? *Proceedings of the National Academy of Sciences of the United States of America,* 97: 7043–7050.

Franklin, J., 2009. *Mapping species distributions: spatial inference and prediction.* Cambridge University Press, Cambridge, UK.

Gaertner, M., Den Bree, A., Hui, C. & Richardson, D. M., 2009. Impacts of alien plant invasions on species richness in Mediterranean–type ecosystems: a meta–analysis. *Progress in Physical Geography,* 33: 319–338.

Hejda, M., Pyšek, P. & Jarosik, V., 2009. Impact of invasive plants on the species richness, diversity and composition of invaded communities. *Journal of Ecology,* 97: 393–403.

Hyman, J. & Pruett–Jones, S., 1995. Natural history of the Monk Parakeet in Hyde Park, Chicago. *Wilson Bull,* 107: 510–517.

Keane, R. M. & Crawley, M. J., 2002. Exotic plant invasions and the enemy release hypothesis. *Trends Ecol. Evol.,* 17: 164–170.

Kolar, C. S. & Lodge, D. M., 2001. Progress in invasion biology: predicting invaders. *Trends in Ecology & Evolution,* 16: 199–204.

Lefebvre, L., Reader, S. M. & Sol, D., 2004. Brains, innovations and evolution in birds and primates. *Brain, Behavior and Evolution,* 63: 233–246.

Leprieur, F., Beauchard, O., Blanchet, S., Oberdorff, T. & Brosse, S., 2008. Fish invasions in the world's river systems: when natural processes are blurred by human activities. *PloS Biology,* 6: 2940–2940.

Lukacs, P., Thompson, W., Kendall, W., Gould, W., Doherty, P., Burnham, K. & Anderson, D. R., 2007. Concerns regarding a call for pluralism of information theory and hypothesis testing. *J. Appl. Ecol.,* 44: 456–460.

Mitchell, C. E. & Power, A. G., 2003. Release of invasive plants from fungal and viral pathogens. *Nature,* 421: 625–627.

Muñoz, A. R. & Real, R., 2006. Assessing the potential range expansion of the exotic monk parakeet in Spain. *Diversity and Distributions,* 12: 656–665.

Navarro, J. L., Martella, M. B. & Bucher, E. H., 1992. Breeding season and productivity of Monk parakeets in Cordoba, Argentina. *Wilson Bulletin,* 104: 413–424.

Petitpierre, B., Kueffer, C., Broennimann, O., Randin, C., Daehler, C., Guisan, A., 2012. Climatic Niche Shifts Are Rare Among Terrestrial Plant Invaders. *Science,* 335: 1344–1348.

Pithon, J., 1998. The status and ecology of Ring–necked Parakeet *Psittacula krameri* in Great Britain. Ph. D. Thesis, Univ. of York.

Pruett–Jones, S. & Tarvin, K. A., 1998. Monk Parakeets in the United States: population growth and regional patterns of distribution. In: *Proccedings, 18th Vertebrate Pest Control Conference:* 55–58 (R. O. Baker & A. C. Crabb, Eds.). Univ. of California, Davis.

Raizada, P., Raghubanshi, A. S. & Singh, J. S., 2008. Impact of invasive alien plant species on soil processes: a review. *Proceedings of the National Academy of Sciences India Section B, Biological Sciences,* 78: 288–298.

Schwab, D. J. & Gwynn III, T. M., 1992. Monk parakeets nesting in Newport News, Virginia. *The Raven,* 63: 34–34.

Senar, J. C., Carrillo–Ortiz, J., Arroyo, L., Montalvo, T., Peracho, V., 2009. Estima de la abundancia de paloma (*Columbia livia* var.) de la ciudad de Barcelona, 2006 y valoración de la efectividad de control por la eliminación de individuos. *Arxius de Miscel·lània Zoològica,* 7: 62–72.

Shields, W. M., 1974. Use of native plants by Monk Parakeets in New Jersey. *Wilson Bulletin,* 86: 172–173.

Soh, M. C. K., Sodhi, N. S., Seoh, R. K. H. & Brook, B. W., 2002. Nest site selection of the house crows (*Corvus splendens*), an urban invasive bird species in Singapore and implications for its management. *Landscape and Urban Planning,* 59: 217–226.

Sol, D., 2007. Do successful invaders exit? Pre–adaptations to novel environments in terrestrial vertebrates. In: *Biological invasions:* 127–141 (W. Nentwing, Eds.). Springer, Heidelberg.

Sol, D., Santos, D. M., Feria, E. & Clavell, J., 1997. Habitat selection by the Monk Parakeet during colonization of a new area in Spain. *The Condor,* 99: 39–46.

Spreyer, M. F. & Bucher, E. H., 1998. Monk Parakeet (*Myiopsitta monachus*). *The Birds of North America,* 322: 1–23.

StatSoft I, 2001. *STATISTICA.* StatSoft, Inc., Tulsa

Stockwell, C. A., Hendry, A. P. & Kinnison, M. T., 2003. Contemporary evolution meets conservation biology. *Trends in Ecology & Evolution,* 18: 94–101.

Strubbe, D. & Matthysen, E., 2009. Establishment success of invasive ring–necked and monk parakeets in Europe. *Journal of Biogeography,* 36: 2264–2278.

– 2011. A radiotelemetry study of habitat use by the ring–necked Parakeet *Psittacula krameri* in Belgium. *IBIS,* 153: 180–184.

Summerour, B., 1990. Monk parakeets (*Myiopsitta monachus*) nesting in Huntsville. *Alabama birdlife,* 37: 9–10.

Taylor, B. W. & Irwin, R. E., 2004. Linking economic activities to the distribution of exotic plants. *Proceedings of the National Academy of Science USA,* 101: 17725–17730.

Temple, S. A., 1992. Exotic Birds: A growing problem with no easy solution. *The Auk,* 109: 395–397.

Torchin, M. E., Lafferty, K. D. & Kuris, A. M., 2002. Parasites and marine invasions. *Parasitology,* 124: S137–S151.

Torchin, M. E., Lafferty, K. D., Dobson, A. P., McKenzie, V. J. & Kuris, A. M., 2003. Introduced species and their missing parasites. *Nature,* 421: 628–630.

Van Bael, S. & Pruett–Jones, S., 1996. Exponential population growth of Monk Parakeets in the United States. *Wilson Bulletin,* 108: 584–588.

Wilcove, D. S., Rothstein, D., Dubow, J., Phillips, A. & Losos, E., 1998. Quantifying threats to imperiled species in the United States. *Bioscience,* 48: 607–615.

Williamson, M., 1996. *Biological invasions.* Chapman & Hall, London.

5

Subjectivism as an unavoidable feature of ecological statistics

A. Martínez–Abraín, D. Conesa & A. Forte

Martínez–Abraín, A. Conesa, D. & Forte, A., 2014. Subjectivism as an unavoidable feature of ecological statistics. *Animal Biodiversity and Conservation*, 37.2: 141–143.

Abstract

Subjectivism as an unavoidable feature of ecological statistics.— We approach here the handling of previous information when performing statistical inference in ecology, both when dealing with model specification and selection, and when dealing with parameter estimation. We compare the perspectives of this problem from the frequentist and Bayesian schools, including objective and subjective Bayesians. We show that the issue of making use of previous information and making *a priori* decisions is not only a reality for Bayesians but also for frequentists. However, the latter tend to overlook this because of the common difficulty of having previous information available on the magnitude of the effect that is thought to be biologically relevant. This prior information should be fed into *a priori* power tests when looking for the necessary sample sizes to couple statistical and biological significances. Ecologists should make a greater effort to make use of available prior information because this is their most legitimate contribution to the inferential process. Parameter estimation and model selection would benefit if this was done, allowing a more reliable accumulation of knowledge, and hence progress, in the biological sciences.

Key words: Ecology, Previous information, Frequentists, Bayesians, Parameter estimation, Hypothesis testing

Resumen

La subjetividad como característica inevitable de los análisis estadísticos en ecología.— En este trabajo abordamos la gestión de información previa al realizar inferencia estadística en ecología, tanto en la especificación y la selección del modelo como en la estimación de los parámetros. Comparamos las perspectivas que aplican a esta problemática la escuela Frecuentista y la Bayesiana, que comprende a los bayesianos objetivos y subjetivos. Mostramos que la problemática de utilizar información previa y tomar decisiones *a priori* no es solo una realidad para los bayesianos, sino también para los frecuentistas. Sin embargo, estos últimos tienden a pasar por alto esta cuestión debido a la dificultad habitual de encontrar información previa sobre la magnitud del efecto que se considera relevante desde el punto de vista biológico. Esta información previa debería utilizarse en las pruebas de potencia *a priori* para buscar los tamaños de muestra óptimos para alcanzar la significación estadística y biológica necesaria. Los ecólogos deberían hacer un mayor esfuerzo por hacer uso de dicha información previa, pues es su contribución más legítima al proceso inferencial. De esta manera, la estimación de parámetros y la selección de modelos se verían beneficiadas, lo que permitiría que el proceso de aprendizaje fuera más fiable y, por tanto, que las ciencias biológicas pudieran progresar.

Palabras clave: Ecología, Información previa, Frecuentistas, Bayesianos, Estimación de parámetros, Contraste de hipótesis

Alejandro Martínez–Abraín, Depto. de Bioloxia Animal, Bioloxia Vexetal e Ecoloxia, Univ. da Coruña, Campus da Zapateira s/n., 15071 A Coruña, España (Spain) and Population Ecology Group, IMEDEA (CSIC–UIB), c/ Miquel Marquès 21, 07190 Esporles, Mallorca, Espanya (Spain).– David Conesa & Anabel Forte, Grup d'Estadística Espacial i Temporal en Epidemilogia i Medi Ambient, Dept. d'Estadística i Investigació Operativa, Univ. de València, c/ Dr, Moliner 50, 46100 Burjassot, Valencia, España (Spain).

Corresponding author: A. Martínez–Abraín, Evolutionary Biology Group (GIBE), Fac. de Ciencias, Univ. da Coruña, Campus da Zapateira s/n., 15071 A Coruña, Spain. E–mail: a.abrain@udc.es

It has become more and more common worldwide in academic departments of biology and several other sciences to question whether frequentist statistics is the right tool to solve our ecological questions or whether Bayesian statistics should be used instead (Efron, 1986; Clark, 2005; Fidler et al., 2006; Hobbs & Hilborn, 2006; Burnham & Anderson, 2014). Apart from the conceptual and methodological differences between these two approaches, a question of great interest in order to make best use of our costly–to–obtain field data is how to handle previous information when performing statistical inference in ecology.

Previous information and model specification

The first scenario in which ecologists are forced to use their previous knowledge, whether they are adopting a Bayesian or a frequentist point of view, is when performing model specification. For any ecological problem, it is important to set an adequate model to work with, and this decision implies some previous knowledge about the problem. Also, when we are selecting a model —and we cannot usually consider every possible model— choosing a set of adequate models from which to select the best one is also a matter of previous information (subjectivism). This step is important because the result can be biased by this decision. For instance, when adopting a frequentist approach, the whole process of model (= hypothesis) selection, based, for example, on Akaike's Information Criteria, could be biased if a wrong set of hypotheses has been previously selected. As a result, the best thing ecologists can do is to trust in their cumulated experience on the study question and system, and define and contrast the set of hypotheses that is most adequate with the information so far available.

Previous information and parameter estimation for objective and subjective Bayesians

Another scenario where it is necessary to handle previous information is when assessing data available for the parameters. In the Bayesian approach, this is done by introducing a prior distribution for the parameters, although in the frequentist approach this can be somewhat trickier, and sometimes cannot be done. Let's review how assessing previous information for the parameters can be performed in both frameworks. Making use of previous information is considered one of the main characteristics of Bayesian statistics. Within this framework, the Bayes' theorem is used to combine the likelihood (a function that measures our trust in a parameter value given our data) and prior information (*i.e.* the probability distribution of our parameter) to obtain a posterior probability distribution of our parameter given our data. Depending when and how we introduce our previous information, there are two approaches within the Bayesian framework. Subjective Bayesians argue that one can define a prior distribution for unknown parameters according to personal experience and impression, recognizing that the opinion of experts has a value and it is better than nothing. This is highly criticized by frequentists, who claim a more objective procedure. It is also one of the

reasons why non–statisticians avoid Bayesian methods. An objective Bayesian approach is then an attempt to unify frequentist and Bayesian statistics, as pointed out by Bayarri & Berger (2004). Objective Bayesians defend the idea that no other information should be considered apart from that introduced during model specification. As stated in Berger (2006), 'the most familiar element of the objective Bayesian school is the use of objective prior distributions, designed to be minimally informative in some sense'. This procedure provides similar results (even exactly the same results under some circumstances) to classical frequentist analyses, although parameters are treated as random variables. In addition to allowing a non–subjective point of view, the objective Bayesian approach can also be useful in scenarios where choosing priors for our parameters is difficult or even unfeasible. For instance, in the presence of a large number of parameters a subjective specification of all priors could be too time–consuming. Alternatively, it would be more practical to invest time in improving model specification (Berger & Pericchi, 2001). However, whenever good previous information about parameters is available, it is important to make use of it, and the Bayesian subjective approach could be a good tool to achieve this goal.

Prior information and parameter estimation in frequentist analyses

In the frequentist framework, assessing previous information for the parameters is different from the Bayesian approach but, importantly, also existent. Many ecologists feel that this problem of dealing with previous information about parameters is not present when using frequentist methods. We think, however, that introducing this sort of previous information is the largest contribution that ecologists can make to the inferential process. The main situation in which one can introduce previous information about parameters in the frequentist world is when seeking the right sample size needed to test informed null hypotheses during the stage of experimental design. In this case, prior information about the magnitude of the effect that is considered to be biologically relevant needs to be introduced. The sample size can be chosen by running a priori power tests (note that here we mean 'a priori' power tests, rather than 'a posteriori' power tests, a highly criticized procedure). For this purpose, the magnitude of interest of the effect is used (Steidl et al., 1997; Thomas, 1997; Peterman, 1990; Nakagawa & Foster, 2004). Note that this sort of prior information is intrinsic to frequentist analyses because it is intended to maximize the power of the tests. In the Bayesian context, as by construction there is not power of tests, the issue of a small sample size is solved by incorporating previous knowledge through more informative prior distributions. The problem is that using a priori tests and the magnitude of effect is seldom done in the real world of ecological statistics, because ecologists seldom know the magnitude of the effect that corresponds to a biologically relevant effect is (*e.g.* Martínez–Abraín, 2007, 2008, 2014). This is easier to know in experimental (Beaumont & Rannala, 2004) and production–based sciences (such as animal breeding science (Blasco, 2001, 2005)) than when dealing with free–ranging wild species. What is the biologically relevant difference between the wing lengths of two bird populations? We simply do not know. So basically, the vast majority of ecologists often cannot inform their analyses better (Martínez–Abraín, 2013). That is why we ecologists end up using uninformed null hypotheses of equality to zero (Anderson et al., 2000), and hence get the feeling that in a frequentist framework, the problem of dealing with previous information is not an issue. It is, however, but we are forced to overlook it and use null hypothesis testing in a poor way because of the types of problems and data we deal with.

Conclusions

Irrespective of which statistical inference approach one decides to use to estimate a parameter or to contrast hypotheses, one cannot avoid the uncomfortable step of making rather subjective decisions in the scientific process of solving questions. All the information from past experiments and observations, performed by ourselves or by others, is of great value, and it is the ecologist's task to make the best possible use of this information. To show a practical example that affects our research with Mediterranean seabirds, the finding by Ruiz et al. (2000) that the first egg in Audouin's gull *Larus audouinii* decreased in size by 5 cm^3 when food was particularly scarce during clutch formation is a valid piece of previous information that can and should be used when assessing differences in egg size between Audouin's gull populations in the future. In this case, we would use this value as a guiding clue when assessing if the difference in the mean egg size between two populations is biologically relevant. Specifically, we would perform a priori power tests, with 5 cm^3 as our effect size of interest (instead of using zero differences between populations as our default reference), and will determine the sample sizes required to obtain statistically significant results if (and only if) the difference between the two study populations is at least of that magnitude (5 cm^3). Alternatively, a 'safe' arbitrary minimum percentage of change in egg size could be established by consensus among researchers if we do not have a concrete reference value of effect size available from previous studies. This is, in fact, our most legitimate contribution to a process that, in all other respects, is in the hands of statisticians. It is good that we acknowledge for once that in the presence of previous information, we do have proper tools to handle it, and that both the precision of our parameter estimates and the results of our hypothesis contrasts can improve substantially if we focus more on this slippery but fundamental compartment of our daily research.

References

Anderson, D. R., Burnham, K. P., Thompson, W. L., 2000. Null hypothesis testing: problems, prevalence, and an alternative. *Journal of Wildlife Management*, 64: 912–923.

Bayarri, M. J. & Berger, J. O., 2004. The interplay between Bayesian and frequentist analysis. *Statistical Science*, 19: 58–80.

Beaumont, M. A. & Rannala, B., 2004. The Bayesian revolution in genetics. *Nature Review Genetics*, 5: 251–261.

Berger, J. O., 2006. The case for objective Bayesian analysis. *Bayesian Analysis,* 1: 385–402.

Berger, J. O. & Pericchi, L. R., 2001. Objective Bayesian Methods for Model Selection: Introduction and Comparison. *Lecture Notes–Monograph Series*, 38: 135–207.

Blasco, A., 2001. The Bayesian controversy in animal breeding. *Journal of Animal Science*, 79: 2023–2046.

– 2005. The use of Bayesian statistics in meat quality analyses: a review. *Meat Science*, 69: 115–122.

Burnham, K. P. & Anderson, D. R., 2014. P values are only an index of to evidence: 20th– vs. 21st–century statistical science. *Ecology*, 95: 627–630.

Clark, J. S., 2005. Why environmental scientists are becoming Bayesians. *Ecology Letters*, 8: 2–14.

Efron, B., 1986. Why isn't everyone a Bayesian? *American Statistician*, 40: 1–5.

Fidler, F., Burgman, M. A., Cumming, G., Buttrose, R. & Thomason, N. R., 2006. Impact of criticism of null–hypothesis significance testing on statistical reporting practices in conservation biology. *Conservation Biology*, 20: 1539–1544.

Hobbs, N. T. & Hilborn, R., 2006. Alternatives to statistical hypothesis testing in ecology: A guide to self teaching. *Ecological Applications*, 16: 5–19.

Martínez–Abraín, A., 2007. Are there any differences? A non–sensical question in ecology. *Acta Oecologica*, 32: 203–206.

– 2008. Statistical significance and biological relevance: A call for a more cautious interpretation of results in ecology. *Acta Oecologica*, 34: 9–11.

– 2013. Why do ecologists aim to get positive results? Once again, negative results are necessary for better knowledge accumulation. *Animal Biodiversity and Conservation*, 36.1: 33–36.

– 2014. Is the 'n = 30 rule of thumb' of ecological field studies reliable? A call for greater attention to the variability in our data. *Animal Biodiversity and Conservation,* 37.1: 95–100.

Nakagawa, S. & Foster, T. M., 2004. The case against retrospective statistical power analyses with an introduction to power analysis. *Acta Ethologica*, 7: 103–108.

Peterman, R. M., 1990. The importance of reporting statistical power: The forest decline and acidic deposition example. *Ecology*, 71: 2024–2027.

Ruiz, X., Jover, L., Pedrocchi, V., Oro, D. & González–Solís, J., 2000. How costly is clutch formation in the Audouin's gull Larus audouinii? *Journal of Avian Biology*, 31: 567–575.

Steidl, R. J., Hayes, J. P. & Schauber, E., 1997. Statistical power analysis in wildlife research. *Journal of Wildlife Management*, 61: 270–279.

Thomas, L., 1997. Retrospective power analysis. *Conservation Biology*, 11: 276–280.

6

A new endemic species of *Bryconamericus* (Characiformes, Characidae) from the Middle Cauca River Basin, Colombia

C. Román–Valencia, R. I. Ruiz–C., D. C. Taphorn B. & C. García–Alzate

Román–Valencia, C., Ruiz–C, R. I., Taphorn B., D. C. & García–Alzate, C., 2014. A new endemic species of *Bryconamericus* (Characiformes, Characidae) from the Middle Cauca River Basin, Colombia. *Animal Biodiversity and Conservation*, 37.2: 107–114.

Abstract

A new endemic species of Bryconamericus *(Characiformes, Characidae) from the Middle Cauca River Basin, Colombia.*— *Bryconamericus caldasi*, a new species, is described from the Middle Cauca River drainage, Andean versant of Colombia. The new species is distinguished from all congeners by: the number of predorsal scales (15–17 *vs.* 9–14); a wide anterior maxilla tooth, at least twice as wide as the posterior tooth, both of which are pentacuspid (*vs.* maxilla teeth of same size); a dark lateral stripe overlaid by a peduncular spot; and a reticulated pattern on the sides of body (*vs.* peduncular spot and other body pigments not superimposed over a dark lateral stripe). We found several differences that distinguish the new species from *B. caucanus,* the only sympatric congener: number of predorsal median scales (15–17 *vs.* 12–13); convex predorsal profile (*vs.* oblique); scale size and number of scale rows at caudal–fin base (small scales arranged in two or more rows vs. large scales in just one row); pectoral fins not or just reaching pelvic fin insertions (*vs.* pectoral fins reaching posterior to pelvic–fin insertions); and dorsal–fin origin position (at vertical through posterior tip of pelvic–fin *vs.* at a vertical anterior to pelvic–fin tip).

Key words: Biodiversity, Taxonomy, Tropical Fish, New Taxon

Resumen

Una nueva especie endémica de Bryconamericus *(Characiformes, Characidae) de la cuenca media del río Cauca, en Colombia.*— Se describe una nueva especie, *Bryconamericus caldasi*, en la cuenca media del río Cauca en los Andes de Colombia. La nueva especie se distingue de todos sus congéneres por el número de escamas predorsales (15–17 *vs.* 9–14) y también por poseer el diente maxilar anterior ancho, al menos dos veces más ancho que el diente posterior; ambos son pentacúspides (*vs.* dientes del maxilar de igual tamaño) y por una banda lateral oscura que se solapa con la mancha peduncular y que tiene un dibujo reticulado a ambos lados del cuerpo (*vs.* mancha peduncular y otros pigmentos no solapados sobre la banda lateral oscura). Se observaron diferencias que distinguen a la nueva especie de *B. caucanus,* el único congénere simpátrico: el número de escamas predorsales (15–17 *vs.*12–13), el perfil dorsal convexo (*vs.* oblicuo), el tamaño de la escama y el número de filas de escamas en la base de la aleta caudal (escamas pequeñas y ordenadas en dos o más filas *vs.* escamas largas y ordenadas en una sola fila), las aletas pectorales que no llegan a la inserción de las aletas pélvicas o llegan muy justo (*vs.* aletas pectorales que llegan a las inserciones de las aletas pélvicas) y la posición del origen de la aleta dorsal (en la vertical del extremo posterior de las aletas pélvicas *vs.* en la vertical del extremo anterior de las aletas pélvicas.

Palabras clave: Biodiversidad, Taxonomía, Pez tropical, Nuevo taxón

C. Román–Valencia, R. I. Ruiz–C, D. C. Taphorn B. & C. García–Alzate, Lab. de Ictiología, ·Univ. del Quindío, A. A. 2639, Armenia, Colombia.– D. C. Taphorn B, 1822 N. Charles St., Belleville, 62221 Illinois, USA.– C. García–Alzate, Programa de Biología, Univ. del Atlántico, km 7 antigua vía a Puerto Colombia, Barranquilla, Colombia.

Corresponding author: C. Román–Valencia. E–mail: ceroman@uniquindio.edu.co

Introduction

Currently, 79 species are assigned to the genus *Bryconamericus* (Román–Valencia et al., 2008; Eschmeyer & Fricke, 2013). Of these, 22 valid species occur in Colombian watersheds (Román–Valencia et al., 2008, 2009a, 2009b, 2010, 2011, 2013), but only one species, *B. caucanus*, has been reported from the Río Cauca (Román–Valencia et al.,2009a), and it is known to have a wide geographic distribution in Colombia (Román–Valencia, 2003; Román–Valencia et al., 2009a). The known distribution of *Bryconamericus* in Colombia suggests that this genus is more diverse in Transandean than in Cisandean drainages, probably because of several geographic isolation events (Román–Valencia et al., 2009b).The description of this new species of *Bryconamericus* from the Middle Cauca River Basin is a result of our ongoing revision of the genus, and is further evidence of the undocumented biodiversity of *Bryconamericus*.

Material and methods

Measurements were taken with digital calipers, recorded to tenths of millimeters and usually expressed as percentages of standard (SL) or head length (HL) (table 1). Counts were made using a stereoscope with a dissection needle to extend the fins. In count ranges, values for the holotype are indicated with an asterisk (*). Counts and measurements were taken from the left side of specimens when possible and following the guidelines in Vari & Siebert (1990) and Armbruster (2012). The 21 morphometric characters analyzed in this study (table 1) were evaluated by Principal Component Analysis (PCA) using the Burnaby method to eliminate the influence of size with the PAST program, version 1.81 for Windows (Hammer et al., 2008).

Observations of bones and cartilage were made on cleared and stained specimens (C&S), prepared according to techniques outlined in Taylor & Van Dyke (1985) and Song & Parenti (1995). Bone nomenclature follows Weitzman (1962), Vari (1995), and Ruiz–C. & Román–Valencia (2006). In the lists of paratypes, the number of individuals is given immediately after the catalog number, which is followed by the range of standard length in mm (SL) for that lot; for example: IUQ 3223 (7) 51.1–66.9 mm SL, indicates seven individuals in lot IUQ 3223, with the smallest fish measuring 51.1 mm SL and the largest 66,9 mm SL. All collections were made in Colombia. Acronyms used follow Sabaj–Pérez (2010). Meters above sea level is abbreviated as m a.s.l., Municipio is translated as County.

Comparative material

Bryconamericus andresoi (see Román–Valencia, 2003). *B. caucanus* (see Román–Valencia, 2003; Román–Valencia et al., 2009a, 2009b); IUQ 3715 (3), Caldas, Arauca, middle Cauca River, La Margarita Creek, at La Margarita on the road to Manizales city, 100 m upstream from bridge, 05° 06' 26.4" N, 5° 45' 30" W,

1,101 m a.s.l., 2 VI 2012. IUQ 3223 (7), Caldas, Viterbo County, vereda el Guamo, Middle Cauca River, El Guamo Creek, 05° 03' 04" N, 75° 45' 20" W, 1,119 m a.s.l., 11 VII 2011; IUQ 3771 (2C&S), Caldas, Viterbo County, vereda el Guamo, Middle Cauca River, El Guamo Creek, 05° 03' 04" N, 75° 45' 20" W, 1,119 m a.s.l. 11 VII 2011; IUQ 3227 (3), Caldas, river at La Marina farm, 200 m from the bridge on the road to San Juan Arauca, 04° 03' 46" N, 76° 18' 44,9" W, 2 VI 2012; IUQ 3228 (24), Caldas, Viterbo County, El Guamo Creek, vereda el Guamo, 05° 03' 04" N, 75° 49' 02" W; IUQ 3716 (4),Caldas, Viterbo County, Apia River, a tributary of Guarne River, 05° 05' 46" N, 75° 52' 04" W, 1,009 m a.s.l., 20 I 2014; IUQ 3717 (2), Caldas, Viterbo County, La Isla Creek on the Viterbo Medellín road, 05° 07.7' 82" N, 75° 49.9' 22" W, 1,028 m a.s.l.; IUQ 3718 (4), Caldas, Anserma County, Lázaro Creek at the outskirts of Anserma–Risaralda on the road to Medellín, 05° 07.3' 42" N, 75° 42' 95" W, 1,044 m a.s.l., 20 I 2014; IUQ 3719 (6), Caldas, Arauca County, La María Creek, vereda La María, 100 m on the road to Arauca, 05° 13.5' 15" N, 75° 42' 35" W, 1,044 m a.s.l. 19 I 2014; IUQ 3720 (2), Caldas, San José County, Los Caimos Creek, tributary of Risaralda River, 200 m. upstream from Asia on the road to Medellín, 05° 13.5' 5" N, 75° 50' 14" W, 995 m a.s.l., 20 I 2014; IUQ 3723 (6), Caldas, San José County, Creek on the road 200 m from La Libertad Creek, vereda La Libertad on the road between San José and Arauca, 05° 06' 26.4" N, 75° 45' 30" W, 1,101 m a.s.l. *B. galvisi* (see Román–Valencia, 2000). *B. huilae* (see Román–Valencia, 2003). *B. plutarcoi* (see Román–Valencia, 2001). *B. foncensis* (see Román–Valencia, 2009b). *B. oroensis* (see Román–Valencia et al., 2013). *B. brevirostris*: IUQ 3215 (19), Ecuador, Loja,on international bridge at Macará, 30 III 1979; MUSM 3393 (3), Perú, Tumbes, San Jacinto, La Peña, Tumbes River, spillway, 6 VII 1992; MUSM 3306 (3), Perú, Tumbes, Tumbes River, 500 m from bridge, 03° 29' 33.60" S, 80° 27' 26.39" W, 0 m a.s.l., 5 VII 1992; MUSM 6889 (50), Ecuador, Guayas, Guayas River Basin, Cotimes, Daule River; MUSM 5732 (7), Perú, Tumbes, San Jacinto Bocatoma, Tumbes River, 16 VIII 1994; MUSM 3394 (9), Perú, Tumbes, Tumbes River in irrigation channel, 6 II 1992; MUSM 3058 (1), Perú, Tumbes, Zarumilla, Matapalo, Zarumilla River, 03° 9' 39" S, 80° 14' 15.35" W, 7 m a.s.l., 11 XII 1990; MUSM 5765 (26), Perú, Tumbes, Zarumilla, Lepanga, Zarumilla River, 15 VIII 1994; MUSM 2582 (20), Perú, Tumbes,Tumbes River, near irrigation channel 10 VIII 1986; MUSM 1983 (20), Tumbes, Zarumilla River and shallow pools, 12 VIII 1986. *B. diaphanus*: All from Perú: ANSP 21216 (Paralectotype), Huallaga River en Moyabamba; MUSM 33475 (40), San Martín, Moyobamba, Indeche River, 826 m a.s.l., *B. osgoodi*: All from Perú: CAS 40828 (Holotype), Moyabamba, Huallaga River basin. *B. pachacuti*: All from Peru: MCZ 31563 (Holotype) Santa Ana, Urumbamba River; CAS 40829 (22) (Paratype), Cusco, Amazon River, Urubamba River; IUQ 3155 (1C&S), Ecuador, Morona–Santiago, Yapapa River tributary Santiago River, 9 V 1991; MUSM 29947 (3), Pasco, Okapampa, Icozacia, Mayo River, 20 V 2004; MUSM 32466 (4), Perú,

Table 1. Morphometry of *Bryconamericus caldasi* n. sp. (standard and total length in mm; average in parentheses).

Tabla 1. Morfometría de Bryconamericus caldasi sp. n. (longitud estándar y longitud total en mm; promedio entre paréntesis).

	Holotype	Paratypes
Standard length	63.7	47.4–73.1 (65.6)
Total length	79.1	57.4–89.9 (80.2)
Percentages of SL		
Body depth	30.6	24.3–34.4 (29.4)
Snout–dorsal fin distance	54.1	43.1–58.2 (50.3)
Snout–pectoral fin distance	23.3	20.4–27.8 (24.2)
Snout–pelvic fin distance	42.9	36.6–52.2 (44.0)
Snout–anal fin distance	57.9	51.9–66.6 (57.0)
Dorsal fin–hypural distance	52.1	46.9–59.4 (52.0)
Dorsal–fin length	22.6	20.4–32.8 (23.6)
Pectoral–fin length	21	16.4–22.3 (19.6)
Pelvic–fin length	13.5	10.7–16.8 (14.0)
Caudal peduncle depth	13.9	10.3–14.9 (12.6)
Caudal peduncle length	12.3	11.4–18.6 (15.8)
Head length	20.5	19.1–24.8 (21.6)
Dorsal–anal fin distance	31.4	24.4–36.0 (30.9)
Dorsal–pectoral fin distance	40.1	35.3–45.6 (39.8)
Anal–fin length	18.2	9.4–18.3 (13.3)
Percentages of HL		
Snout length	26.2	19.9–44.4 (26.1)
Orbital diameter	35.4	28.9–39.5 (33.4)
Postorbital distance	48	41.1–55.6 (48.8)
Maxilla length	35.9	35.8–48.9 (43.7)
Interorbital distance	37.3	31.2–39.7 (34.7)
Mandible superior distance	32.4	26.4–38.9 (30.6)

Cusco, La Convenciòn, Echarate, Iherimpituari Creek, Paratori River, 16 III 2008; MUSM 11120 (24), Perú, Puno, Sandia, Candamo River, 358 m a.s.l., 2 IV 1997; MUSM 30199 (3), Perú, Pasco, Oxapampa, Pto. Bera tributary Apunmacayali River, 26 V 2004; MUSM 35771 (9), Perù,Ucayali, Atalaya, Sepahua,Huayashi Creek, 26 VII 2007; MUSM 12329 (1), Perú, Junin, Perené River road to Satipo, 21 IX 1995; MUSM 37348 (22), Perú, Ucayali, PadreAbad, Aguaytia River, Shamabo River, 8° 50' 03" S, 75° 34' 10" W, 258 m a.s.l., 26 V 2009; MUSM 16144 (30), Perú,Ucayali, Padre Abad, Aguaytia River, Huiango Creek km 18 road Curimaná, 14 V 1997; MUSM 30363 (50), Perú, Pasco, Oxampa, Villa Rica, San Pedro de Pichanos village, Pichanos Creek, 3 VI 2004; MUSM 18017 (50), Peru, Huanuco, CCA, Pachitea River, Honoris, Sargento Lores islands, 4 VII 2005; MUSM 20562 (50), Perú, Pasco, Oxapampa Pto. Bermudez, Ataz Creek, 9 VIII 2002; MUSM 34324 (27), Perú, Cusco, Convención, Echarate, CCNN Camanà, Alto Uru-bamba River, 29 IX 2008; MUSM 29125 (1), Perú, Madre de Dios, Tambopata,Tambopata River, Botafogo beach, 12° 17' 05.52" S, 71° 52' 07.01" W, 276 m a.s.l., 13 VI 2006; MUSM 25433 (2), Peru, Madre de Dios, Tambopata Mazuko, Inambari River, Quenque Creek, 8 IX 2009. *B. phoenicopterus*: ANSP 8093 (holotype), Perú,Ambyacu Loreto River, Maynas, near Pebas, tributary of the Amazonas, (07° 46' 32.62" S, 77° 53' 46.49" W, 2,297 m a.s.l.; MEPN 2120 (200), Ecuador, Zamora, Chinchipe, beach near military

Table 2. Physicochemical variables in the habitat of *Bryconamericus caldasi* n. sp., middle Cauca River Basin, Colombia: 1. La Libertad Creek; 2. La Libertad Creek; 3. La Libertad Creek in La Marina; 4. La Libertad Creek in El Gril; 5. La Libertad Creek in Marandua; Rd. Rocks and detritus; Rs. Rocks and sand.

Tabla 2. Variables fisicoquímicas del hábitat de Bryconamericus caldasi *sp. n., cuenca media del río Cauca, Colombia: 1. Riachuelo La Libertad; 2. Riachuelo La Libertad; 3. Riachuelo La Libertad en La Marina; 4. Riachuelo La Libertad en El Gril; 5. Riachuelo La Libertad en Marandua; Rd. Rocas y detritos; Rs. Rocas y arena.*

	Locality				
	1	2	3	4	5
m a.s.l.	1,124	1,101	1,003–1,078	1,203	1,095
Water temperature (°C)	17	18.9	17.4–24.0	23.5	23
Air temperature (°C)	24	26	23.5–26	27	28
Dissolved oxygen (mg/l)	5.4	5.3	4.2–7.2	5.7	5.4
pH	7.9	8	7–8.3	7.6	7.7
Width (m)	3–4	1–2	5–6	1–2	0.2–0.5
Depth (m)	0.5–1.0	0.5–1	0.5–1.0	0.5–1.0	0.2–0.5
Color	clear	clear	clear	clear	clear
Substrate	Rd, Rs	Rs	Rs	Rd, Rs	Rs

base, lower Mayaicu, 03° 58' 15" S, 78° 41' 15" W, 18 VIII 1993; MEPN 44 (11), Ecuador, Zamora Chinchipe, Nangantza River beach near military base, Mayaicu, 18 VII 1993; IUQ 3135 and 3153 (2C&S), Ecuador, Morona Santiago, Gualaquiza River, 22 IX 1978; MUSM 20722 (4), Perú, San Martin, Tarapota, Morales, San Antonio, Cuerabraza River, 18 IX 1998. *B. pectinatus*: All from Peru, Madre de Dios: MUSM 3821 (1), Calli Creek, 05 IX 1988; MUSM 3809 (1), Manu National Park, Manu River beach near Cucha, 12° 17.05' 52" S, 7° 52' 07.01" W, 8 IX 1988; MUSM 3825, Manu National Park, 2 II 2002. *B. peruanus* MUSM 5752 (80), Perú, Piura, Sullana, Mallares, Saman bridge, side channel, Chira River, 18 VIII 1994; (see also Román–Valencia et al., 2011).

B. thomasi: all from Bolivia: ANSP 68740 (Holotype), Paraná–Paraguay system, Río Lipeo, tributary of the Bermejo River at the Argentina–Bolivia border, 22° 44' 51.57" S, 64° 20' 30.51" W, 393 m a.s.l., VIII 1936; Chuquisaca, 2 km in a straight line SE of Monteagudo, 29 IX 1998; CBF 01228 (10), Tarija, Gran Chaco County, 1.5 km in a straight line SO of Villamontes, 2 X 1988; CBF 01198 (3), Chuquisaca, H. Stiles, Bermejo Ichilo–Mamoré River Basin, 2 km in a straight line SE of Monteagudo; UMSS 00806 (2), del Plata/ Pilcomayo, Pilaya River, 12 VII 2005; UMSS 00740 (12), del Plata / Bermejo, Emborozu River, 12 VII 2005; UMSS 00805 (35), del Plata / Bermejo, Orosas River, 12 VII 2005; UMSS 03131 (14), del Plata/Bermejo, Guadalquivir River, 10 VII 2006; UMSS 04945 (1), del Plata/Bermejo, Bermejo/ Gran de Tarija, Tarija River, 21 XI 2006; UMSS 04530 (3), del Plata/Bermejo, Grande de Tarija/Tarija, Salinas

River, 5 X 2004; UMSS 5106 (11), del Plata / Bermejo, Arroyo Toro, 1 VII 2006; UMSS 00719 (1), Amazon/ Mamoré, Salado River, 17° 30' 46.20" S, 64° 48' 30.56" W, 1,862 m a.s.l., 11 VII 2005; UMSS 00891 (8), Amazon/ Itenez, San Pablo/ Parapeti, Heredia River, 23 X 2005; UMSS 04968 (3), del Plata / Bermejo, Bermejo/ Gran de Tarija, Tarija River, Saycan River, 6 X 2004. *Bryconamericus* sp.1 CBF 06023 (10); Bolivia, Santa Cruz, Ichilo PN–AMI, Amboro, San Juan del potrero, Amazonas, Ichilo–Mamoré River Basin, 17° 48' 59.82" S, 64° 13' 00,00" W, 2,289 m.a.s.l., *Bryconamericus* sp. 2 UMSS 00699 (25) Bolivia, Salado River, Amazonas/Mamorè,–22.426168, 59.617998, 12 VII 2005.

Bryconamericus sp. 3: UMSS 01227 (50); Bolivia, Santa Cruz, Amazonas, Itènez, Izozog/Parapeti, Parapeti River Basin, upstream from Camiri (–20.017632,–63.560430).

Bryconamericus sp. 4: MUSM 31598 (30); Perú, Costa Pacífica, Lambayeque, Tenerife, Lañamis, Huancdocinibe River, Cañariaco River, 19 IX 2007. *Hemibrycon boquiae*; IUQ 3226 (15), Caldas, San José County, Middle Cauca River, road to San José, Arauca, finca El Gril, La Libertad Creek, 04° 03' 46" N, 76° 18' 44,9" W, 2 I 2012; IUQ 3692 (3), Caldas, Viterbo County, El Guamo Creek nearl Guamo, 05° 03' 04" N, 75° 49' 02" W,2 VII 2011; IUQ 3721 (4); Caldas, Lazaro Creek at Anserma–Risaralda borde ron road to Medellin, 05° 07' 34" N, 75° 42.9' 75" W, 1,044 m a.s.l., 20 I 2014; IUQ 3722 (6); Caldas, Viterbo County, Los Caimos Creek tributary Los Caimos River, tributary Risaralda River 200 m from Asia, road to Medellín, 05° 05.43' 08" N, 75° 50.0' 14" W, 995 m a.s.l.,

Fig. 1. *Bryconamericus caldasi* n. sp., holotype, male, IUQ 3714, 63.7 mm SL, San José County, La Libertad Creek, 200 m from La Libertad school on the San José, Arauca Road, Caldas, Colombia.

Fig. 1. Bryconamericus caldasi *sp. n., holotipo, macho, IUQ 3714, 63,7 mm de longitud estándar, condado de San José, riachuelo La Libertad, a 200 m de la escuela La Libertad, situada en la carretera de San José a Arauca, en Caldas, Colombia.*

20 I 2014. IUQ 3725 (13), Caldas, Belalcazar county, Middle Cauca River, Creek on the road Belalcazar–Arauca, 5° 03.6' 05" N, 75° 46.1' 75" W, 1,025 m a.s.l., 5 II 2014.

Bryconamericus caldasi n. sp. (tables 1–2, figs. 1–2)

Holotype: IUQ 3714, 63.7 mm SL, Colombia, Caldas, San José County, Middle Cauca River Basin, La Libertad Creek, two hundred meters from La Libertad School, on the San José–Arauca road, 75° 45' 53.6" W, 05° 63' 07.4" N, 1,124 m a.s.l.

Paratypes: all from Colombia, Caldas, middle Cauca River Basin: IUQ 3225 (9), 54.8–73.5 mm SL, San José County, creek on the road 200 m from La Libertad Creek, La Libertad on the road San José–Arauca, 05° 06' 26.4" N, 75° 45' 30" W, 1,101 m.a.s.l.; IUQ 3772 (2C&S), 54.6–69.2 mm SL, San José County, creek on the road, 200 m from La Libertad Creek, La Libertad on the road San José–Arauca, 05° 06' 26.4" N, 75° 45' 30" W, 1,101 m a.s.l.; IUQ 3229 (17), 47.4–67.1mm SL, same locality as holotype. IUQ 3691 (1), 67.1 mm SL, Arauca County, La Libertad Creek, El Gril Ranch, on the San José–Arauca road, 05° 6' 9.22" N, 75° 46' 27.8" W, 1,203 m a.s.l.; IUQ3723 (14), 58.8–70.51 mm SL, Arauca–San José County, La Libertad Creek, La Marina–El Eden Ranch, 05° 875' N, 75° 52.037' W, 1,007 m a.s.l.

Diagnosis
Bryconamericus caldasi is distinguished from most congeners by having a dark lateral stripe overlaid by a peduncular spot and reticulated pattern on the sides of the body (*vs.* peduncular spot and other body pigments not superimposed over a dark lateral stripe, except in *B. oroensis* which has a dark lateral stripe on body from posterior edge of opercle to base of caudal fin; see Romàn–Valencia et al., 2013); by the predorsal scale counts (15–17 *vs.* 9–14, except *B. andresoi* with 13–15, *B. galvisi* with 12–17, *B. huilae* with 14–19, *B. plutarcoi* and *B. foncensis* with 11–16); and a wide anterior maxilla tooth, at least twice as wide as the posterior tooth, both of which are pentacuspid (*vs.* maxilla teeth of same size). The new species differs from *B. oroensis* (Román–Valencia et al., 2013) by: pectoral–fin length (16.4–22.3% SL *vs.* 32.0–39.3% SL), pelvic–fin length (10.7–16.8% SL *vs.* 22.0–25.9% SL), caudal peduncle depth (9.3–14.9% SL *vs.* 16.7–23.8% SL), dorsal–anal fin distance (24.4–36.0% SL *vs.* 10.8–13.0% SL), dorsal–pectoral distance (34.4–45.6% SL *vs.*9.4–13.5% SL), anal–fin length (9.4–18.3% SL *vs.* 23.4–27.8% SL), length of maxilla (35.8–48.9% HL *vs.* 22.3–36.2% HL), and by number of scale rows between pelvic–fin and lateral lines (4–5 *vs.* 6–8). We found the following differences that distinguish the new species from the sympatric *B. caucanus*: number of predorsal median scales (15–17 *vs.*12–13 see Román–Valencia et al., 2009a, table 2); pectoral–fins not reaching or just reaching pelvic–fin insertions (*vs.* pectoral fins reaching posterior to pelvic–fin insertions); convex predorsal profile (*vs.* oblique); scale size and number of scale rows at caudal–fin base (small scales arranged in two or more rows, vs. large scales in just one row); dorsal–fin origin position (at vertical through posterior tip of pelvic–fin *vs.* at a vertical anterior to pelvic–fin tip).

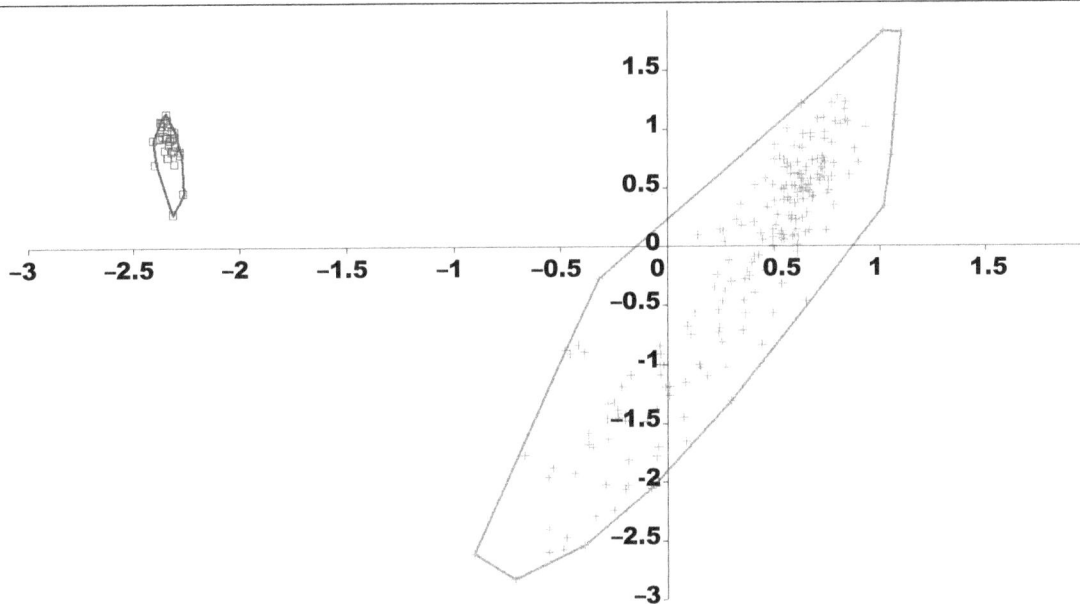

Fig. 2. Schematic diagram of the first two principal components of morphometric data for *Bryconamericus caucanus* (*n* = 217, +) and *B. caldasi* n. sp. (*n* = 35, □). First and second component in the x and y axis, respectively.

Fig. 2. Representación de los primeros dos componentes principales de los datos morfométricos de Bryconamericus caucanus (n = 217, +) y B. caldasi sp. n. (n = 35, □). El primer y el segundo componente se representan en los ejes x e y, respectivamente.

Description

Table 1 shows the study morphometrics. Greatest body depth near dorsal–fin origin (mean maximum body depth about 29.4% SL). Area above orbits convex. Dorsal profile of head and body curved from supraoccipital to dorsal–fin origin and from last dorsal–fin ray to base of caudal–fin. Ventral profile of body curved from snout to anal–fin base. Caudal peduncle laterally compressed. Head and snout short, mandibles equal, mouth terminal, lips soft and flexible and covering outer row of premaxilla teeth; ventral border of upper jaw not straight; posterior edge of maxilla reaching anterior edge of orbit; opening of posterior nostrils vertically ovoid; opening of anterior nostrils with a membranous flap. Distal tip of pectoral–fin not or just reaching pelvic–fin insertions. Distal tip of pelvic–fin not reaching anal–fin origin. Caudal–fin not scaled, except at its base, forked with short pointed lobes, principal caudal rays 1/18/1 with 9/10 procurrents. Lateral line usually complete, with 34–43 pored scales that extend, in a downward curved line, from supracleithrum back towards the hypurals.

Premaxilla with two rows of teeth and large rounded lateral process. Four to six teeth of outer row tricuspid. Inner row with four pentacuspid teeth of equal size. Maxilla long, exceeding two–quarter lengths of the second infraorbital, anterior margin with notches, with four tricuspid teeth, at least twice as wide as the posterior tooth. Dentary with four large pentacuspid teeth,

gradually diminishing in size laterally. Supraoccipital spine short, covering anterior axis of neural complex.

Lateral line complete, perforated scales 34(5), 35(3), 36(3), 37(7), 38*(5), 39(4), 40(2), 41(2) or 43(2). Scale rows between dorsal–fin origin and lateral line 4(1), 5(34) or 6*(1); scale rows between lateral line and anal–fin origin 4(16), 5*(18) or 6(1); scale rows between lateral line and pelvic–fin insertion 4(31), 5*(3) or 6(1). Predorsal scales 15*(9), 16(24) or 18(1), arranged in regular series. Anal–fin rays iv (25) or v*(11), 23(1), 24(3), 25(12), 26(9), 27(6) or 28*(2). Pectoral–fin rays ii, 10*(16), 11(18) or 12(1). Dorsal–fin rays ii, 8*; Pelvic–fin rays i, 6, i*; last ray simple; first unbranched ray approximately one–half length of second ray, its tip reaching first bifurcation of first branched ray. Total number of vertebrae 38–41.

Eight to nine supraneurals present between head and anterior part of dorsal–fin, located between sixth and thirteenth dorsal spine and the first dorsal pterygiophores. Proximal pterygiophores (26 to 28) of the anal fin completely ossified. Cleithrum with pointed dorsal process that does not surpass entire supracleithrum, which is joined to post temporal.

Secondary sexual dimorphism

Sexually mature males have 15–22 hooks on anterior branched anal–fin rays, including first to ninth, and on one simple anal–fin ray with eight to ten hooks. They also have twelve to nineteen hooks along the ventral surface of branched pelvic–fin rays and one simple ray.

Color in alcohol
Dorsum dark, greenish. Body with very dark lateral band from posterior edge of opercle to base of caudal fin. Humeral spot round with faint ventral and dorsal projections. Peduncular spot rounded, extending beyond caudal peduncle, continuing on to middle caudal–fin rays. Ventro–lateral region of body from snout tip to caudal peduncle light yellow. All fins gray, on the anal fin with dark bands on distal portions of rays.

Live colors
Dorsum of body and head and postventral region dark greenish, with black pigment. Body silvery white with yellow lateral stripe. Middle caudal–fin rays covered by a narrow band of melanophores that forms a slender arc or half–moon shaped spot on caudal peduncle. There is a small purple spot between the fifth and sixth infraorbitals and the opercle. The opercle has melanophores concentrated on the posterior portion. Humeral spot dark and rounded with disperse pigments; dark caudal spot elongate and continued on to middle caudal–fin rays. Pectoral, and pelvic–fins hyaline, dorsal, anal and caudal– fins yellow, but distal tips of caudal–fin rays white, and anal–fin yellowish on anterior rays but with posterior rays and distal tips of anterior rays intense white; dispersed melanophores present on interradial membranes. Opercular area silvery blue covering posterior margin of eye, opercular bone series and extending on to ventral region of body.

Distribution and ecological notes

This species is known from the middle Cauca River Basin in La Libertad Creek, San José County, Caldas, Colombia. *B. caldasi* was captured in La Libertad creek, a clear creek that is characterized by a relatively rapid water current, running over rocky and sandy bottoms. The pH was near basic values (7.0–8.3), and dissolved oxygen values were high (table 2), typical of oligotrophic environments.

The new species is syntopic with *Andinoacara* sp., *Astroblepus* sp., *Brycon henni*, *Hemibrycon boquiae*, *H. rafaelense*, *Chaetostoma fischeri*, *Poecilia caucana*, *Xiphophorus hellerii*, and *Trichomycterus caliensis*. The analysis of stomach contents of three specimens revealed the presence of adults and larvae of different species of Diptera (12.5%, 4.0 mm^3), Diptera: Simulidae (12.5%, 3.0 mm^3), Diptera: Chironomidae (12.5%, 3.0 mm^3), Diptera: Dixidae (12.5%, 2.0 mm^3), Trichoptera: Leptoceridae (12.5%, 14.0 mm^3), Ephemeroptera: Baetidae (25%, 5.0 mm^3), Hymenoptera: Vespidae (12.5%, 2.0 mm^3). The presence of both autochthonous and some allochthonous items suggests that this species is insectivorous with a considerable plasticity in its diet.

Etymology
Bryconamericus caldasi named to honor the memory of the Colombian naturalist Francisco José de Caldas, who devoted his life to the study of Neotropical nature, and whose intellectual merit lies in having embraced the incipient patriotic fervor in the struggle for Colombian independence in the first half of the nineteenth century.

Comments
Principal component analyses performed on all species examined were not informative for most of them, but did distinguish *Bryconamericus caldasi* from the sympatric *B. caucanus* a by differences in pectoral–fin length, pelvic–fin length, caudal peduncle depth and dorsal–pectoral fin distance along the axis x, and snout length, postorbital distance and mandible superior distance along the axis y. The first component explained 87.5% of the total variability and the second 5.51%, for a total of 93.01% of the variation (fig. 2).

Although we made several attempts to collect this new species in different tributaries of the middle Cauca and Risaralda rivers (see comparative examined material), it was only collected from La Libertad Creek. This is the first record of such a narrowly endemic species of *Bryconamericus* in South America.

Discussion

In most species of *Bryconamericus*, there are one or two large, rounded scales located at the base of the caudal lobes. Furthermore, squamation does not extend beyond one–third of the length of the caudal–fin rays, and when well preserved, scales do not cover the procurrent caudal–fin rays (Romàn–Valencia et al., 2013). In *B. caldasi* n. sp., these scales are smaller than in most species of *Bryconamericus* and arranged in two rows, showing an intermediate condition between other species of *Bryconamericus* and *Knodus*, in which there are more than two scale rows that cover more than the proximal third of the caudal–fin.

Román–Valencia et al. (2009a) reported that the populations of *B. caucanus* from the middle Cauca River Basin were fairly uniform morphologically as analyzed using Principal Component Analysis, and in their osteology and meristic characters. However, in this study of the new species, we found that it differs from *B. caucanus* by the number of predorsal median scales, the length and position of the pectoral–fins, shape of the predorsal profile, number of scale rows at the base of the caudal–fin and position of the dorsal–fin.

Acknowledgements

We thank the University of Quindío, Vicerrectoría de Investigaciones, for financial assistance to carry out this study (to C. R.–V. & R. I. R.–C.) grant 594, Soraya Barrera, Jaime Sarmiento (CBF), Mabel Maldonado (UMSS), Hernan Ortega (MUSM) and Ramiro Barriga (EPN) for the loan and donation of specimens. We also thank Oscar A. Duque (IUQ) for elaborating figure 1 and Cristian Román–P. (IUQ and UV) for identifying stomach contents. We are also grateful to the editors of ABC, Jonathan W. Armbruster (AUM), and Juan Marcos Mirande for their comments and corrections that helped to improve this paper.

A new endemic species of Bryconamericus (Characiformes, Characidae) from the Middle Cauca...

65

References

Armbruster, J. W., 2012. Standardized measurements, landmarks, and meristic counts for cypriniform fishes. *Zootaxa*, 3586: 8–16.

Eschmeyer, W. N. & Fricke, R. (Eds.), 2013. *Catalog of Fishes*. Electronic version (22 Aug 2013). http://research.calacademy.org/research/ichthyology/catalog/fishcatmain.asp

Hammer, O., Harper, D. A. & Ryan, P. D., 2008. *PAST-Paleontological Statistics*, ver.1.81: 1–88.

Román–Valencia, C., 2000. Tres nuevas especies de *Bryconamericus* (Ostariophysi: Characidae) de Colombia y diagnóstico del género. *Revista de Biología Tropical*, 48: 449–464.

– 2001. Descripción de una nueva especie de *Bryconamericus* (Ostariophysi, Characidae) del alto río Suárez, cuenca del Magdalena, Colombia. *Bolletino Museum Regionalli Science Naturali, Torino*, 18: 469–476.

– 2003. Sistemàtica de las especies Colombianas de *Bryconamericus* (Characiformes, Characidae). *Dahlia (Rev. Asoc. Colomb. Ictiol.)*, 6: 17–58.

Román–Valencia, C., García, M. D. & Ortega, H., 2011. Revisión taxonómica y geográfica de *Bryconamericus peruanus* (Teleostei, Characidae). *Revista Mexicana de Biodiversidad*, 82: 844–853.

Román–Valencia, C., García–Alzate, C. A., Ruiz–C., R. I. & Taphorn, D. C., 2010. *Bryconamericus macarenae* n. sp. (Characiformes, Characidae) from the Güejar River, Macarena mountain range, Colombia. *Animal Biodiversity and Conservation*, 33.2: 195–203.

Román–Valencia, C., Ruiz–C., R. I., Taphorn B., D. C. & Garcia–Alzate, C., 2013. Three new *Bryconamericus* (Characiformes, Characidae), with key to Ecuador species and a discussion on validez of *Knodus*. *Animal Biodiversity and Conservation*, 36.1: 123–139.

Román–Valencia, C., Vanegas–Ríos, A. & García, M. D., 2009a. Análisis comparado de las especies del genero *Bryconamericus* (Teleostei: Characidae) en la cuenca de los ríos Cauca–Magdalena y Ranchería, Colombia. *Revista Mexicana de Biodiversidad*, 80: 465–482.

Román–Valencia, C., Vanegas–Ríos, A. & Ruiz–C., R. I., 2008. Una nueva especie de pez del género *Bryconamericus* (Ostariophysi: Characidae) del río Magdalena, con una clave para las especies de Colombia. *Revista de Biología Tropical*, 56: 1749–1763.

– 2009b. Especie nueva del género *Bryconamericus* (Teleostei: Characidae) del río Fonce, sistema río Magdalena, Colombia. *Revista Mexicana de Biodiversidad*, 80: 455–463.

Ruiz–C., R. I. & Román–Valencia, C., 2006. Osteología de *Astyanax aurocaudatus* Eigenmann, 1913 (Pisces: Characidae), con notas sobre la validez de *Carlastyanax* Géry, 1972. *Animal Biodiversity and Conservation*, 29.1: 49–64.

Sabaj–Pérez, M. H. (Ed.), 2010. *Standard symbolic codes for institutional resource collections in herpetology and ichthyology: an online reference*. Version 2.0 (8 November 2010). Electronically accessible at http://www.asih.org/, American Society of Ichthyologists and Herpetologists, Washington, D.C.

Song, J. L. & Parenti, R., 1995. Clearing and staining whole fish specimens for simultaneous demonstration of bone, cartilage and nerves. *Copeia*, 1995: 114–118.

Taylor, W. R. & Van Dyke, G. C., 1985. Revised procedures for staining and clearing small fishes and other vertebrates for bone and cartilage study. *Cybium*, 9: 107–119.

Vari, R., 1995. The Neotropical fish family Ctenoluciidae (Teleostei: Ostariophysi: Characiformes): supra and intrafamilial phylogenetic relationships, with a revisionary study. *Smithsonian Contribution to Zoology*, 564: 1–96.

Vari, R. P. & Siebert, D. J., 1990. A new, unusually dimorphic species of *Bryconamericus* (Pisces: Ostariohysi: Characidae) from the Peruvian Amazon. *Proceeding Biological Society Washington*, 103: 516–524.

Weitzman, S. H., 1962. The osteology of *Brycon meeki*, a generalized characid fish, with an osteological definition of the family. *Stanford Ichthyological Bulletin*, 8: 1–77.

7

Is the 'n = 30 rule of thumb' of ecological field studies reliable? A call for greater attention to the variability in our data

A. Martínez–Abraín

Martínez–Abraín, A., 2014. Is the 'n = 30 rule of thumb' of ecological field studies reliable? A call for greater attention to the variability in our data. *Animal Biodiversity and Conservation*, 37.1: 95–100.

Abstract

Is the 'n = 30 rule of thumb' of ecological field studies reliable? A call for greater attention to the variability in our data.— A common practice of experimental design in field ecology, which relies on the Central Limit Theorem, is the use of the 'n = 30 rule of thumb'. I show here that papers published in *Animal Biodiversity and Conservation* during the period 2010–2013 adjust to this rule. Samples collected around this relatively small size have the advantage of coupling statistically–significant results with large effect sizes, which is positive because field researchers are commonly interested in large ecological effects. However, the power to detect a large effect size depends not only on sample size but, importantly, also on between–population variability. By means of a hypothetical example, I show here that the statistical power is little affected by small–medium variance changes between populations. However, power decreases abruptly beyond a certain threshold, which I identify roughly around a five–fold difference in variance between populations. Hence, researchers should explore variance profiles of their study populations to make sure beforehand that their study populations lies within the safe zone to use the 'n = 30 rule of thumb'. Otherwise, sample size should be increased beyond 30, even to detect large effect sizes.

Key words: Sample size, Variance, Statistical power, Effect size, Field ecology, Reliability.

Resumen

¿Es fiable la regla de oro de n = 30 de los estudios ecológicos de campo? Se debe prestar más atención a la variabilidad de nuestros datos.— La utilización de la regla de oro de n = 30 es una práctica común del diseño experimental en ecología de campo que se fundamenta en el teorema del límite central. A continuación se muestra que los artículos publicados en A*nimal Biodiversity and Conservation* durante el período comprendido entre los años 2010 y 2013 se ajustan a esta regla. Las muestras recogidas cuyo tamaño se aproxima a esta cifra relativamente pequeña tienen la ventaja de relacionar resultados estadísticamente significativos con efectos de gran magnitud, lo cual es positivo porque por lo general los investigadores de campo están interesados en los efectos ecológicos de gran magnitud. No obstante, la posibilidad de detectar un efecto de gran magnitud no solo depende del tamaño de la muestra, sino también en gran medida de la variabilidad existente entre las poblaciones. Mediante un ejemplo hipotético, a continuación se muestra que la potencia estadística se ve poco afectada por los cambios pequeños o medios de varianza que pueda haber entre las poblaciones. Sin embargo, la potencia se reduce bruscamente a partir de un determinado límite, que nosotros establecemos aproximadamente en una diferencia de cinco veces en la varianza entre poblaciones. Por consiguiente, los investigadores deberían analizar los perfiles de varianza de sus poblaciones de estudio con el fin de asegurarse de antemano de que sus poblaciones en estudio se encuentran en la zona de seguridad en que puede emplearse la regla de oro de n = 30. De lo contrario, será necesario aumentar el tamaño de la muestra a más de 30, incluso para detectar efectos de gran magnitud.

Palabras clave: Tamaño de la muestra, Varianza, Potencia estadística, Magnitud del efecto, Ecología de campo, Fiabilidad.

Alejandro Martínez–Abraín, Depto. de Bioloxía Animal, Bioloxía Vexetal e Ecoloxía, Univ. da Coruña, Fac. de Ciencias, Campus da Zapateira s/n., 15071 A Coruña, España (Spain); Population Ecology Group, IMEDEA (CSIC–UIB), Miquel Marquès 21, 07190 Esporles, Mallorca, España (Spain).

E–mail: amartinez@imedea.uib–csic.es

It is common to read in introductory books on biostatistics that working with a sample size of at least 30 is safe for the design of field studies (*e.g.* pg. 43 in Cohen & Cohen, 1995). This recommendation relies on the Central Limit Theorem, according to which if random samples of size n are drawn from a normal population the means of these samples will conform to a normal distribution (Zar, 1999). Supposedly random samples of a minimum size = 30 allow recovery of a normal distribution of the mean even if samples are non–normal. A common consequence is that field researchers tend to use samples adjusted to this minimum. To verify this tendency, I analyzed the sample size used in the papers published in Animal Biodiversity and Conservation from 2010 to 2013 (n = 4 years), and the results suggest that this rule holds for ecological and conservation field studies, since the arithmetic mean of both the medians and averages of sample sizes used in each paper was very close to 30 (see table 1).

This is in contrast with experimental design recommendations, where sample size is known to be directly dependent on the variance in the population (as estimated by the sample standard deviation), and indirectly dependent on the maximum allowable absolute difference between the estimated population parameter and the true population parameter (*d*) from equation 1,

$$n \geqslant \frac{z^2 \, \sigma^2}{d^2}$$

where *z* is 1.96, the value for a 95% confidence interval from a normal distribution (Quinn & Keough, 2002). That is, the minimum required sample size to accurately estimate a parameter will be higher if a) population variance is high, in order to minimize the risk of our sample not being representative of the statistical population, and b) if the desired accuracy is high, for a given confidence level, since increasing the confidence level also increases the required sample size. Of course, deciding the value of *d* means that we have previous knowledge about the true magnitude of the study parameter, which is not usually the case in field ecological studies (*e.g.* Martínez–Abraín, 2008, 2013).

Sample size and null hypothesis testing

This is not only the theoretical basis of experimental design for parameter estimation, but also for a priori or prospective power tests within the framework of null hypothesis statistical testing (Zar, 1999; Schneir & Gurevitch, 2001). The required sample size to couple biological and statistical significance (and hence to make sense of statistically non–significant results) is determined after providing alpha (*i.e.* the type–I error rate, typically fixed at 5%), power (*i.e.* 1–the type II error rate or probability of correctly failing to reject a null hypothesis, typically fixed at 80%), and effect size (the minimum magnitude of the difference between two populations that is considered to be biologically relevant if dealing with mean comparison, or the amount of variance of each variable that is explained by other variables considered, if dealing with regression problems). Again, this implies that we have some a priori knowledge of what represents a biologically–relevant effect in our study system, which unfortunately is seldom the case in field ecological studies.

Empirically, sample size could be obtained by plotting the standard deviation against sample size until a plateau is reached, provided that our sample correctly represents the variability in the population. However, this is mostly viable for experimental studies (preferentially lab studies, although also some field studies), where sample size can be modified along a large range of possible values. A better–suited option for field studies could be to perform this plotting by applying resampling with repetition (bootstrap) to our data, if our sampling protocol includes samples of different size. Moreover, this exercise would be necessary for each variable under study, because each variable has its own profile. This means that the usual procedure of measuring many variables from the same sample of individuals —taking advantage of having captured them, for example— does not respect the prerequisite of accounting for variance to determine the right sample size, because variances for each trait of an individual do not necessarily co–vary in a strong way. Different sample sizes would thus be necessary to study different traits, something that seems impracticable in field studies where information

Table 1. Sample size extracted from n = 76 papers published in *Animal Biodiversity and Conservation* from 2010 to 2013: V. Journal volume; I. Journal isue; i. Initial page; f. Final page; n. Sample size; M1. Median sample size; M2. Average sample size; E. Cause of exclusion (1. Species description; 2. Ring recoveries; 3. Essay; 4. Survey, hunting–bag data, bioacoustics/RADAR data; 5. Species atlas or similar; 6. Review papers; 7. Genetics; 8. Museum collections; 9. Simulated data).

Tabla 1. Tamaño de la muestra extraída de n = 76 artículos publicados en Animal Biodiversity and Conservation *entre 2010 y 2013: V. Volumen de la revista; I. Número de la revista; i. Página inicial; f. Página final; n. Tamaño de la muestra; M1. Mediana del tamaño de la muestra; M2. Media del tamaño de la muestra; E. Motivo de exclusión (1. Descripción de especies; 2. Recuperación de anillas; 3. Ensayo; 4. Encuesta, datos de caza, datos de bioacústica y radar; 5. Atlas de especies o similar; 6. Artículos de revisión; 7. Genética; 8. Colecciones museísticas; 9. Datos simulados).*

Year	V	I	i	f	n	n	n	n	n	n	n	n	n	n	n	n	M1	M2	E
2010	33	1	1	13	22	19	11										13	13.2	
2010	33	1	15	18															3
2010	33	1	19	29	18	4	17	39	23	26	18	27	24	21	11	27	22	21.6	
2010	33	1	31	45	5	10	20	30	53								30	27.7	
2010	33	1	47	51															1
2010	33	1	53	61	7												53	40.3	
2010	33	1	63	87	13	16	11	20	11	9	10	10	12	14	10	19	12.5	21.8	
2010	33	1	89	115															2
2010	33	1	117	117															3
2010	33	2	119	129	50	50	50	50	50								50	50.0	
2010	33	2	131	142															1
2010	33	2	143	150	16	16	21	25	23								21	20.2	
2010	33	2	151	185															6
2010	33	2	187	194	30	30											30	30.0	
2010	33	2	195	203															1
2010	33	2	205	208															3
2011	34	1	1	10															1
2011	34	1	11	21	39	41	19	31	20	10							25.5	26.7	
2011	34	1	23	29	60	60	60	60									60	60.0	
2011	34	1	31	34	8												8	8.0	4
2011	34	1	35	45	20	14	36	24	16	11	16	36					18	21.6	
2011	34	1	47	66															1
2011	34	2	229	247	17	3											10	10.0	5
2011	34	2	249	256	13	132											72.5	72.5	
2011	34	2	257	264															1
2011	34	2	265	272															1
2011	34	2	273	285															8
2011	34	2	287	294	4	17											10.5	10.5	
2011	34	2	295	308															1
2011	34	2	309	317															6
2011	34	2	319	330	10												10	10.0	
2011	34	2	331	340	8												8	8.0	
2011	34	2	341	353															3
2011	34	2	355	361	37	25	22										25	28.0	
2012	35	1	1	11															5

Table 1. (Cont.)

Year	V	l	i	f	n	n	n	n	n	n	n	n	n	n	n	n	n	n	n	n	M1	M2	E
2012	35	1	13	21																			1
2012	35	1	23	26																			7
2012	35	1	27	50																			1
2012	35	1	51	58																			8
2012	35	1	59	69																			7
2012	35	1	71	94																			1
2012	35	1	107	117	73	51															62	62.0	
2012	35	1	119	124	29	26	39	35	36	33											34	33.0	
2012	35	1	125	139	30																30	30.0	
2012	35	1	141	150	10	4	2	14													7	7.5	
2012	35	2	151																				3
2012	35	2	153	154																			3
2012	35	2	155																				3
2012	35	2	159	161																			3
2012	35	2	163	170																			4
2012	35	2	171	174																			3
2012	35	2	175	188																			4
2012	35	2	189	196	18	10	13	15	54	44	33	28									23	26.9	
2012	35	2	197	207	27	64	60	6	48	12	5										27	31.7	
2012	35	2	209	217																			
2012	35	2	219	220																			3
2012	35	2	221	233	27																27	27.0	
2012	35	2	235	246	5	29	42	9	21	15											18	20.2	
2012	35	2	247	252																			5
2012	35	2	253	265	5	20	49	158	42												42	54.8	
2012	35	2	267	275																			7
2012	35	2	277	283																			5
2012	35	2	285	293	48	28	16	16	16	16	16	16	16	4							16	19.2	
2012	35	2	295	306																			3
2013	36	1	1	11	12	12	29	31	32	3											20.5	19.8	
2013	36	1	13	31	25	9															17	17.0	
2013	36	1	33	36																			3
2013	36	1	37	46																			9
2013	36	1	47	57																			4
2013	36	1	59	67																			1
2013	36	1	69	78																			2
2013	36	1	79	88																			4
2013	36	1	89	99	50	51	41	50	50	50	45	49	59	22	51	30	30	21	30	6	47	39.7	
2013	36	1	101	111																			1
2013	36	1	113	121	3	30	14	91	30	21	45	11	13	20	30	28					24.5	28.0	
2013	36	1	123	139																			
Mean																					27.3	28.0	
SD																					17.2	16.7	

Table 2. Change in statistical power as between–population variance increases, using the n = 30 rule of thumb of field ecological statistics. Exercise using fictitious data in open software G*Power 3.1.3: n. Sample size; M1, M2. Means of both populations; Sd1, Sd2. Standard deviations of both populations; Ratio. Ratio Sd1/Sd2; Es. Effect size (Cohen's d); Po. Statistical power.

*Tabla 2. Cambio en la potencia estadística a medida que aumenta la varianza entre poblaciones, utilizando la regla de oro de n = 30 de los estadísticos de la ecología de campo. Ejercicio que utiliza datos ficticios en el programa informático abierto G*Power 3.1.3: n. Tamaño de la muestra; M1, M2. Medias de ambas poblaciones; Sd1, Sd2. Desviaciones estándar de ambas poblaciones; Ratio. Coeficiente Sd1/Sd2; Es. Magnitud del efecto (d de Cohen); Po. Potencia estadística.*

ID	n	M1	M2	Sd1	Sd2	Ratio	Es	Po
1	30	1.5	1.0	0.2	0.2	1.0	2.50	1.00
2	30	1.5	1.0	0.3	0.2	1.5	1.96	1.00
3	30	1.5	1.0	0.4	0.2	2.0	1.58	1.00
4	30	1.5	1.0	0.5	0.2	2.5	1.31	1.00
5	30	1.5	1.0	0.6	0.2	3.0	1.12	0.99
6	30	1.5	1.0	0.7	0.2	3.5	0.97	0.96
7	30	1.5	1.0	0.8	0.2	4.0	0.86	0.90
8	30	1.5	1.0	0.9	0.2	4.5	0.77	0.83
9	30	1.5	1.0	1.0	0.2	5.0	0.69	0.75
10	30	1.5	1.0	1.1	0.2	5.5	0.63	0.67
11	30	1.5	1.0	1.2	0.2	6.0	0.58	0.60
12	30	1.5	1.0	1.3	0.2	6.5	0.54	0.53
13	30	1.5	1.0	1.4	0.2	7.0	0.50	0.48
14	30	1.5	1.0	1.5	0.2	7.5	0.47	0.43

for each individual is exploited as much as possible given the difficulty in obtaining it.

The 'rule of thumb of n = 30' in field ecological studies comes from the fact that we are usually interested in "large" effect sizes (of unknown absolute magnitude). The reasoning follows that if we are able to obtain a statistically–significant result (this only meaning that the properties of our sample can be applied to the whole statistical population, and hence that the desired inference from particular to general can be done) with a small sample size such as 30, the effect we are dealing with is probably large, and hence, most likely a biologically–relevant effect (Martínez–Abraín, 2007).

However, this approach of reasoning around n = 30 in relation to the magnitude of the effects (in the denominator of equation 1) is influenced by variance (in the numerator of equation 1). Power decreases with increasing between–population variance, when sample size is kept constant at n = 30 (table 2, fig. 1). However, this decrease proceeds in a non–linear fashion, indicating a strong resilience of statistical power to small–to–medium changes in between–population variance. Only when the change in between–populations variance is large (around a

five–fold difference in the variance between groups, which corresponds to our study case #9) does power drop abruptly below the desired minimum value of 0.8 (fig. 1). In this hypothetical example, it is necessary to increase our sample size to n = 34 in case study #9, and to n = 73, in case study #14, to allow the recovery of a 0.80 power.

The n = 30 rule of thumb also overlooks the possibility that small or medium effect sizes can be biologically relevant in some cases (Igual et al., 2005). Since we typically do not know when that is the case, we are forced or limited to work with large effect sizes. On the contrary, working with too large a sample size (as is commonplace among theoretical ecologists) could even be counter–productive at times because we could be focusing on small effects which could be biologically irrelevant.

Hence, it seems reasonable to use the n = 30 rule in ecological field studies to make inferences on parameter values or to test null hypotheses, owing to our usual lack of knowledge on *d* or effect size of interest, but we should make an effort to explore beforehand the variance profile of our study populations in order to be able to detect large effects with that sample size and with a high power. Populations with

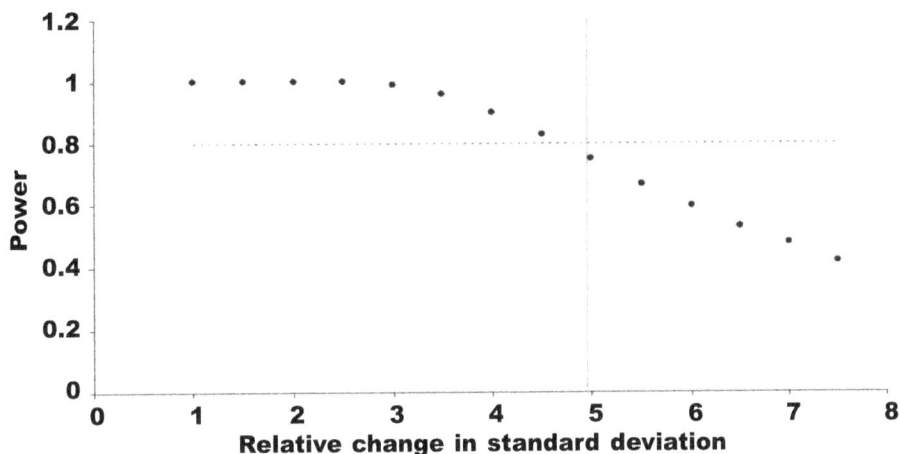

Fig. 1. Non–linear decrease of statistical power with increasing variance between two populations (ratio of standard deviations between two groups) while keeping means and sample sizes constant (n = 30), for data in table 2. The reference line for a desired power of 0.80 is shown as well as the line for a five–fold difference in between–populations variance.

Fig. 1. Para los datos que figuran en la tabla 2, la potencia estadística disminuye de forma no lineal a medida que aumenta la varianza entre dos poblaciones (coeficiente de las desviaciones estándar entre dos grupos) y con las medias y los tamaños de las muestras (n = 30) constantes. Se muestran también la línea de referencia de una potencia deseada de 0,80 y la línea que representa una diferencia de cinco veces en la varianza entre poblaciones.

high differences between them in variance will require larger sample sizes (compared to 30) to detect even large differences.

Rules of thumb exist for a reason, but they should be used with great caution and as an approximation. The exercise of thinking, typical of the scientific enterprise, cannot and should not be set aside during the process of experimental design, despite the added complexities of ecological field studies compared to lab studies. Reduced power, such as increases in between–populations variance, can lead to increased prevalences of Type II errors (*i.e.* incorrectly failing to reject a null hypothesis of equality to zero), resulting in serious problems regarding decision–making in conservation, when evaluating the effect of human activities. We may conclude that nothing happens when indeed it does. Let's hence give more attention to the variability of our field data for the benefit of proper knowledge acquisition and correct decision making.

Acknowledgements

I would like to thank the editor and two anonymous referees who helped improve the manuscript with their critical comments. Daniel Oro commented on an early version of the manuscript.

References

Cohen, J. & Cohen, L., 1995. *Statistics for ornithologists*. BTO Guide 22.

Igual, J. M., Forero, M. G., Tavecchia, G., González–Solís, J., Martínez–Abraín, A., Hobson, K. A., Ruiz, X., Oro, D., 2005. Short–term effects of data–loggers on Cory's shearwater (*Calonectris diomedea*). *Marine Biology*, 146: 619–624.

Martínez–Abraín, A., 2007. Are there any differences? A non–sensical question in ecology. *Acta Oecologica*, 32: 203–206.

– 2008. Statistical significance and biological relevance: A call for a more cautious interpretation of results in ecology. *Acta Oecologica*, 34: 9–11.

– 2013. Why do ecologists aim to get positive results? Once again, negative results are necessary for better knowledge accumulation. *Animal Biodiversity and Conservation*, 36.1: 33–36.

Quinn, G. P. & Keough, M. J., 2002. Experimental design and data analysis for biologists. Cambridge University Press, Cambridge.

Schneir, S. M. & Gurevitch, J. (Eds.), 2001. *Design and analysis of ecological experiments*. Oxford University Press, Oxford.

Zar, J. H., 1999. *Biostatistical analysis*. Prentice Hall, Upper Saddle River, New Jersey.

8

Pets becoming established in the wild: free–living Vietnamese potbellied pigs in Spain

M. Delibes–Mateos & A. Delibes

Delibes–Mateos, M. & Delibes, A., 2013. Pets becoming established in the wild: free–living Vietnamese potbellied pigs in Spain. *Animal Biodiversity and Conservation*, 36.2: 209–215.

Abstract

Pets becoming established in the wild: free–living Vietnamese potbellied pigs in Spain.— Vietnamese potbellied (VPB) pigs (*Sus scrofa*) are a common pet in North America and Europe, but their recent decrease in popularity has increased their abandonment. Our main aim was to identify potential cases of free–living VPB pigs in Spain through an in–depth Google search. We identified 42 cases of free–living VPB pigs distributed throughout the country. The number of free–living VPB pigs reported increased by year but the species abundance still seems to be low. Signs of VPB pig reproduction and possible hybrids between VPB pigs and wild boar or feral pigs have been also reported. Free–living VPB pigs could erode the gene pool of the Spanish wild boar population and exacerbate the damage (*e.g.* crop damage or spread of diseases) already caused by wild board. Urgent evaluation and adequate management of wild VPB pig sightings is needed to prevent their establishment in natural habitats.

Key words: Feral pig, Google search, Pet trade, Wild boar.

Resumen

Mascotas que se establecen en la naturaleza: cerdos vietnamitas que viven en libertad en España.— Los cerdos vietnamitas (*Sus scrofa*) son una mascota habitual en Norteamérica y Europa; sin embargo, su popularidad ha disminuido recientemente y esto ha provocado que se abandonen cada vez más. El objetivo principal de este trabajo es identificar casos de cerdos vietnamitas que viven en libertad en España a través de una búsqueda exhaustiva en Google. Se han identificado 42 casos de cerdos vietnamitas que viven en libertad distribuidos por todo el país. El número de casos aumenta cada año, aunque la abundancia de la especie aún parece ser baja. También se han observado indicios de que los cerdos vietnamitas se reproducen en libertad y de posibles híbridos de éstos con jabalíes o cerdos asilvestrados. El hecho de que los cerdos vietnamitas vivan en libertad podría reducir el patrimonio genético de la población española de jabalí, así como agravar los daños que este ya causa en España (como los daños a cultivos o los accidentes de tráfico). Con vistas a evitar que se establezcan en hábitats naturales, es urgente evaluar y gestionar debidamente las observaciones de cerdos vietnamitas salvajes.

Palabras clave: Cerdo asilvestrado, Búsqueda en Google, Comercio de mascotas, Jabalí.

Miguel Delibes–Mateos, Instituto de Investigación en Recursos Cinegéticos (IREC, CSIC–UCLM–JCCM), Ronda de Toledo s/n., 13071 Ciudad Real, España (Spain).– Adolfo Delibes, Junta de Castilla y León, c/ Rigoberto Cortejoso 14, 41014, Valladolid, España (Spain).

Corresponding author: Miguel Delibes–Mateos. E–mail: mdelibesmateos@gmail.com

Introduction

The deliberate or accidental release of pets has frequently resulted in the successful establishment of new wild populations (*e.g.* Bertolino & Lurz, 2013). Over the last years, Vietnamese potbellied pigs (*Sus scrofa*; hereafter VPB pigs; fig. 1) have become a popular pet species in many regions of the world (Braun & Casteel, 1993), including Spain (Jarandilla, 2011). However, their popularity has recently declined, leading to their frequent abandonment (Tynes, 1997). Given that this is a very opportunistic species, it could be expected that released or abandoned VPB pigs eventually establish free–living populations. Nevertheless, to our knowledge this has not yet been demonstrated. In this paper our main objective was to report, for the first time, the existence of free–living VPB pigs in Spain. In addition, we aimed to gather complementary information regarding these potential introductions, concerning the year, locality and habitat type in which they were observed (table 1). We also investigated whether pregnant VPB female pigs, piglets and/or hybrids between VPB pigs and wild boar (or feral pigs) were reported, as such sightings could indicate that VPB pigs are successfully establishing wild populations (see an example in Magalhaes & Jacobi, 2013). Finally, we briefly discuss the potential risks associated with the successful establishment of free–living VPB pigs.

Material and methods

We identified cases of free–living VPB pigs by means of a thorough Google search. We searched for terms that stemmed from the following words (in Spanish) in the following combinations: Vietnamese pig or Vietnamese boar, combined with wild boar or hybrid or wild or hunting or stalking or harvested or escaped or abandoned or abandonment (*i.e.* 9*2 = 18 combinations). We identified all the cases in which VPB pigs (or their reported hybrids with wild boar or feral pigs; see below) were unequivocally reported as roaming freely. We stopped the search for a word combination when 100 consecutive results did not identify any additional valid record, *i.e.*, when we retrieved only repeated cases, results that did not exactly relate to free–living VPB pigs (*e.g.* web pages offering VPB pigs for sale), and/or cases in which it was impossible to unequivocally identify free–living VPB pigs (*e.g.* unclear observations, inconsistent rumours, etc.). Most valid cases were reported through searches in Internet forums (mainly in hunting websites) and in online newspapers and television channels. In total, more than 1,800 records were checked (18 word combinations and at least 100 records per search). All the searches were conducted between mid–February and mid–March 2013.

When possible, we recorded the year, locality (and province), and number of individuals for each reported case (table 1). We differentiated between cases in urban and rural areas because the former would correspond to momentary VPB pig escapes or abandonments, whereas the latter would be more closely associated with individuals potentially becoming established in the wild. All the records from cities/towns, urban parks, highways close to the city/town, and so on were included in the urban area category, and those from natural environments (scrublands, woodlands, crops interspersed with natural vegetation, etc.) in the rural area category (table 1). We also recorded the presence of pregnant VPB pig females, piglets and/or potential VPB pig hybrids with wild boar or feral pig (table 1), as such sightings could indicate the establishment of viable free–living populations. Reported hybrids were described as individuals showing mixed morphological characteristics between VPB pigs and wild boar or feral pigs (fig. 1B). Furthermore, we differentiated between killed, captured or observed individuals, and we recorded the source of information: reported by hunters, reports in the news (online newspapers or televisions), and reports provided by forest rangers, hikers, bikers, and similar (table 1).

Results

We identified 42 cases of free–living VPB pigs (table 1, fig. 2). A very low percentage (< 1%) of the total records obtained through a Google search reported valid cases of free–living VPB pigs. Cases of free–living VPB pigs were relatively well distributed throughout Spain (fig. 2). Many occurred close to main cities, especially those identified near the coast (fig. 2). Free–living VPB pigs were firstly reported in Spain in 2007, and since then the number of cases has increased considerably. VPB pigs were recorded slightly more frequently in rural than in urban areas (57.15% *vs* 42.85%, *n* = 42). The first hybrid between VPB pig and wild board (or feral pig) was reported in 2010, and since then, the proportion of records of potential hybrids has increased substantially (table 1). Hybrids were reported in > 25% of the total cases (table 1). In addition, pregnant females or females taking care of piglets were also sporadically observed, captured or killed (table 1).

Table 1 also shows that: 1) on most occasions, only one individual was observed, captured or killed (70%, n = 34), although some pairs and larger groups were also reported; 2) the number of records of observed, captured and killed individuals was similar; captured and killed animals were better sources for the identification of VBP pigs and hybrids (fig. 1); 3) most cases were reported either by hunters or in newspapers; and 4) nearly all the cases recorded in urban areas were reported in newspapers, and all the records provided by hunters came from rural landscapes.

Discussion

In this study we provide the first evidences of free–living VPB pigs in Spain, and, to the best of our knowledge, elsewhere in Europe, although they are also common pets in other countries. According to

Fig. 1. A. A docile Vietnamese potbellied pig captured by a hunter in a wild environment within northern Spain (ID 3 in table 1 and fig. 2); B. Four potential hybrids between Vietnamese potbellied pigs and wild boar (on the left of the picture) killed during a hunting day in northern Spain (ID 1 in table 1 and fig. 2). The morphological aspect of reported hybrids differs notably to that of young wild boar (on the right of the picture).

Fig. 1. A. Un cerdo vietnamita dócil capturado por un cazador en un entorno natural del norte de España (ID 3 en la tabla 1 y en la fig. 2); B. Cuatro posibles híbridos entre cerdos vietnamitas y jabalíes (a la izquierda de la imagen) cazados en el norte de España (ID 1 en la tabla 1 y en la fig. 2). El aspecto morfológico de los híbridos observados difiere notablemente del de los jabalíes jóvenes (a la derecha de la imagen).

our findings, free–living VPB pigs have been reported throughout the country. Although the number of cases identified was not high, it has increased in recent years. Importantly, our results indicate that VPB pigs might have successfully established wild populations because signs of VPB pig reproduction (*i.e.* pregnant females and females taking care of piglets, and/or hybrids of VPB pigs and wild boar or feral pigs) have been frequently observed in the wild. Reasons for the successful establishment, spread and adaptation of VPB pigs to wild habitats are probably related to their highly adaptive and opportunistic behaviour, the scarcity of natural predators, and the favourable climatic conditions in Spain.

Internet search engines are increasingly used in ecological studies (*e.g.* Maccallum & Bury, 2013), and in particular to monitor the escapes/releases of non–native pets into the wild (see Kikillus et al., 2012). From this perspective, Google searching could be an appropriate tool to monitor the presence of free–liv-

Table 1. Cases of free–living Vietnamese potbellied pigs in Spain identified through an in–depth Google search (see Methods for details): ID. Locality numbers plotted in fig. 2; N. Number of pigs observed, captured or killed in each case; A. Area (R. Rural, U. Urban); H. Hybrid; P/N. Pregnant/nursing (* females that were taking care of piglets); T. Type (K. Killed, C. Captured, O. Observed); S. Source (H. Hunters, N. Newspaper, O. Others: O[1] Forest rangers, O[2] Conservationists).

Tabla 1. Casos de cerdos vietnamitas que viven en libertad en España identificados mediante una búsqueda exhaustiva en Google (en el apartado Methods se ofrece información más detallada): ID. Números de las localidades indicadas en la fig. 2; N. Número de cerdos observados, capturados o cazados en cada caso; A. Área (R. Rural, U. Urbana); H. Híbrido; P/N. Embarazada/lactante (* hembras con crías); T. Tipo (K. Muerto, C. Capturado, O. Observado); S. Fuente (H. Cazadores, O. Otros: O[1] Guardabosques, O[2] Conservacionistas).

ID	Locality	Province	N	A	H	P/N	Year	T	S
1	Castillo de Onielo	Palencia	4	R	Yes	Yes	2013	K	H
2	Olivares de Duero	Valladolid	1	R	–	–	2013	O	H
3	Mucientes	Valladolid	1	R	No	No	2013	C	H
4	Maella	Zaragoza	6	R	–	No	2012	C	N
5	Unknown	Alicante	–	R	Yes	–		O	H
6	Mallorca	Baleares	16	U	Yes	No	2012	C	N
7	Puig de Santa Magdalena de Inca	Baleares	1	R	Yes	No	2012	K	N
8	Cartagena	Murcia	1	U	No	–	2012	O	N
9	Elche	Alicante	1	U	No	No	2007	C	N
10	Santander	Cantabria	4	U	Yes	Yes*	2011	C	N
11	Argentona	Barcelona	1	R	No	No	2011	K	H/N
12	Pontons	Barcelona	1	R	No	No	2010	K	H
13	San Antonio	Baleares	1	U	No	No	2011	C	N
14	Olesa de Montserrat	Barcelona	2	R	Yes	No	2010	K	H
15	Valencia	Valencia	1	U	No	No	2008	C	N
16	András	Pontevedra	1	U	No	No	2008	C	N
17	Unknown	Huesca	–	R	–	–	2012	K	H
18	Unknown	Coruña	1	R	–	–	2012	O	H
19	Concejo de Siero	Asturias	1	R	No	No	2012	O	H
20	Murcia	Murcia	1	U	No	No	2012	K	O[1]
21	Cuenca río Verdugo	Pontevedra	–	R	Yes	–	2012	O	H
22	Unknown	Valencia	2	R	Yes	No	2012	K	H
23	Unknown	Almería	–	R	No	–	2012	O	H
24	Retuerta del Bullaque/ Los Yébenes	Ciudad Real/ Toledo	4	R	No	No	2012	K	H
25	Villaviciosa y Sariego	Asturias	–	R	No	No	2012	O	H
26	Valencia	Valencia	1	R	No	No	2012	O	O[2]
27	Elda	Alicante	4	U	No	Yes*	2012	O	N
28	Parque Almijara	Málaga/ Granada	–	R	Yes	–	2011	O	O
29	La Seu de Urgell	Lleida	1	R	Yes	No	2013	O	O
30	Badajoz	Badajoz	1	U	No	No	2010	O	N

Table 1. (Cont.)

ID	Locality	Province	N	A	H	P/N	Year	T	S
31	Baños de la Encina	Jaén	–	R	Yes	No	2012	K	H
32	Zaragoza	Zaragoza	1	U	No	No	2012	C	N
33	Vélez–Málaga	Málaga	1	U	No	No	2008	C	N
34	Rojales	Alicante	2	U	No	No	2011	C	N
35	Santiago de Compostela	Coruña	1	U	No	No	2010	C	N
36	Valencia	Valencia	1	U	No	No	2012	C	N
37	Zaragoza	Zaragoza	1	U	No	No	2007	C	N
38	Campo Lameira	Pontevedra	1	R	No	No	2010	C	H
39	La Solana	Ciudad Real	1	U	No	No	2012	C	N
40	Parque Natural del Montgó	Alicante	1	R	No	No	2007	C	N
41	San Vicente del Raspeig	Alicante	4	U	No	No	2009	C	N
42	Southern Alicante	Alicante	–	Rural	No	No	2013	O	H

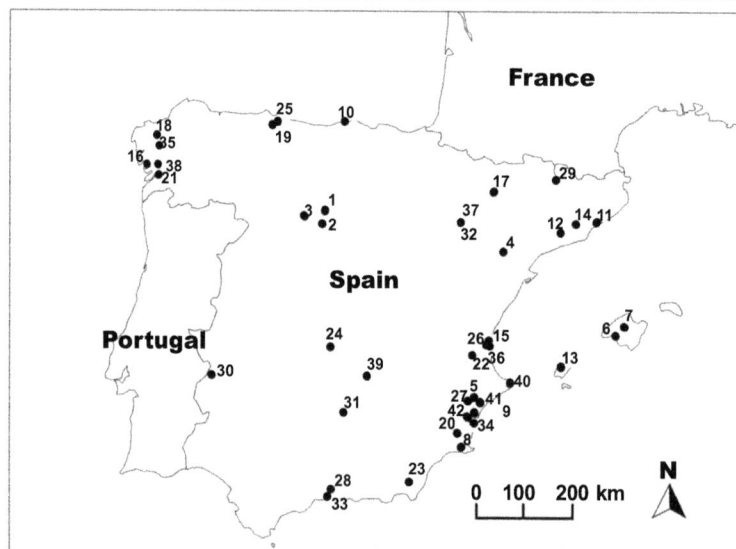

Fig. 2. Localities in which free–living Vietnamese potbellied pigs or reported hybrids between Vietnamese potbellied pigs and wild boar or feral pigs were observed, captured or killed. Dots generally represent approximate locations, as on most occasions we could not obtain the exact coordinates where free–living Vietnamese potbellied pigs were observed. Numbers correspond to the ID of each locality in table 1.

Fig. 2. Localidades en las que se observaron, capturaron o cazaron cerdos vietnamitas en libertad o híbridos entre cerdos vietnamitas y jabalíes o cerdos asilvestrados. Por lo general, los puntos representan las localidades aproximadas, puesto que en la mayoría de las ocasiones no fue posible obtener las coordenadas exactas en las que se observaron los cerdos vietnamitas en libertad. Los números son los que figuran en la columna ID de cada localidad en la tabla 1.

ing VPB pigs in Spain. However, the use of Internet searches in scientific studies presents several limitations that must be taken into account. For example, in our study it was almost impossible to check the reliability of most of the observations of free–living VPB pigs; nevertheless, pictures of killed or captured VPB pigs were frequently shown on the websites returned from our searches. Similarly, identification of hybrids between VPB pigs and wild boar (or feral pigs) was not unequivocal through Google searching, and therefore genetic studies are needed to accurately confirm their presence in the wild. In addition, Google searching excludes communities that use Internet less frequently (Maccallum & Bury, 2013), such as older people, or residents in small villages (Proulx et al., in press). Such communities may be closer to wildlife, as these people frequently work in the field (e.g. farmers and gamekeepers); it could therefore be expected that their knowledge about the presence of free–living VPB pigs would be higher than that of other people. Taking this into consideration, we could have underestimated the number and distribution of wild VPB pigs. In contrast, given that striking news is especially highlighted in websites, the presence of free–living VPB pigs could also have been overestimated in our study. In any case, Internet search engines offer some advantages over conventional field–monitoring programs. For example, they permit cost–effective and rapid assessments to detect the recent introduction of invasive animals (Proulx et al., in press), such as the case addressed in the present manuscript.

The introduction of VPB pigs into the wild is a result of their deliberate or accidental release, and seems to have been favoured by the huge development of the online pet–trade (Magalhaes & Jacobi, 2010). Thus, the number of Spanish websites advertising VPB pigs for sale estimated through a Google search increased from less than 10 in 2006 to more than 1,300 in 2012. As a result, the prize paid for VPB pigs in Spain has decreased drastically, dropping from several hundred euros to as little as 20 euros, a fact that facilitates their abandonment (Lord & Wittum, 1998). According to our results, VPB pigs observed were mostly solitary, suggesting that the number of individuals released per event was low. Nevertheless, as observed with other introduced pets (e.g. Alda et al., 2013), only a few founders may have been sufficient to establish free–living populations of VPB pigs across Spain. Furthermore, free–living VPB pigs were frequently documented in parks and gardens in urban areas, indicating these areas could play a key role in the establishment of free–living populations. This closely resembles the manner in which racoons were introduced (*Procyon lotor*) in central Spain (García et al., 2012).

As VPB pigs have become established in wild ecosystems in Spain only recently, we can only speculate about the risk they may pose. Given the notable genetic divergence between European and Asian *S. scrofa* (Fernández et al., 2010), free–living VPB pigs could cause erosion of the gene pool of Spanish wild boar populations. In addition, they could

create problems similar to those already caused by the wild boar in large areas of Spain (e.g. negative effects on crops and/or natural vegetation, Herrero et al., 2006; Gómez & Hódar, 2008).

The fact that we found only a few cases of apparently successful establishment of VPB pigs in the wild suggests they are not yet abundant in our setting. However, previous experiences suggest that calls for early management action are required, because by the time released exotic animals are publicly recognised as a problem, it is often too late for effective action, due to logistic, economic, or scale factors (Bertolino & Genovesi, 2003). We therefore recommend the situation should be suitably managed before it runs out of control. An in–depth monitoring plan is needed to determine the current distribution and abundance of free–living VPB pigs in Spain, and to accurately assess the risk they may cause. To prevent further releases of exotic pets in the wild, we urgently need stricter trade regulations in Spain (see for a wider explanation on this topic Bertolino & Lurz, 2013). At the same time, we need to increase public awareness about the risks of free–living pets and the benefits of their prevention and mitigation for native biodiversity.

Acknowledgements

We are very grateful to Drs. M. Delibes, M. Díaz and P. White, and an anonymous reviewer for helpful comments on previous drafts, and to all the people who kindly provided information about free–living Vietnamese potbellied pigs in Spain. M. Delibes–Mateos was supported by a JAE–doc contract (Programa Junta para la Ampliación de Estudios), funded by CSIC and the European Social Fund.

References

Alda, F., Ruiz–López, M. J., García, F. J., Grompper, M. E., Eggert, L. S. & García, J. T., 2013. Genetic evidence for multiple introduction events of racoons (*Procyon lotor*) in Spain. *Biological Invasions* 15: 687–698.

Bertolino, S. & Genovesi, P., 2003. Spread and attempted eradication of the grey squirrel (*Sciurus carolinensis*) in Italy, and consequences for the red squirrel (*Sciurus vulgaris*) in Eurasia. *Biological Conservation,* 109: 351–358.

Bertolino, S. & Lurz, P. W. W., 2013. *Callosciurus* squirrels: worldwide introductions, ecological impacts and recommendations to prevent the establishment of new invasive populations. *Mammal Review,* 43: 22–33.

Braun, W. F. & Casteel, S. W., 1993. Potbellied pigs: miniature porcine pets. *Veterinary Clinics of North America: Small Animal Practice,* 23: 1149–1161.

Fernández, A. I., Alves, E., Óvilo, C., Rodríguez, M. C. & Silió, L., 2010. Divergence time estimates of East Asian and European pigs based on multiple near complete mitochondrial DNA sequences.

Animal Genetics, 42: 86–88.

García, J. T., García, F. J., Alda, F., González, J. L., Aramburu, M. J., Cortés, Y., Prieto, B., Pliego, B., Pérez, M., Herrera, J. & García–Román, L., 2012. Recent invasion and status of the racoon (*Procyon lotor*) in Spain. *Biological Invasions,* 14: 1305–1310.

Gómez, J. M. & Hódar, J. A., 2008. Wild boars (*Sus scrofa*) affect the recruitment rate and spatial distribution of holm oak (*Quercus ilex*). *Forest Ecology and Management,* 256: 1384–1389.

Herrero, J., García–Serrano, A., Couto, S., Ortuño, V. M. & García–González, R., 2006. Diet of wild boar *Sus scrofa* L. and crop damage in an intensive agroecosystem. *European Journal of Wildlife Research,* 52: 245–250.

Jarandilla, L., 2011. *Cerdos vietnamitas.* Editorial Hispano Europea, Barcelona.

Kikillus, K. H., Hare, K. M. & Hartley, S., 2012. Online trading tools as a method of estimating propagule pressure via the pet–release pathway. *Biological Invasions,* 14: 2657–2664.

Lord, L. & Wittum, T., 1998. Survey of humane organizations and slaughter plants regarding experiences with Vietnamese potbellied pigs. *The Journal of the American Veterinary Medical Association,* 211: 562–565.

Magalhaes, A. L. B. & Jacobi, C. M., 2010. E–commerce of freshwater aquarium fishes: potential disseminator of exotic species in Brazil. *Maringá,* 32: 243–248.

– 2013. Asian aquarium fishes in a Neotropical biodiversity hotspot: impeding establishment, spread and impacts. *Biological Invasions,* 15: 2157–2163.

Mccallum, M. L. & Bury, G. W., 2013. Google search patterns suggest declining interest in the environment. *Biodiversity and Conservation,* 22: 1355–1367.

Proulx, R., Massicotte, P. & Pépino, M., in press. Googling trends in conservation biology. *Conservation Biology.*

Tynes, V. V., 1997. Potbellied pig husbandry and nutrition. *Veterinary Clinics of North America: Exotic Animal Practice,* 2: 193–208.

9

The impact of non–local birds on yellow–legged gulls (*Larus michahellis*) in the Bay of Biscay: a dump–based assessment

O. Jordi, A. Herrero, A. Aldalur, J. F. Cuadrado & J. Arizaga

Jordi, O., Herrero, A., Aldalur, A., Cuadrado, J. F. & Arizaga, J., 2014. The impact of non–local birds on ye–llow–legged gulls (*Larus michahellis*) in the Bay of Biscay: a dump–based assessment. *Animal Biodiversity and Conservation*, 37.2: 183–190.

Abstract

*The impact of non–local birds on yellow–legged gulls (*Larus michahellis*) in the Bay of Biscay: a dump–based as–sessment.*— Understanding how animals exploit non–natural feeding sources such as garbage dumps is necessary from many perspectives, including conservation, and population dynamics and management. Several large predatory gulls (*Larus* spp.) are among the species which most clearly benefit from using dumps. The yellow–legged gull (*L. michahellis*) is the most abundant gull in the southwestern Palaearctic, and its fast population increase until at least the 2000s was partly due large waste dumps becoming more numerous. The Bay of Biscay is an area that hosts resident local and also wintering non–local yellow–legged gulls. Using data collected over a period of eight years (bird counts, identification of colour–ringed individuals) at four dumps situated within a 60–km radius from the colonies of Gipuzkoa (southwestern Bay of Biscay), we aimed to answer: (1) the origin of gulls using dumps at the Bay of Biscay; (2) the impact of local and non–local gulls at these dumps; (3) the possible age–dependent use of these sites; and (4) the possible seasonal fluctuations in the use of dumps by gulls. Gulls in our area (study dumps) came from nearby colonies in Gipuzkoa, Atlantic Iberia, the Mediterranean region, and other areas such as Atlantic France and inland colonies (Navarra, Germany). Our study dumps seemed to be used mostly by local gulls.

Key words: Bird counts, Colour–ring, Generalist foragers, Gipuzkoa, Food availability, Trophic ecology

Resumen

*El impacto de los individuos no locales en la gaviota patiamarilla (*Larus michahellis*) en el Golfo de Vizcaya: una estimación a partir de vertederos.*— Es necesario comprender la forma en que los animales explotan los recursos tróficos de origen no natural, como es el caso de los vertederos, desde múltiples perspectivas como la conservación, la dinámica de poblaciones y la gestión. Son varias las especies de gaviotas depredadoras de gran tamaño (*Larus* sp.) las que indudablemente se benefician de utilizar los vertederos. La gaviota patiamarilla (*L. michahellis*) es la especie de gaviota más abundante del Paleártico sudoccidental y el rápido crecimiento de sus poblaciones hasta al menos la primera década del siglo XXI se debe, parcialmente, al aumento de vertederos. El Golfo de Vizcaya es una zona que alberga gaviotas locales residentes y gaviotas invernantes procedentes de otras zonas. A partir de los datos obtenidos en censos y avistamientos de gaviotas marcadas con anillas de color que se recopilaron durante un periodo de ocho años en cuatro vertederos situados en un radio de 60 km desde las colonias de cría en Gipuzkoa, se trató de responder a las siguientes cuestiones: (1) el origen de las gaviotas que usan los vertederos en el Golfo de Vizcaya; (2) el impacto de los individuos locales y no locales en estos vertederos; (3) la posibilidad de que exista un uso distinto según la edad y (4) la posibilidad de que haya fluctuaciones estacionales en el uso de los vertederos. Las gaviotas en los vertederos estudiados provienen de las colonias costeras cercanas de Gipuzkoa, la zona atlántica de la península ibérica, la región mediterránea y otras zonas como la costa atlántica de Francia y las colonias continentales (Navarra y Alemania). Parece que los vertederos de nuestro estudio fueron utilizados, principalmente, por aves locales.

Palabras clave: Censos, Anilla de color, Consumidores generalistas, Gipuzkoa, Disponibilidad de alimento, Ecología trófica

Olga Jordi, Alfredo Herrero, Asier Aldalur, Juan F. Cuadrado, Juan Arizaga, Dept. of Ornithology, Aranzadi Sciences Society, Zorroagagaina 11, E20014 Donostia (San Sebastián), España (Spain).

Corresponding author: J. Arizaga. E–mail: jarizaga@aranzadi–zientziak.org

Introduction

Human activity often produces a super–abundance of food that is exploited by generalist animal foragers (Oro et al., 1995; Giaccardi & Yorio, 2004; Oro et al., 2013; Heath et al., 2014). Dumps constitute a paradigmatic case of this phenomenon. The availability of huge amounts of organic waste attracts multiple species of animals, so some dumps can give rise to concentrations of up to many thousands of individuals (Donázar, 1992; Pons, 1992; Tortosa et al., 2002; Admasu et al., 2004).

Dumps promote large changes in several wildlife aspects, such as demography (Newton, 2013), dispersal and migration (Newton, 2008), trophic ecology (Ramos et al., 2009), and diseases (Monaghan et al., 1985). In parallel, animal concentrations around particular dumps often generate socio–economic (Belant, 1997; Raven & Coulson, 1997; Rock, 2005), sanitary (Monaghan et al., 1985; Ramos et al., 2010), and ecological problems (Rusticali et al., 1999; Vidal et al., 2000; Oro et al., 2005). In attempts to solve this situation, managers have tried to control over–population using a variety of methods, such as culling. These approaches are often of doubtful efficiency (Bosch et al., 2000; Álvarez, 2008) and can even promote undesired effects (Newton, 2013). Alternatively, or complementarily, managers have used methods such as falconry to deter gulls from sites such as dumps (Arizaga et al., 2013a).

Several large predatory gulls (*Larus* spp.) are among the species that benefit most from dumps (Olsen & Larson, 2004). As opportunistic foragers, they exploit a feeding source that has promoted rapid growth rates in their populations (Duhem et al., 2002; Skorka et al., 2005; Duhem et al., 2008). The yellow–legged gull (*L. michahellis*) is the most abundant gull in the southwestern Palaearctic (Olsen & Larson, 2004). Its fast population increase until, at least, the 2000s (Arizaga et al., 2009; Molina, 2009) was partly due to the generalization of large dumps (Duhem et al., 2008) and some colonies have been strongly linked to this type of food (Ramos et al., 2009; Ramos et al., 2011). Other colonies, that depend more on marine prey or other types of natural feeding sources, also forage, to a greater or lesser extent, on waste from dumps (Moreno et al., 2009; Arizaga et al., 2013b). Dumps, in consequence, play a key role for the species.

The yellow–legged gull population is divided into several subspecies that have different migratory behaviour (Olsen & Larson, 2004). Populations from Atlantic Iberia (mostly attributed to belong to *L. m. lusitanius*) are resident, and populations from the Mediterranean (belonging to *L. m. michahellis*) are partially migratory (Munilla, 1997; Arizaga et al., 2010; Galarza et al., 2012). The latter overwinter in part within the Bay of Biscay (Martínez–Abraín et al., 2002). Dumps within this region offer a great foraging opportunity to gulls but the use of these sites by local and non–local gulls is still poorly understood (Álvarez, 2008; Galarza et al., 2012). Here we aimed to determine (1) the origin of gulls using dumps in the Bay of Biscay, (2) the impact of local and non–local gulls at these dumps, (3) the possible age–dependent use of these sites,

and (4) the possible seasonal fluctuations in the use of dumps by gulls.

Using data collected over a period of eight years at four dumps in the south–eastern Bay of Biscay area, we aimed to answer these questions. We accordingly increased our understanding of dump use and the population structure of the yellow–legged gull within this region, where local and non–local individuals coexist for several months each year.

Material and methods

Study area and data collection

We considered the dumps situated within a radius of 60 km from the colonies of Gipuzkoa province (north of Spain). These colonies are situated in the east–most distribution range of the yellow–legged gull, subspecies *L. m. lusitanius*, in the Bay of Biscay (Olsen & Larson, 2004).

From January 2006 to February 2014, the species was surveyed foraging at four dumps within this 60–km radius: S. Marcos, Urteta, Zaluaga and Sasieta (fig. 1). There were two other dumps within this radius (Igorre, Lemoiz) where the species was known to occur, but they were not included in the analyses due to the lack of surveys. The use of the four study dumps by the yellow–legged gull varied during the study period, in accordance with dump management and the amount of food (waste) available at each site (Arizaga et al., 2013a).

At each dump, the yellow–legged gull population size was assessed by means of visual counts. These were always done from the same site at each dump and by the same observer. The time invested to count gulls at each dump was also constant so, overall, the sampling effort at each dump remained constant. Counts from days when gulls were flying around the dump and/or when we observed that they were continuously moving/ flying, due to the use of falconry or other dissuasive methods, were not considered for our analyses.

The yellow–legged gull was the dominant gull among the white–headed gull species at all dumps, and therefore the occurrence of other species could be considered marginal. The second gull in terms of numbers was the Lesser Black–backed Gull (*L. fuscus*) but it comprised ca. < 5% of the counts. Total gull counts were therefore considered to provide a good estimate of the yellow–legged gull population at each dump.

Apart from counts, our databank also contained sightings of colour–ringed gulls seen alive by us or reported by birdwatchers. These included data from colour–ringed gulls seen at both the study dumps and in sites outside these dumps (*e.g.* rivers, harbors, beaches, etc.). We only considered data from individuals ringed as chicks. Sighting data were used to determine the origin of the gulls and to quantify their relative amount with regard to the entire population. Finally, we compiled the number of chicks ringed at the colonies from which ringed gulls were seen at our study dumps.

Overall, data were collected from January of 2006 to February of 2014.

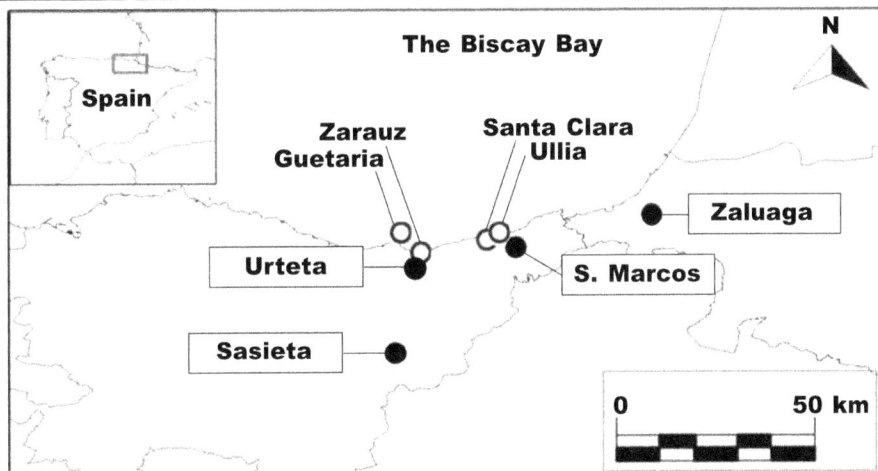

Fig. 1. Location of the study dumps (dark dots) situated at less than 60 km from the reference colonies (open dots) in Gipuzkoa.

Fig. 1. Localización de los vertederos estudiados (puntos negros) situados en un radio de 60 km desde las colonias de referencia (puntos en blanco) en Gipuzkoa.

Data analyses

We pooled years and dumps for all the analyses due to the relatively low sample size (number of counting days) at most dumps (table 1).

To examine the origin sites of gulls visiting our study dumps, we built a table with the number of individual colour–ringed gulls and the total number of origin colonies detected at each dump. Colonies were grouped into four areas of origin: Gipuzkoa (colonies situated at < 60 km), Atlantic, Mediterranean, and others (Atlantic France, inland Iberia, central–western Europe).

We also checked whether the use of the dumps varied between age groups and in relation to the regions of origin. To do this, we considered data obtained both at and outside the study dumps, within a radius of 60 km. We considered five age groups: 1st–year, 2nd–year, 3rd–year, 4th–year, and older (> 4 year) birds. An age category was considered as the year elapsing from July (when chicks fledge) through to June the following year. Groups from two origins were considered: local gulls (Gipuzkoa colonies) and gulls of Mediterranean origin. The gulls from other origins (Atlantic Iberia, Others) were not included in this analysis due to low sample size (< 10 gulls per age class). For each category of origin (Gipuzkoa or Mediterranean), we conducted a chi–square test to see whether the relative number of gulls at and outside the dumps varied between age classes. Standardized residual values from this test were used to identify significant biases from a distribution assuming the same proportion of counts between zones and group. Values > 3 indicate significant differences (Agresti, 2002).

To estimate the yellow–legged gull population size at each dump, we divided the year into two periods, the breeding (January to June) period, and the non–breeding period (July to December). The breeding period corresponded to the time when the occurrence of yellow–legged gulls of Mediterranean origin is minimal (Galarza et al., 2012), while the non–breeding period corresponded to a period when local resident gulls (Arizaga et al., 2010) live in sympatry with yellow–legged gulls from other origins (Galarza et al., 2012). To analyse whether the population size of yellow–legged gulls varied between these periods and between dumps, we conducted a generalized linear model (GLM) on bird counts (log–transformed) with dump and period as factors. Bird counts were log–transformed to fit the normal distribution (K–S test: $P > 0.05$). A linear–link function was used for the GLM.

All analyses were run using the software SPSS v.21.0.

Results

A total of 1226 colour–ringed gulls were observed. We detected 38 origin colonies: four in Gipuzkoa, nine in Atlantic Iberia, 22 in the Mediterranean region and three at other sites (Atlantic France, inland Iberia, central–western Europe) (table 2; fig. 2).

Considering the number of chicks ringed at the origin colonies (table 3), we observed that 39.8% of the chicks ringed at the colonies in Gipuzkoa were seen at our study dumps (all the year is considered here). This proportion was lower for the other origin zones: Atlantic Iberia, 5.6%; Mediterranean, 1.8%; others: 4.2%.

Regarding the use of our dumps between age classes in relation to their origin region, we obser-

Table 1. Number of survey (counting) days at each dump. We show the total number of visits and also those when birds were present (data from 2014 collected only until February).

Tabla 1. Número de días de censo (conteo) en cada vertedero. Mostramos tanto el total de visitas como las visitas en que se detectaron gaviotas (los datos de 2014 se recogieron solo hasta febrero).

Dump	Coordinates	Year	No. counts (> 0)		No. counts (all)	
			Jan–Jun	Jul–Dec	Jan–Jun	Jul–Dec
S. Marcos	43° 18' N – 01° 56' W	2006–2007	3	2	3	2
Urteta	43° 15' N – 02° 10' W	2006–2009	30	15	25	15
Zaluaga	43° 23' N – 01° 34' W	2009–2014	44	55	33	54
Sasieta	43° 02' N – 02° 13' W	2012–2014	8	4	2	4

ved that the proportion of each age category within and outside the dumps did not vary for any of the origin categories considered (Gipuzkoa: $\chi^2 = 7.896$, $P = 0.095$; Mediterranean: $\chi^2 = 7.896$, $P = 0.095$). Overall (data obtained at and outside the study dumps pooled), we detected that the number of 4th year gulls seen at our dumps was proportionally lower for birds of Mediterranean origin. This finding was reversed for older (> 4 years) birds ($\chi^2 = 49.887$, $P < 0.001$; fig. 3)

The population size did not vary between periods but differed between dumps (Period: Wald $\chi^2 = 0.126$, $P = 0.723$; Dump: Wald $\chi^2 = 21.642$, $P < 0.001$; Period × Dump: Wald $\chi^2 = 2.463$, $P = 0.482$; fig. 4). This difference was due to the higher population at Sasieta (> 3,000 gulls) than at the other three dumps (1,000–2,000 gulls) (table 4; fig. 4).

Discussion

The origin of yellow–legged gulls at four dumps near the southeastern Bay of Biscay was diverse. It ranged from Gipuzkoa (local resident gulls; *L. m. lusitanius*) and other colonies along the Bay of Biscay from nor-

thwestern Iberia (also *L. m. lusitanius*) to northwestern France (*L. m. michahellis*; Yésou, 1991), to the Mediterranean and a few inland colonies, including inland Iberia and central–western Europe (*L. m. michahellis*) (Bermejo & Mouriño, 2003; Olsen & Larson, 2004).

Overall, the results are in accordance with the migration patterns described for these two yellow–legged gull subspecies (Munilla, 1997; Olsen & Larson, 2004; Arizaga et al., 2010; Galarza et al., 2012). Thus, while *L. m. lusitanius* is mostly resident, with only a slight fraction moving > 60 km from their natal sites (Arizaga et al., 2010), *L. m. michahellis* migrates to overwinter mostly within the Bay of Biscay (*e.g.*, Galarza et al., 2012). However, considering only the latter subspecies, we did not detect gulls from south–western Iberia, northern Africa (except Algeria), or the central–eastern Mediterranean. Although in some of these areas (*e.g.* northern Africa) few gulls are ringed, this is not the case in others (*e.g.* Italy) (Spina & Volponi, 2008). Therefore, it can be reasonably stated that the central–eastern Mediterranean and the south–western Iberian gulls are rare visitors to our dumps and hence in the south–eastern Bay of Biscay. The occurrence of sufficient food in these two extensive regions would prevent local birds from having the need to move north to the

Table 2. Number of individually colour–ringed yellow–legged gulls (each bird considered only once) detected at the study dumps. We show how many of these gulls were ringed in each origin: * Atlantic France (Ré island), inland Iberia (Navarra), central–western Europe (Germany).

*Tabla 2. Número de gaviotas marcadas con una anilla de color (cada ejemplar solo se tuvo en cuenta una vez) que se detectaron en los vertederos estudiados. Se muestra cuántas de estas gaviotas se anillaron en cada región de origen: * Costa atlántica de Francia (isla de Ré), interior de la península ibérica (Navarra) y Europa centrooccidental (Alemania).*

Individual rings	Origin colonies			
	Gipuzkoa (< 60 km)	Atlantic Iberia	Mediterranean	Others*
1,226	930	127	166	3

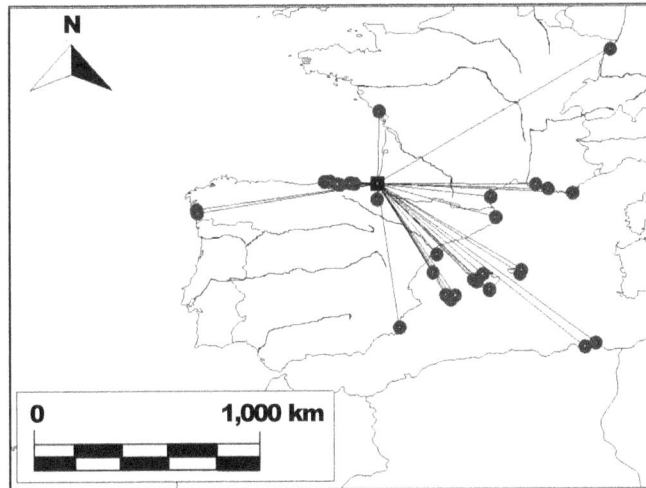

Fig. 2. Origin (dots) of yellow–legged gulls at dumps shown in fig. 1 (square) situated less than 60 km from the colonies in Gipuzkoa. Origins reported using individuals colour–ringed as chicks. The administrative limits are shown in order to facilitate the location of the colonies. We also show the main rivers.

Fig. 2. Origen (puntos) de las gaviotas patiamarillas que se observaron en los vertederos de la fig. 1 (cuadrado) situados en un radio inferior a 60 km desde las colonias de Gipuzkoa. Los orígenes se determinaron a partir de aves marcadas cuando eran pollos con una anilla de color. Se muestra el límite administrativo de los estados con el fin de facilitar la localización de las colonias. También se muestra el cauce de los ríos más importantes.

Biscay Bay area. For instance, the areas surrounding Cádiz Bay, and the Guadalquivir and other nearby river mouths are among the most nutrient productive areas in south–western Europe (Huertas et al., 2006).

The presence of gulls at dumps as compared to sites at a distance from the dumps did not vary between age classes. This was independent of the region of origin and suggests that the use of the dumps was not age–dependent. The use of refuse tips as a food resource was general for all age classes within the region. This result contrasts with findings from earlier studies carried out in the Bay of Biscay, where adult Mediterranean yellow–legged gulls were observed to be proportionally more abundant at dumps than young, sub–adult gulls (Galarza et al., 2012). A possible reason for this difference is a bias associated with local conditions close to our study dumps. We considered a relatively small survey area, so it is possible that the presence of gulls outside the dumps but still rather close to them may be conditioned by the use of these dumps. Thus, some sighting points around or close to dumps may be used as resting areas by the same gulls that have fed in the dumps.

We also observed that, up to the 4th year, yellow–legged gulls of Mediterranean origin become progressively less abundant than local yellow–legged gulls, indicating that older gulls of Mediterranean origin tend to disappear from our area. This is likely due to the fact that adult Mediterranean yellow–legged gulls may remain near their breeding sites during the non–breeding period (Martínez–Abrain et al., 2002; Ramos et al., 2011). The proportionally higher

Table 3. Number of individually colour–ringed yellow–legged gulls (each bird considered only once) detected at our study dumps and number of chicks ringed in the origin colonies of these gulls. We show in brackets the number of colour–ringed gulls coming from colonies from which the total number of chicks ringed was provided: [1] See caption of table 2; [2] During the years in which the gulls seen at our dumps hatched.

Tabla 3. Número de gaviotas patiamarillas marcadas con una anilla de color (cada ejemplar solo se tuvo en cuenta una vez) que se detectaron en los vertederos estudiados y número de pollos anillados en las colonias de origen. En paréntesis, se indica el número de gaviotas con anilla de color provenientes de colonias para las que se pudo saber el total de pollos anillados: [1] Ver cabecera de la tabla 2; [2] Durante los años en que nacieron los pollos que fueron vistos en los vertederos estudiados.

Origin colonies	Ringed gulls seen	Chicks ringed [2]
Gipuzkoa (< 50 km)	930 (930)	2,339
Atlantic Iberia	127 (126)	2,267
Mediterranean	166 (141)	7,586
Others [1]	3 (2)	48

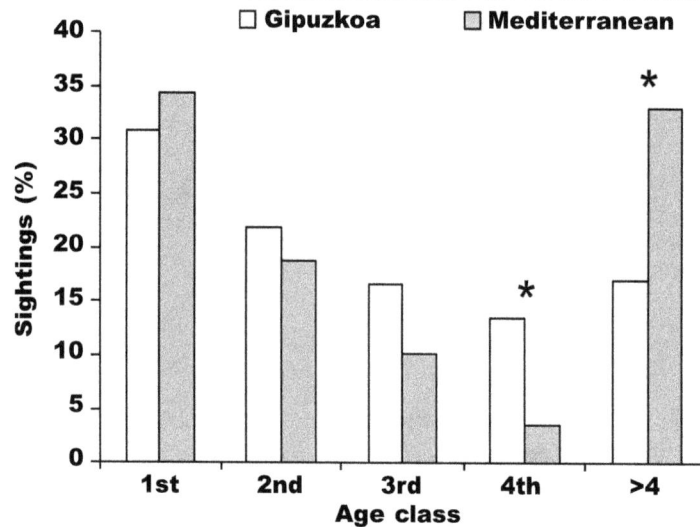

Fig. 3. Relative abundance (percentage) of gulls at the study dumps and their surroundings (within a 60–km radius) in relation to their age class and origin. The symbol (*) indicates significant differences between the two origins for each age class in relation to an expected distribution similar for the two regions.

Fig. 3. Abundancia relativa (porcentaje) de gaviotas en los vertederos estudiados y su entorno (en un radio de 60 km) en relación con la edad y el origen. El símbolo (*) indica la existencia de diferencias significativas entre ambos orígenes para cada edad en relación con una distribución esperada similar para ambas regiones.

percentage of adult Mediterranean gulls, compared to those from Gipuzkoa, is likely associated to the fact that, overall, ringing at the Mediterranean colonies has been done for longer and, therefore, a higher number of adult ringed birds of Mediterranean origin were still alive when the study was carried out.

Finally, we found no statistical evidence to support relevant fluctuations of gull abundances bet-

ween dumps (except at Sasieta, where more birds were detected) or between seasons. With counts ranging between 1,000 and 2,000 individuals, and considering a breeding population at Gipuzkoa of ca. 1,000 pairs (Arizaga et al., 2009; Molina, 2009), which is known to depend on refuse tips to a relevant extent (Arizaga et al., 2013a, 2013b), it can be deduced that most gulls at our dumps were local. The

Table 4. B–parameters from a GLM used to test if the number (population size) of gulls varied between dumps and periods: D. Dump; P. Period; Br. Breeding; NBr. Non–breeding; * Reference values.

*Tabla 4. Parámetros B de un modelo lineal general empleado para comprobar si el número (tamaño de la población) de gaviotas varió entre vertederos y periodos: D. Vertedero; P. Período; Br. Crianza; NBr. No crianza; * Valores de referencia.*

Parameters	B	SE(B)	P	Parameters	B	SE(B)	P
D: Sasieta	+ 0.425	0.183	0.020	Sasieta × NBr	0*		
D: Zaluaga	+ 0.136	0.095	0.152	Zaluaga × Br	+ 0.185	0.128	0.150
D: S. Marcos	+ 0.080	0.241	0.741	Zaluaga × NBr	0*		
D: Urteta	0*			S. Marcos × Br	+ 0.123	0.310	0.693
P: Breeding (Br)	− 0.186	0.106	0.081	S. Marcos × NBr	0*		
P: Non–breeding (NBr)	0*			Urteta × Br	0*		
Sasieta × Br	+ 0.285	0.301	0.343	Urteta × NBr	0*		

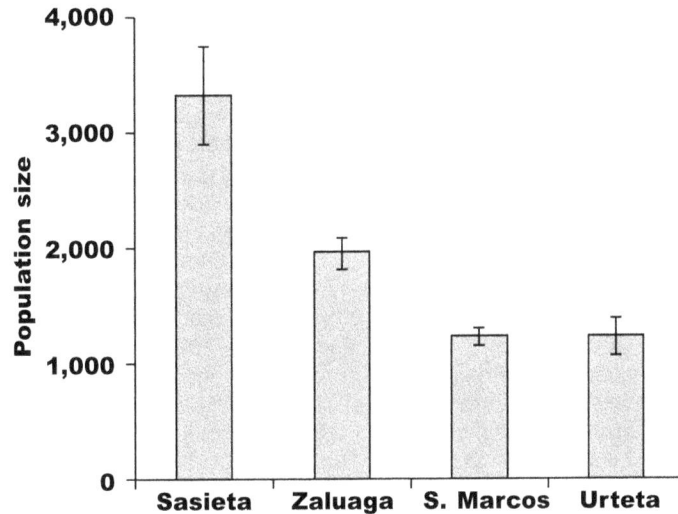

Fig. 4. Population size (mean ± SE) of the yellow–legged gull at four dumps situated within a 60 km radius of the colonies in Gipuzkoa.

Fig. 4. Tamaño de la población (media ± EE) de las gaviota patiamarillas que utilizaron los cuatro vertederos situados en un radio de 60 km desde las colonias en Gipuzkoa.

higher number of gulls at Sasieta was probably due to the fact that there were no other dumps nearby during the survey period.

Acknowledgements

This research was funded by the Basque Government and the Gipuzkoa Regional Council. Many thanks to the people who provided the sighting data and those who provided data on the number of chicks ringed at various colonies, especially: N. Baccetti, A. Galarza, M. McMinn, J. Mouriño, G. Orizaola, B. Samraoui, B. Sarzo, V. Saravia, and the institutions Asociación Naturalista del Sudeste, Catalan Institute of Ornithology, Columbretes Islands Natural Reserve, LPO–Réserve Naturelle Lilleau des Niges, Equipo de Anillamiento Milvus–GOES, Ebro Delta Natural Park. J. A. Donázar and two referees provided very valuable comments that helped us to improve an earlier version of this work.

References

Admasu, E., Thirgood, S. J., Bekele, A. & Karen Laurenson, M., 2004. Spatial ecology of golden jackal in farmland in the Ethiopian Highlands. *African Journal of Ecology*, 42: 144–152.

Agresti, A., 2002. *Categorical Data Analysis.* Wiley. London.

Álvarez, C. M., 2008. *La problemática de las gaviotas en Asturias. El caso del Vertedero Central de CO-GERSA.* Principado de Asturias/COGERSA. Gijón.

Arizaga, J., Aldalur, A., Herrero, A., Cuadrado, J., Díez, E. & Crespo, A., 2013a. Foraging distances of a resident yellow–legged gull (*Larus michahellis*) population in relation to refuse management on a local scale. *European Journal of Wildlife Research* in press.

Arizaga, J., Galarza, A., Herrero, A., Hidalgo, J. & Aldalur, A., 2009. Distribución y tamaño de la población de la Gaviota Patiamarilla *Larus michahellis lusitanius* en el País Vasco: tres décadas de estudio. *Revista Catalana d'Ornitologia*, 25: 32–42.

Arizaga, J., Herrero, A., Galarza, A., Hidalgo, J., Aldalur, A., Cuadrado, J. F. & Ocio, G., 2010. First–year movements of Yellow–legged Gull (*Larus michahellis lusitanius*) from the southeastern Bay of Biscay. *Waterbirds*, 33: 444–450.

Arizaga, J., Jover, L., Aldalur, A., Cuadrado, J. F., Herrero, A. & Sanpera, C., 2013b. Trophic ecology of a resident Yellow–legged Gull (*Larus michahellis*) population in the Bay of Biscay. *Marine Environmental Research*, 87–88: 19–25.

Belant, J. L., 1997. Gulls in urban environments: landscape–level management to reduce conflict. *Landscape and Urban Planning*, 38: 245–258.

Bermejo, A. & Mouriño, J., 2003. *Gaviota Patiamarilla, Larus cachinnans.* In: Atlas de las aves reproductoras de España: 272–273 (R. Martí & J. C. del Moral, Eds.). DGCN–SEO/BirdLife, Madrid.

Bosch, M., Oro, D., Cantos, F. J. & Zabala, M., 2000. Short–term effects of culling on the ecology and population dynamics of the yellow–legged gull. *J.*

Appl. Ecol., 37: 369–385.

Donázar, J. A., 1992. Muladares y Basureros en la biología de la conservación de las aves en España. *Ardeola,* 39: 29–40.

Duhem, C., Bourgeois, K., Vidal, E. & Legrand, J., 2002. Food resources accessibility and reproductive parameters of Yellow–legged Gull *Larus michahellis* colonies. *Revue D Ecologie–La Terre Et La Vie,* 57: 343–353.

Duhem, C., Roche, P., Vidal, E. & Tatoni, T., 2008. Effects of anthropogenic food resources on yellow–legged gull colony size on Mediterranean islands. *Population Ecology,* 50: 91–100.

Galarza, A., Herrero, A., Domínguez, J. M., Aldalur, A. & Arizaga, J., 2012. Movements of Mediterranean Yellow–legged Gulls *Larus michahellis* to the Bay of Biscay. *Ringing and Migration,* 27: 26–31.

Giaccardi, M. & Yorio, P., 2004. Temporal patterns of abundance and waste use by kelp gulls (Larus dominicanus) at an urban and fishery waste site in northern coastal Patagonia, Argentina. *Ornitologia Neotropical,* 15: 93–102.

Heath, M. R., Cook, R. M., Cameron, A. I., Morris, D. J. & Speirs, D. C., 2014. Cascading ecological effects of eliminating fishery discards. *Nature Communications,* 5: 3893.

Huertas, I. E., Navarro, G., Rodríguez–Gálvez, S. & Lubián, L. M., 2006. Temporal patterns of carbon dioxide in relation to hydrological conditions and primary production in the northeastern shelf of the Gulf of Cadiz (SW Spain). *Deep Sea Research Part II: Topical Studies in Oceanography,* 53: 1344–1362.

Martínez–Abrain, A., Oro, D., Carda, J. & Del Señor, X., 2002. Movements of Yellow–Ledged Gulls *Larus [cachinnans] michahellis* from two small western Mediterranean colonies. *Atlantic Seabirds,* 4: 101–108.

Molina, B. E., 2009. *Gaviota reidora, sombría y patiamarilla en España. Población en 2007–2009 y método de censo.* SEO/BirdLife, Madrid.

Monaghan, P., Shedden, C. B., Ensor, K., Fricker, C. R. & Girdwood, R. W. A., 1985. *Salmonella* carriage by Herring gulls in the Clyde area of Scotland in relation to their feeding ecology. *J. Appl. Ecol.,* 22: 669–680.

Moreno, R., Jover, L., Munilla, I., Velando, A. & Sanpera, C., 2009. A three–isotope approach to disentangling the diet of a generalist consumer: the yellow–legged gull in northwest Spain. *Marine Biology,* 157: 545–553.

Munilla, I., 1997. Desplazamientos de la Gaviota Patiamarilla (*Larus cachinnans*) en poblaciones del norte de la Península Ibérica. *Ardeola,* 44: 19–26.

Newton, I., 2008. *The migration ecology of birds.* Academic Press. London.

– 2013. *Bird populations.* Collins New Naturalist Library, London.

Olsen, K. M. & Larson, H., 2004. *Gulls of Europe, Asia and North America.* Christopher Helm, London.

Oro, D., Bosch, M. , Ruiz, X. 1995. Effects of a trawling moratorium on the breeding success of the Yellow–legged Gull *Larus cachinnans. Ibis,* 137: 547–549.

Oro, D., De Leon, A., Minguez, E. & Furness, R. W., 2005. Estimating predation on breeding European storm–petrels (*Hydrobates pelagicus*) by yellow–legged gulls (*Larus Michahellis*). *Journal of Zoology,* 265: 421–429.

Oro, D., Genovart, M., Tavecchia, G., Fowler, M. S. & Martínez–Abraín, A., 2013. Ecological and evolutionary implications of food subsidies from humans. *Ecol. Lett.,* 16: 1501–1514.

Pons, J. M., 1992. Effects of changes in the availability of human refuse on breeding parameters in a Herring Gull *Larus argentatus* population in Brittany, France. *Ardea,* 80: 143–150.

Ramos, R., Cerda–Cuellar, M., Ramirez, F., Jover, L. & Ruiz, X., 2010. Influence of Refuse Sites on the Prevalence of *Campylobacter* spp. and *Salmonella Serovars* in Seagulls. *Applied and Environmental Microbiology,* 76: 3052–3056.

Ramos, R., Ramírez, F., Carrasco, J. L. & Jover, L., 2011. Insights into the spatiotemporal component of feeding ecology: an isotopic approach for conservation management sciences. *Diversity and Distributions,* 17: 338–349.

Ramos, R., Ramirez, F., Sanpera, C., Jover, L. & Ruiz, X., 2009. Diet of Yellow–legged Gull (*Larus michahellis*) chicks along the Spanish Western Mediterranean coast: the relevance of refuse dumps. *Journal of Ornithology,* 150: 265–272.

Raven, S. J. & Coulson, J. C., 1997. The distribution and abundance of *Larus* gulls nesting on buildings in Britain and Ireland. *Bird Study,* 44: 13–34.

Rock, P., 2005. Urban gulls: problems and solutions. *British Birds,* 98: 338–355.

Rusticali, R., Scarton, F. & Valle, R., 1999. Habitat selection and hatching success of Eurasian Oystercatchers in relation to nesting Yellow–legged Gulls and human presence. *Waterbirds,* 22: 367–375.

Skorka, P., Wojcik, J. D. & Martyka, R., 2005. Colonization and population growth of Yellow–legged Gull Larus cachinnans in southeastern Poland: causes and influence on native species. *Ibis,* 147: 471–482.

Spina, F. & Volponi, S., 2008. *Atlante della migrazione degli uccelli in Italia. Vol. 1: non–Passeriformi.* ISPRA–MATTM, Roma.

Tortosa, F. S., Caballero, J. M. & Reyes–López, J., 2002. Effect of Rubbish Dumps on Breeding Success in the White Stork in Southern Spain. *Waterbirds,* 25: 39–43.

Vidal, E., Medail, F., Tatoni, T. & Bonnet, V., 2000. Seabirds drive plant species turnover on small Mediterranean islands at the expense of native taxa. *Oecologia,* 122: 427–434.

Yésou, P., 1991. The sympatric breeding of *Larus fuscus, L. cachinnans* and *L. argentatus* in western France. *Ibis,* 133: 256–263.

10

Effects of migrations on the nestedness structure of bird assemblages in cays of the Jardines de la Reina archipelago, Cuba

A. García–Quintas & A. Parada Isada

García–Quintas, A. & Parada Isada, A., 2014. Effects of migrations on the nestedness structure of bird assemblages in cays of the Jardines de la Reina archipelago, Cuba. *Animal Biodiversity and Conservation*, 37.2: 127–139.

Abstract

Effects of migrations on the nestedness structure of bird assemblages in cays of the Jardines de la Reina archipelago, Cuba.— The nested subset hypothesis states that species in fragmented, less species–rich biotas are non–random subsets of those inhabiting richer sites. The effect of migration on these models has not been yet fully addressed. We compared the phenological stages of the community during the spring and fall migrations. Presence–absence data of bird species occurring at 43 cays of the Jardines de la Reina archipelago was compiled and two incidence matrices were built for fall and spring periods. The degree of nestedness was estimated based on the overlap and decreasing fill, and its significance was assessed by means of 1,000 replicates of four null models. Bird assemblages showed a higher number of species during fall (67) than they did in spring (51). They also showed a significant and stable pattern of nestedness, although this was slightly higher in spring. Seasonal fluctuations caused by migratory movements thus barely affected the nested structure of bird assemblages.

Key words: Community organization, Nested subset, Fragmented biota, Selective extinction, Differential colonization, Null model

Resumen

Efecto de las migraciones sobre la estructura de anidamiento de los ensamblajes de aves en los cayos del archipiélago de los Jardines de la Reina, Cuba.— La hipótesis del subgrupo anidado plantea que, en biotas fragmentadas, las especies de los sitios empobrecidos constituyen subconjuntos no aleatorios de las especies de los sitios con mayor riqueza. El efecto de las migraciones sobre estos modelos aún no ha sido abordado plenamente. Se compararon los estados fenológicos de la comunidad durante las migraciones primaveral y otoñal. Se recogieron datos sobre las presencias y ausencias de las especies de aves en 43 cayos del archipiélago de los Jardines de la Reina. Luego se construyeron dos matrices de incidencia para los periodos otoñal y primaveral. El grado de anidamiento de las matrices se calculó mediante el índice de anidamiento basado en el relleno superpuesto y decreciente, y se evaluó su significación mediante 1.000 réplicas de cuatro modelos nulos. Los ensamblajes de aves presentaron un mayor número de especies en el periodo otoñal (67) que en el primaveral (51). También manifestaron un modelo de anidamiento significativo y estable, que fue ligeramente mayor durante el periodo primaveral. Así, las fluctuaciones estacionales debidas a los movimientos migratorios prácticamente no alteraron la estructura anidada de los ensamblajes de aves.

Palabras claves: Organización comunitaria, Subgrupo anidado, Biota fragmentada, Extinción selectiva, Colonización diferencial, Modelo nulo

Antonio García–Quintas & Alain Parada Isada, Centro de Investigaciones de Ecosistemas Costeros (CIEC), Cayo Coco, Ciego de Ávila, 69400 Cuba.

Corresponding author: Antonio García Quintas. E–mail: antonio@ciec.fica.inf.cu

Introduction

One of the best studied and most controversial subjects within the ecological context is how communities are assembled (Patterson, 1990; Gotelli & McCabe, 2002; Bloch et al., 2007) because the fundamental disjunctive question regarding this topic is based on whether such structuring is deterministically or stochastically originated. Diamond's assembly rules, published in 1975, were supported by the idea that interspecific competition was the basic causative factor shaping community structure. These rules are still considered to be among the most remarkable assumptions explaining the natural organization of communities but their validity has been subjected to much discussion during the past quarter of the last century (Gotelli & McCabe, 2002). The hypothesis of the nested subgroup (Patterson & Atmar, 1986) stands out among the most widely known rules (Rohde et al., 1998; Bloch et al., 2007) as no causal factors are assumed *a priori*. The remaining assembly rules, despite being based on varying criteria, assume that interspecific competition is the crucial factor for the structuring of natural communities. In this regard, analysis of the nested subgroup provides methodological advantages for studies on community organization over many other approaches.

The core of the nested subgroup hypothesis rests on the fact that communities exhibit a nested structure if poor species assemblages are non–random subgroups of those with greater species richness (Rohde et al., 1998; Fernández–Juricic, 2000; Bloch et al., 2007). This issue is closely related to studies on fragmented or isolated biotas such as islands, mountaintops, parasite hosts, isolated forests, and caves. In all cases, if structure of communities or assemblages is described by a nested model, this differs significantly from any randomly generated organization.

The nestedness of species assemblage could be generated by one or many factors depending on the taxonomic group and main features of the study area. Basic factors promoting nested structures are selective extinction and differential colonizationof species (*i.e.* Patterson & Atmar, 1986; Patterson, 1990). Selective extinction induces species loss within ecosystems, forming a predictable sequence without replacements by nearby colonizers (species relaxation). This may provoke non–random losses because species requiring large minimum areas or those forming small populations face high risks of extinction. The origin of nestedness caused by differential colonization was based on the idea that the dispersal capability of differential species leads to the occupation of a larger number of sites by stronger dispersers. Further research has since revealed that other factors may influence nestedness patterns (*e.g.* Calmé & Desrochers, 1999; Ulrich et al., 2009), such as passive sampling, habitat nestedness, disturbances, fragmentation, and age and superficial extension of fragments.

Differential colonization of species shows great potential to promote such structural patterns among the causative factors of nestedness. However, its influence has only been studied taking into account the consequences of the species permanently occupying the sites (immigration) (*e.g.* Cook & Quinn, 1995; Rohde et al., 1998). For instance, birds' annual migrations —which may be described as temporal selective colonizations since each species follows its own migratory pathway and determines its wintering ground— await further scientific scrutiny. Such phenological events are known to bring about remarkable annual compositional changes in tropical bird communities within the Caribbean, depending on the migratory movements in question.

Many authors (*e.g.* Cook & Quinn, 1995; Rohde et al., 1998; Calmé & Desroches, 1999; Fernández–Juricic, 2000; Bloch et al., 2007) have referred to the differential or selective colonization as one of the main forces generating nestedness patterns within the species' natural assemblages. This may imply that the differential occupancy experienced by many migratory bird species in several archipelagos during each season should increase the degree of nestedness on the assemblages of which they temporarily form part. On the contrary, Patterson & Atmar (1986), Patterson (1990), and Calmé & Desrochers (1999) disregard the role of colonization as a critical process to unfolding nested structures in species' assemblages.

This scientific paradox can be assessed in the bird communities of the Jardines de la Reina archipelago, off the southern coast of Cuba because the area is a critical site for residence and transit of migrant bird species (Parada & García–Quintas, 2012). One tentative hypothesis to the aforementioned contradiction is that bird assemblages inhabiting the Jardines de la Reina archipelago have a stable nested pattern under the influence of the many migratory species. If such an assertion is true, then annual fluctuations in the composition of bird species in the Jardines de la Reina archipelago do not affect the nestedness degree of assemblages. To test this hypothesis, the bird assemblages of the Jardines de la Reina archipelago should be assessed and compared in different phenological stages, that is to say, during the periods of highest turnover rates when the influx of neotropical migrants from northern latitudes in September–October and of migrants from southern latitudes in March–April take place on an annual basis.

Material and methods

Study area

The present study was conducted in 43 cays of the Jardines de la Reina archipelago which stretches along the southern coast of Cuba from the Ancon peninsula (Sancti Spíritus province) to Cabo Cruz (Granma province) and comprises numerous islets, shoals and reefs. Three main insular groups can be distinguished: the Ana María cays, the central cays of the gulf of Ana María, and the Doce Leguas cays (fig. 1).

Terrestrial landscapes tend to have a relatively small area (table 1), early geological evolution, and high ecological fragility due to exposure to extreme physical–geographic conditions (*i.e.* strong winds and tidal waves, high salinization and evaporation rates,

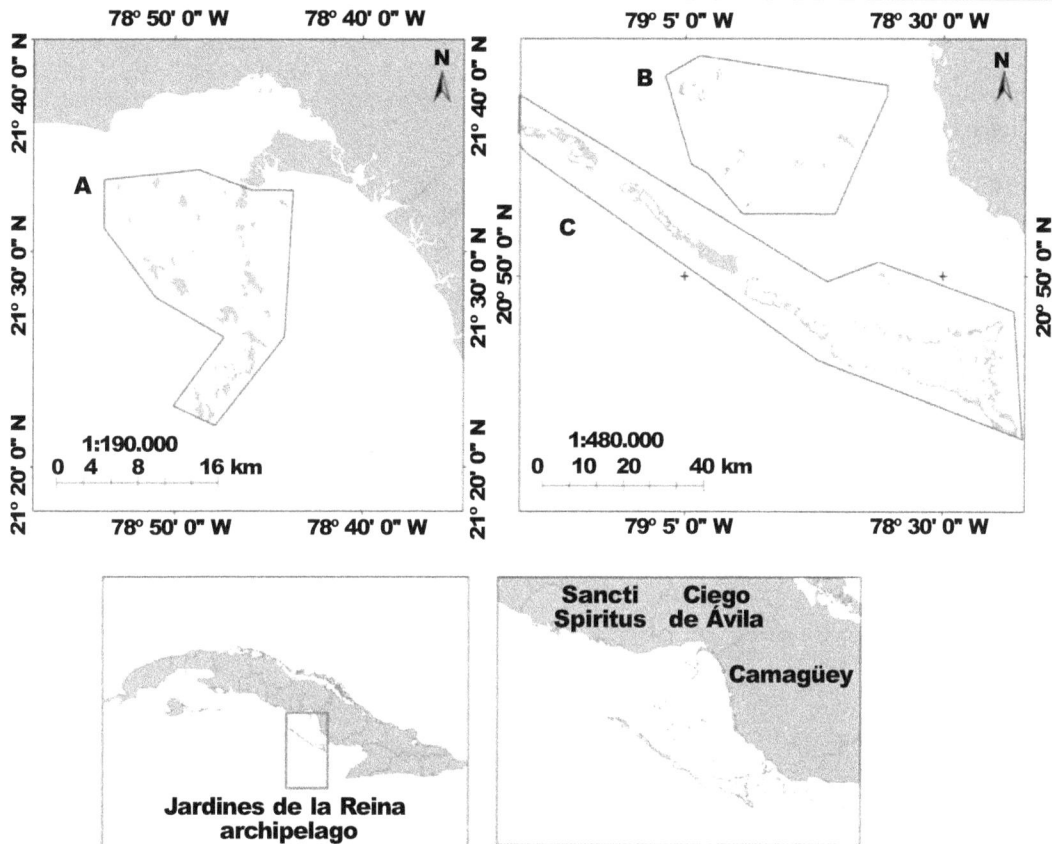

Fig. 1. Study area where the seasonality effect of the avian composition on the structure of bird assemblages in 43 cays of the Jardines de la Reina archipelago, Cuba, was evaluated. A. Ana María cays, B. Central cays of the gulf of Ana María; C. Doce Leguas cays.

Fig. 1. Área de estudio donde se evaluó la influencia de la estacionalidad de la composición de especies sobre la estructura de anidamiento de los ensamblajes de aves en 43 cayos del archipiélago de los Jardines de la Reina, Cuba: A. Cayos de Ana María; B. Cayos del centro del golfo de Ana María; C. Cayos de las Doce Leguas.

seasonal droughts, and intense hydromorphism). The main vegetation forms are mangrove forests, xeromorphic scrub, and sandy and rocky shoreline vegetation, which are best represented in the cays of Doce Leguas (larger and older cays). Lower levels of floral diversity can be found in the cays along the Ana María gulf, where mangrove forests are the prevailing vegetation and may even cover all the emerged land in the Bergantines, Cuervo, Cayuelo and Balandras cays. Some shallow lagoons are also a notable feature in the terrestrial landscapes of many of these cays.

Data source and filtering on the species presence–absence

Presence–absence data was obtained from two different sources. First, we carried out a thorough search encompassing most of the available researches on the study site and extracted many inventory lists. This insular region has been little surveyed and its avifauna is broadly considered among the least studied within the Cuban archipelago. Therefore, most information on the species occurrence was restricted to a few compilation works such as Garrido & García (1975), Buden & Olson (1989), Parada et al. (2012), and Parada & García–Quintas (2012). Second, three surveys conducted in the Caguama (March, July, 2012), and Grande and Caguama (October, 2012) cays were also included to enlarge the final dataset for further analyses. On the March and July field expeditions, visual counts were used to detect bird species inhabiting main vegetation types in Caguama cay. In October, data from visual counts conducted along the south coastline of Grande and Caguama cays during early mornings along with a mist–netting protocol described by Parada et al. (2012) were also used.

Inventory lists per cay were then put together and species exclusively exploiting the ecological resources

Table 1. Main features of the 43 sampled cays of Jardines de la Reina archipelago, Cuba: AMC. Ana María cays; CCG. Central cays of the gulf of Ana María; DLC. Doce Leguas cays; * Information obtained through a classified Landsat image; ** Information gathered from field censuses and specialists´ personal communications.

*Tabla 1. Características principales de los 43 cayos muestreados del archipiélago de los Jardines de la Reina, Cuba: AMC. Cayos de Ana María; CCG. Cayos del centro del golfo de Ana María; DLC. Cayos de las Doce Leguas; * Información obtenida mediante una imagen Landsat clasificada; ** Información obtenida a través de los censos de aves y por comunicaciones personales de especialistas.*

Cays	Insular subgroup	Area* (km^2)	Perimeter* (km^2)	Number of birds per habitat**
Cayuelo	AMC	0.02	0.72	1
Obispito	AMC	0.03	0.84	4
Quitasol	AMC	0.05	1.02	2
La Loma	AMC	0.06	1.50	2
Obispo	AMC	0.09	2.76	4
Guinea	AMC	0.13	1.92	4
La Tea	AMC	0.17	2.46	2
Caoba	AMC	0.26	3.84	3
Flamenco	AMC	0.84	7.50	4
Cana	AMC	0.91	11.34	5
Arenas	AMC	0.97	9.84	4
Tío Joaquín	AMC	1.21	11.07	5
Providencia	AMC	1.29	13.38	5
Guásimas	AMC	1.59	8.70	4
Balandras	AMC	1.62	15.36	2
Punta de Los Machos	AMC	2.14	26.61	3
Cargado	CCG	0.15	3.48	4
Bergantines	CCG	0.22	4.20	4
Palomo	CCG	0.28	6.84	4
Santa María	CCG	0.29	3.12	4
Algodoncito	CCG	0.77	5.82	4
Manuel Gómez	CCG	2.11	34.80	4
Cuervo	CCG	2.16	35.28	5
Algodón Grande	CCG	3.64	32.70	6
Boca Rica	DLC	0.36	6.96	2
Largo	DLC	0.48	7.32	3
Juan Grin	DLC	0.63	16.74	3
Camposanto	DLC	0.82	6.42	4
Alcatracito	DLC	1.34	11.04	4
Boca de la Piedra de Piloto	DLC	1.52	19.14	4
Piedra Grande	DLC	1.53	16.51	5
Boca Seca	DLC	1.76	30.24	2
Alcatraz	DLC	1.84	16.38	4
Cachiboca	DLC	2.44	57.00	6
Boca Piedra Chiquita	DLC	2.88	11.28	5
Las Cruces	DLC	3.64	55.37	4

Table 1. (Cont.)

Cays	Insular subgroup	Area* (km²)	Perimeter* (km²)	Number of birds per habitat**
Cabeza del Este	DLC	6.82	94.44	5
Bretón	DLC	7.51	71.46	4
Caguama	DLC	7.66	87.42	6
Anclitas	DLC	9.06	158.64	6
Grande	DLC	24.29	193.17	5
Caballones	DLC	33.52	73.68	5
Cinco Balas	DLC	43.56	151.20	3

from coastal waters were removed as were all those without any explicit reference of their locality name when first reported. These two simple steps increased the reliability of checklists reliability and avoided associated biases when the numerical analyses were run. The migratory status of bird species occurring in the Jardines de la Reina archipelago was categorized into four main groups: permanent resident (PR), winter resident (WR), summer resident (SR) or transient (Tr), following the criteria of Garrido & Kirkconnell (2011) and those of knowledgeable researchers on the study area avifauna.

Analysis

Transients were excluded from further analyses as they occur at low numbers and exploit ecological niches over only a few days while migrating; the structuring of communities is therefore unlikely to be significantly affected by them. Two matrices containing the remainder of the bird species were then built in order to represent each phenological phase: fall (PR + WR) and spring (PR + SR) migrations. Within each matrix, entries indicated the presence (1) or absence (0) of a species at a site. Typically, matrices were ordered according to the marginal row and column sums. Common species were placed in the upper rows, and species–rich sites placed in the left–hand columns.

Nestedness was calculated by means of two indexes: matrix temperature (T) (Atmar & Patterson, 1993) and a nestedness metric based on overlap and decreasing fill (NODF) (Almeida–Neto et al., 2008) by running the software ANINHADO 3.0.3 (Guimarães & Guimarães, 2006). The former index was solely calculated to allow comparisons with many earlier works on communities' nestedness owing to its vast usage in the specialized literature. The T index values were relativized using the Lomolino (1996) formula to calculate the percentage of perfect nestedness (PN). Nested presence–absence matrices were visualized using the nestedness temperature calculator (Atmar & Patterson, 1995).

Four null models (Er, Ce, Co and Li) provided by the software ANINHADO 3.0.3 were used to assess whether the bird assemblages were nested or randomly structured by generating 1,000 iterations for each one. The calculation of PN was only evaluated using the Er null model. All the randomization algorithms of the null models followed the following rules: (1) Er, presences are randomly assigned to any cell within the matrix; (2) Ce, probability of a cell a_{ij} show the average of the probabilities of occupancy of its row and column (equation 1); (3) Co, presences are randomly assigned within the columns; and (4) Li, presences are randomly assigned within the rows:

$$[(P_i / C) + (P_j / R)] / 2 \quad \text{(equation 1)}$$

where P_i is the number of presences in the i row, P_j is the number of presences in the j column, and C and R are the number of columns and rows, respectively.

Statistical tools were used as null models to compare the degree of nestedness between the spring and fall migrations. To do this, the difference between the values of NODF in the two seasonal stages was calculated (size effect). Afterwards, an overall presence–absence matrix which comprised all the species contained in the two former matrices and all the sampled cays was built up. This matrix was then organized following the general requirements to unfold the nestedness analysis, and randomized twice using the four null models provided by the software ANINHADO 3.0.3. The difference between the simulated values of the NODF generated by each null model was calculated per pairs of iterations/replicates. The statistical significance of the observed difference (size effect) was calculated in terms of its associated probability from the differences simulated by the four null models. To calculate the descriptive statistics of the values generated by the null models, the software *Statistica* 8.0 (StatSoft, 2007) was used and significance level was set at $p < 0.1$.

Results

Avifauna in the study area

The regional avifauna was made up of 120 species, although *Phoenicopterus ruber* (Greater Flamingo), *Anas acuta* (Northern Pintail), *Tringa solitaria* (Solitary Sandpiper), *Antrostomus carolinensis* (Chuck–will's–widow) and *Vermivora chrysoptera* (Golden–winged Warbler) were excluded as their locality names where they were first recorded are unknown (table 2). We also excluded *Fregata magnificens* (Magnificent Frigatebird) as reports of this species were mostly based on individuals in flight exploiting several aerial strata along vast areas, including the surrounding waters. Permanent residents accounted for 37.4% of the species whereas transients were represented by 38 species for the archipelago as a whole. Summer and winter residents were represented by nine and 25 species, respectively (table 2).

Patterns of nestedness

Assemblages were made up by 67 and 51 species during the fall and spring migrations, respectively, and matrices of organized data showed filling values of 28.36% (fall) and 33.01% (spring). The degree of nestedness in spring was greater than in fall (fig. 2), though bird assemblages exhibited patterns of nested structures in both seasons since their NODF values showed significant differences with regard to the simulated values generated by the four null models (table 3).

Comparison between the degrees of nestedness during the two seasonal periods yielded a difference of 1.95. This was not significant if compared to the simulated differences generated by most null models (table 4). The seasonal migrations did not therefore promote any major changes in the degree of nestedness of bird assemblages.

Discussion

Research analyzing the compliance of species' assembly rules can enrich our knowledge of community ecology and help in the planning and implementation of management and conservation efforts. Pianka (1999) and Bloch et al. (2007) stated that nestedness patterns on the community structure provide useful solid grounds for the designing and planning of fragmented protected areas, ecosystem management and meta–communities studies.

Our findings provide additional clues to the growing body of evidence pointing to nestedness as a ubiquitous phenomenon underlying community structure in fragmented biotas. Many authors (*e.g.* Patterson, 1990; Calmé & Desrochers, 1999; Fernández–Juricic, 2000) have found nested structures in numerous bird assemblages, and Méndez (2004) has referred to birds as one of most widely used taxonomic entities in studies focused on community structure. In the study site, a higher degree of in spring could be explained by the lower number of coexisting species and the

ecological relationships among them. Accordingly, nestedness could be produced by the ecological differences among species (Azeria & Kolasa, 2008).

A smaller number of species occurring during spring may have favored many processes, such as spatial segregation, habitat selection and territoriality which are especially remarkable during the breeding season, as in the case of most permanent and summer resident landbirds. These behavioral patterns may promote the species segregation between cays. Thus, the habitat quality, resources availability and intra– and inter–specific hierarchical organizations may have played crucial roles in the differential occupation of species in the cays. For instance, small fragments usually lack sufficient resources to permanently support populations of fruit–eating species (Feeley et al., 2007), causing these birds to move towards larger cays with higher food availability. The numbers of nectarivore (*Chlorostilbon ricordii* only) and frugivore (*Spindalis zena* not currently reported) species in the Jardines de la Reina, for example, is extremely low as they rely on food items that have marked spatial and temporal availability. These species tend to be patchily distributed and may face higher risks of extinction.

Lastly, the increase in nestedness of bird assemblages during spring may be due to the fact that the study site is inhabited by a lower overall number of species, made up only of the breeding populations of summer and permanent residents. In turn, these populations tend to be relatively large and widely distributed across the archipelago´s mangroves and coastal ponds. Aggressive displays by large breeding colonies of cormorants, egrets, pelicans, terns and gulls, and also the presence of *Buteogallus gundlachii* (Cuban Black Hawk) and *Tyrannus caudifasciatus* (Loggerhead Kingbird), which do not breed colonially, may force other birds to shift to cays capable of supporting higher numbers, possibly increasing the degree of nestedness during this period. Likewise, when a larger set of species coexist during fall, the degree of nestedness could decrease given the possible increase in competitive interactions among species, as suggested by Albrecht & Gotelli (2001). Nonetheless, Bloch et al. (2007) pointed out that competitive exclusion reduces the nestedness by preventing the coexistence of species that could otherwise could share the same habitats and resources. On this point, Méndez (2004) believes that nested structures may be ambiguously influenced by the interspecific competition.

In addition, species segregation caused by the ecological dominance among species may lead to weak species being displaced towards resource–poorer habitats (Mac Nally & Timewell, 2005). The density of generalist species may have strong effects on the dynamics of local communities, and therefore the suitability of species assemblages is differentially modulated (Azeria & Kolasa, 2008). Extreme environmental conditions, or the introduction of exotic or invasive species (strong competitors), for example, may have more profound effects on specialist species than those exploiting a much broader range

Table 2. Bird species reported in 43 cays of the Jardines de la Reina archipelago, Cuba: Tr. Transient; PR. Permanent resident; WR. Winter resident; SR. Summer resident; * Number of cays where each species was reported.

*Tabla 2. Especies de aves registradas en 43 cayos del archipiélago de los Jardines de la Reina, Cuba: Tr. Transeúnte; PR. Residente permanente; WR. Residente invernal; SR. Residente veraniego; * Número de cayos en los que fue registrada cada especie.*

Species	Common name	Permanence status	Cays*
Anas discors	Blue–winged Teal	Tr	2
Mergus serrator	Red–breasted Merganser	Tr	1
Fregata magnificens	Magnificent Frigatebird	PR	32
Sula leucogaster	Brown Booby	PR	1
Phalacrocorax auritus	Double–crested Cormorant	PR	30
Anhinga anhinga	Anhinga	PR	12
Pelecanus occidentalis	Brown Pelican	PR	22
Ardea herodias	Great Blue Heron	PR	32
Ardea alba	Great Egret	PR	19
Egretta thula	Snowy Egret	PR	7
Egretta caerulea	Little Blue Heron	PR	13
Egretta rufescens	Reddish Egret	PR	19
Egretta tricolor	Tricoloured Heron	PR	17
Bubulcus ibis	Cattle Egret	Tr	6
Butorides virescens	Green Heron	PR	20
Nyctanassa violacea	Yellow–crowned Night–heron	PR	5
Eudocimus albus	White Ibis	PR	14
Platalea ajaja	Roseate Spoonbill	PR	10
Cathartes aura	Turkey Vulture	PR	18
Pandion haliaetus	Osprey	PR	23
Buteogallus gundlachii	Cuban Black Hawk	PR	14
Buteo jamaicensis	Red–tailed Hawk	Tr	2
Falco peregrinus	Peregrine Falcon	Tr	4
Falco columbarius	Merlin	Tr	6
Rallus longirostris	Clapper Rail	PR	13
Pluvialis squatarola	Grey Plover	WR	10
Charadrius wilsonia	Wilson's Plover	SR	27
Charadrius semipalmatus	Semipalmated Plover	WR	3
Charadrius vociferus	Killdeer	PR	5
Himantopus mexicanus	Black–necked Stilt	PR	5
Actitis macularius	Spotted Sandpiper	WR	11
Tringa melanoleuca	Greater Yellowlegs	WR	3
Tringa semipalmata	Willet	PR	5
Numenius phaeopus	Whimbrel	Tr	1
Arenaria interpres	Ruddy Turnstone	WR	19
Calidris minutilla	Least Sandpiper	WR	10

Table 2. (Cont.)

Species	Common name	Permanence status	Cays*
Calidris mauri	Western Sandpiper	Tr	3
Limnodromus griseus	Short–billed Dowitcher	Tr	1
Leucophaeus atricilla	Laughing Gull	PR	13
Sternula antillarum	Least Tern	SR	7
Thalasseus maximus	Royal Tern	PR	27
Hydroprogne caspia	Caspian Tern	WR	4
Thalasseus sandvicensis	Sandwich Tern	SR	15
Geotrygon montana	Ruddy Quail–dove	Tr	1
Patagioenas squamosa	Scaly–naped Pigeon	PR	3
Patagioenas leucocephala	White–crowned Pigeon	PR	27
Zenaida asiatica	White–winged Dove	PR	23
Zenaida aurita	Zenaida Dove	PR	2
Zenaida macroura	Mourning Dove	PR	19
Columbina passerina	Common Ground–dove	PR	4
Coccyzus americanus	Yellow–billed Cuckoo	SR	5
Coccyzus minor	Mangrove Cuckoo	PR	1
Crotophaga ani	Smooth–billed Ani	PR	4
Tyto alba	Barn Owl	Tr	1
Chordeiles minor	Common Nighthawk	Tr	1
Chordeiles gundlachii	Antillean Nighthawk	SR	18
Chlorostilbon ricordii	Cuban Emerald	PR	18
Megaceryle alcyon	Belted Kingfisher	WR	12
Sphyrapicus varius	Yellow–bellied Sapsucker	Tr	4
Xiphidiopicus percussus	Cuban Green Woodpecker	PR	8
Contopus caribaeus	Greater Antillean Pewee	PR	19
Contopus virens	Eastern Wood–pewee	Tr	1
Myiarchus sagrae	La Sagra's Flycatcher	PR	14
Tyrannus dominicensis	Grey Kingbird	SR	22
Tyrannus caudifasciatus	Loggerhead Kingbird	PR	21
Vireo griseus	White–eyed Vireo	WR	2
Vireo olivaceus	Red–eyed Vireo	Tr	3
Vireo altiloquus	Black–whiskered Vireo	SR	20
Progne cryptoleuca	Cuban Martin	SR	6
Petrochelidon fulva	Cave Swallow	SR	5
Hirundo rustica	Barn Swallow	Tr	12
Catharus minimus	Grey–cheeked Thrush	Tr	1
Catharus fuscescens	Veery	Tr	1
Turdus plumbeus	Red–legged Thrush	PR	3
Dumetella carolinensis	Grey Catbird	WR	5
Mimus polyglottos	Northern Mockingbird	PR	3

Table 2. (Cont.)

Species	Common name	Permanence status	Cays*
Seiurus aurocapilla	Ovenbird	WR	7
Helmitheros vermivorum	Worm–eating Warbler	WR	2
Parkesia noveboracensis	Northern Waterthrush	WR	19
Mniotilta varia	Black–and–white Warbler	WR	9
Protonotaria citrea	Prothonotary Warbler	Tr	1
Oreothlypis peregrina	Tennessee Warbler	Tr	1
Geothlypis trichas	Common Yellowthroat	WR	11
Setophaga citrina	Hooded Warbler	Tr	2
Setophaga ruticilla	American Redstart	WR	16
Setophaga tigrina	Cape May Warbler	WR	4
Setophaga americana	Northern Parula	WR	10
Setophaga castanea	Bay–breasted Warbler	Tr	1
Setophaga fusca	Blackburnian Warbler	Tr	1
Setophaga petechia	Yellow Warbler	PR	40
Setophaga caerulescens	Black–throated Blue Warbler	WR	10
Setophaga palmarum	Palm Warbler	WR	15
Setophaga dominica	Yellow–throated Warbler	WR	8
Setophaga discolor	Prairie Warbler	WR	21
Icteria virens	Yellow–breasted Chat	Tr	1
Tiaris olivaceus	Yellow–faced Grassquit	PR	1
Piranga rubra	Summer Tanager	Tr	1
Piranga olivacea	Scarlet Tanager	Tr	2
Pheucticus ludovicianus	Rose–breasted Grosbeak	Tr	2
Passerina caerulea	Blue Grosbeak	Tr	1
Passerina cyanea	Indigo Bunting	Tr	5
Dolichonyx oryzivorus	Bobolink	Tr	1
Agelaius humeralis	Tawny–shouldered Blackbird	PR	14
Quiscalus niger	Greater Antillean Grackle	PR	31
Icterus galbula	Baltimore Oriole	Tr	1
Asio dominguensis	Short–eared Owl	Tr	1
Sula dactylatra	Masked Booby	Tr	1
Icterus melanopsis	Cuban Oriole	Tr	1
Catharus ustulatus	Swainson's Thrush	Tr	1
Mycteria americana	Wood Stork	Tr	1
Haematopus palliatus	American Oystercatcher	WR	1
Tringa flavipes	Lesser Yellowlegs	WR	1
Calidris alba	Sanderling	WR	5
Polioptila caerulea	Blue–grey Gnatcatcher	Tr	1
Oreothlypis ruficapilla	Nashville Warbler	Tr	1

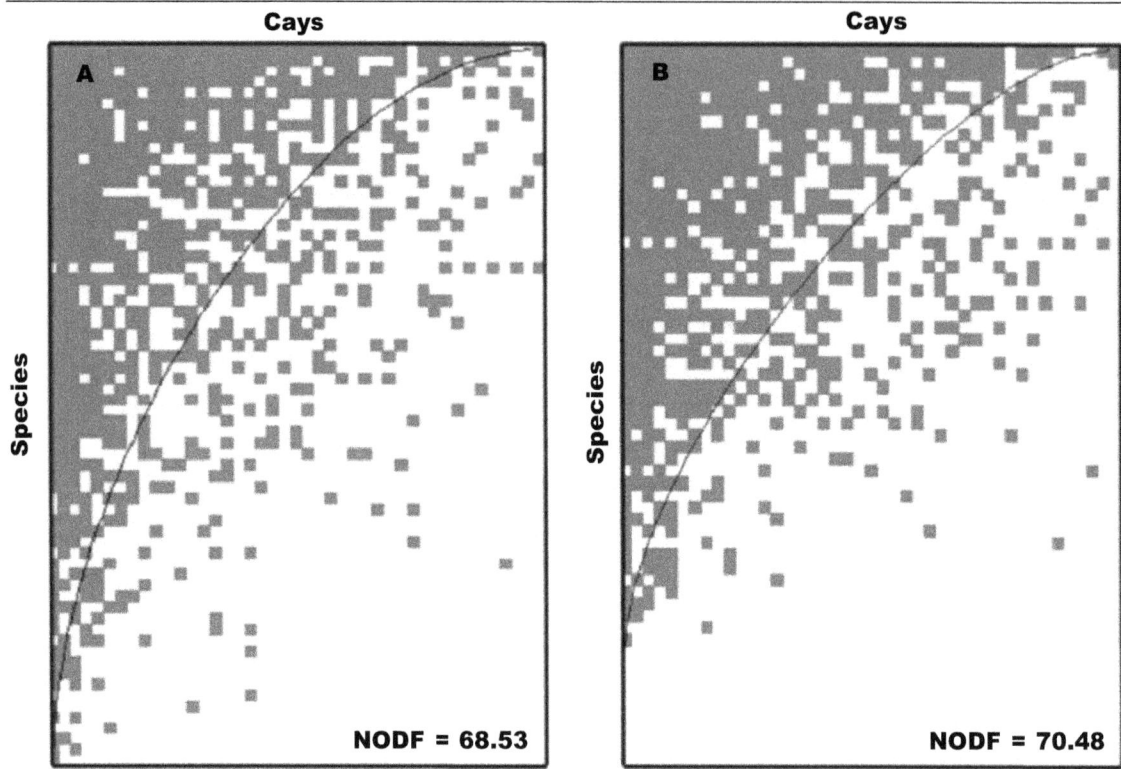

Fig. 2. Nestedness degree of the bird assemblages inhabiting 43 cays of the Jardines de la Reina archipelago, Cuba, during fall (A) and spring (B) migrations. Assemblages made up by 67 and 51 species in A and B, respectively: grey squares, presence; white squares, absence; central line, isocline of perfect nestedness; NODF. Nestedness metric based on overlap and decreasing fill.

Fig. 2. Grado de anidamiento de los ensamblajes de aves presentes en 43 cayos del archipiélago de los Jardines de la Reina, Cuba, durante las migraciones otoñal (A) y primaveral (B). Ensamblajes compuestos por 67 y 51 especies en A y B, respectivamente: cuadros grises, presencias; cuadros blancos, ausencias; línea central, isoclina de anidamiento perfecto; NODF. Índice de anidamiento basado en el relleno superpuesto y decreciente.

of ecological niches. These effects could change species composition, and therefore nested patterns. Generalist species such as herons and doves, both widely distributed populations across the Jardines de la Reina archipelago, could favor the generation of nested structures. Azeria & Kolasa (2008) highlight the importance of thoroughly assessing the relative role of the ecological differences among species (*e.g.* niche breadth) as a cause of nestedness and its temporal stability.

It is plausible to consider that many species may have developed behavioral and morphological adaptations to tolerate higher levels of niche overlapping from other species during fall migrations. Morphological differences (*i.e.* bill size and structure) in similar–sized bird species occupying the same habitat might promote coexistence through feeding niche differentiation as referred in many earlier works (Conant, 1988; Grant, 1999). Such adaptive strategies aiming to minimize the effects of ecological competition upon the arrival of

numerous neartic–neotropical migrants to the Jardines de la Reina archipelago may, to some extent, facilitate species coexistence across this insular region. This would in turn lead to homogenization of the compositional features of bird communities and thus decrease the degree of nestedness on the assemblages.

We agree with Almeida–Neto et al. (2008) and Ulrich et al. (2009) on considering the NODF metric as the most sensible index to evaluate the degree of nestedness given by its properties, which are deemed to be more statistically suitable than the T and D (nestedness discrepancy) indexes (Hu et al., 2011). In fact, this index not only far exceeds the remainder of conventional metrics, but also functions adequately within the null models framework (Almeida–Neto et al., 2008). On the other hand, the use of the T index may yield not only overestimated degrees of nestedness but also contrasting results compared to those obtained by the NODF metric. Such discrepancy was also verified in the present study since bird assemblages exhibited a

Table 3. Assessment of the nestedness degree (NODF index) in the bird assemblages of 43 cays of the Jardines de la Reina archipelago, Cuba, during fall and spring migrations, by comparing the observed (obs.) and the four null models (Er, Ce, Co and Li) simulated values (N = 1,000) of NODF (nestedness metric based on the overlap and decreasing fill): T. Matrix temperature; PNEr. % of perfect nestedness with respect to Er model. All values are expressed as mean ± SD (min–max). There was a significant nestedness (< 0.01) for all the cases.

Tabla 3. Evaluación del grado de anidamiento (índice NODF) en los ensamblajes de aves de 43 cayos del archipiélago de los Jardines de la Reina, Cuba, durante las migraciones otoñal y primaveral, mediante la comparación de los valores observados (obs.) y los simulados (N = 1.000) de los cuatro modelos nulos (Er, Ce, Co y Li) del NODF (Índice de anidamiento basado en el relleno superpuesto y decreciente): T. Temperatura de la matriz; PNEr. % de anidamiento perfecto respecto al modelo Er. Todos los valores se expresan como la [media ± DE (mín–máx)]. En todos los casos el anidamiento fue significativo (< 0,01).

	Sampling periods	
Index	Fall	Spring
T (PNEr)	12.68 (82.57)	13.06 (82.38)
NODFobs.	68.53	70.48
NODFEr	29.41 ± 1.04 (25.43–32.55)	34.00 ± 1.21 (30.27–37.50)
NODFCe	39.01 ± 1.38 (34.56–44.35)	43.55 ± 1.51 (38.67–48.23)
NODFCo	43.11 ± 0.90 (39.22–46.10)	46.37 ± 0.87 (43.30–49.56)
NODFLi	40.74 ± 0.62 (38.78–42.84)	46.18 ± 0.81 (43.40–48.56)

higher degree of nestedness in fall and spring according to the T and NODF indexes, respectively.

A limitation of the present study is that data on species' occurrence was not homogeneous across the study area due to the differences in the number of visits and census techniques used per cay. This may have underestimated the numbers of species along the sampled sites, and may therefore have generated false nested structures. For this reason published compilation works providing the most comprehensive and thorough species checklists were targeted as the ideal currently available information sources (*i.e.* Parada & García–Quintas, 2012; Parada et al., 2012). The inclusion of

Table 4. Comparisons of nestedness degree of the bird assemblages in 43 cays of the Jardines de la Reina archipelago, Cuba, between fall and spring migrations. Differences in the simulated values of NODF index (N = 1,000) are shown as mean ± SD (min–max): NODF. Nestedness metric based on the overlap and decreasing fill; * Statistical significance.

*Tabla 4. Comparación del grado de anidamiento de los ensamblajes de aves en 43 cayos del archipiélago de los Jardines de la Reina, Cuba, entre las migraciones otoñal y primaveral. Las diferencias de los valores simulados del NODF (N = 1.000) se muestran como media ± DE (mín–máx): NODF. Índice de anidamiento basado en el relleno superpuesto y decreciente; * Significación estadística.*

Null models	Differences between pairs of NODF values of replicates	Size effect	Probability
Er	1.11 ± 0.85 (0.00–5.46)	1.95	0.17
Ce	1.39 ± 1.06 (0.00–6.26)	1.95	0.25
Co	0.92 ± 0.71 (0.00–4.20)	1.95	0.10
Li	0.59 ± 0.46 (0.00–2.42)	1.95	0.01*

inventory datasets from over 50 years ago (*i.e.* data from September 1930 published in Buden & Olson, 1989) could influence the nested patterns, if we consider the colonization–extinction dynamics of natural communities. However, we believe that this did not pose a serious problem because most of the species reported by Buden & Olson (1989) and Garrido (1978) are still abundant and widely distributed throughout Jardines de la Reina archipelago. The exceptions are a few vagrant species (*Antrostomus carolinensis* and *Chordeiles minor*) reported on a single occasion some decades ago and *Turdus plumbeus* whose formerly scarce and locally distributed populations may have become extinct as far back as the late 1990s.

Furthermore, the significant differences expressed by the null model Li in the degree of nestedness between migratory seasons may reflect the drawbacks of the randomization algorithm in this model to generate many replicates. Indeed, it produced a narrower range of simulated values than the other models by randomizing the presence of the matrix with row totals kept constant. Thus, the variability of the simulated size effect may have been small enough to bias the detection probabilities of significant differences.

Overall, studies of the temporal changes on nested structures have been little studied to date (Bloch et al., 2007) even though temporal–spatial variations in distribution patterns of non–random species have long been a cornerstone in the ecology of community (Longo–Sánchez & Blanco, 2009). In this regard, no significant differences between the degree of nestedness of assemblages in the two distinct phenological stages (spring and fall seasons) was detected. This indicates that bird communities inhabiting the Jardines de la Reina archipelago showed a stable nested structure despite the influence of many migrants causing seasonal changes on species composition, with higher values of nestedness recorded in spring migrations.

We believe that causative factors such as the isolation of the cays and the habitat nestedness may have a significant role in the generation of nestedness patterns of bird communities in Jardines de la Reina. However, the effects of passive sampling should not be ruled out a priori taking into account the limitations and biases of the currently available datasets. Further identification and evaluation of the factors that promote nested structures among the avifauna in the study site may become relevant to establish conservation priorities and goals in the archipelago and thus maintain the stability of its bird assemblages.

Acknowledgments

We want to thank all those researchers and technicians at CIEC who contributed to this research and we are also grateful to Dennis Denis Ávila and Vicente Osmel Rodríguez Cárdenas for their valuable comments and corrections which greatly improved the work.

References

Albrecht, M. & Gotelli, N. J., 2001. Spatial and temporal niche partitioning in grassland ants. *Oecologia*, 126: 134–141.

Almeida–Neto, M., Guimarães, P., Guimarães, P. R. Jr., Loyola, R. D. & Ulrich, W., 2008. A consistent metric for nestedness analysis in ecological systems: reconciling concept and measurement. *Oikos*, 117: 1227–1239.

Atmar, W. & Patterson, B. D., 1993. The measure of order and disorder in the distribution of species in fragmented habitat. *Oecologia*, 96: 373–382.

– 1995. *The nestedness temperature calculator: a visual basic program, including 294 presence–absence matrices*. AICS Res., Inc., University Park, New Mexico, and The Field Mus., Chicago (http://aicsresearch.com/nestedness/tempcalc.html).

Azeria, E. T. & Kolasa, J., 2008. Nestedness, niche metrics and temporal dynamics of a metacommunity in a dynamic natural model system. *Oikos*, 117: 1006–1019.

Bloch, C. P., Higgins, C. L. & Willing, M. R., 2007. Effects of large–scale disturbance on metacommunity structure of terrestrial gastropods: temporal trends in nestedness. *Oikos*, 116: 395–406.

Buden, D. W. & Olson, S. L., 1989. The avifauna of the cayerías of southern Cuba, with the ornithological results of the Paul Bartsch expedition of 1930. *Smithsonian Contributions to Zoology*, 477: 1–34.

Calmé, S. & Desrochers, A., 1999. Nested bird and micro–habitat assemblages in a peatland archipelago. *Oecologia*, 118: 361–370.

Conant, S., 1988. Geographic variation in the Laysan Finch (*Telespiza cantans*). *Evolutionary Ecology*, 2: 270–282.

Cook, R. R. & Quinn, J. F., 1995. The influence of colonization in nested species subsets. *Oecologia*, 102: 413–424.

Feeley, K. J., Gillespie, T. W., Lebbin, D. J. & Walter, H. S., 2007. Species characteristics associated with extinction vulnerability and nestedness rankings of birds in tropical forest fragments. *Animal Conservation*, 10: 493–501.

Fernández–Juricic, E., 2000. Bird community composition patterns in urban parks of Madrid: The role of age, size and isolation. *Ecological Research*, 15: 373–383.

Garrido, O. H., 1978. Nuevo Bobito Chico (Aves: Tyrannidae) para Cuba. *Academia de Ciencias de Cuba, Instituto de Zoología, Informe Científico–Técnico. La Habana, Cuba*, 68: 1–6.

Garrido, O. H. & García, F., 1975. *Catálogo de las Aves de Cuba*. Academia de Ciencias de Cuba, La Habana, Cuba.

Garrido, O. H. & Kirkconnell, A., 2011. *Aves de Cuba*. Cornell University Press, Ithaca, Nueva York, USA.

Gotelli, N. J. & Graves, G. R., 1990. Body size and the occurrence of avian species on landbridge islands. *Journal of Biogeography*, 17: 315–325.

Gotelli, N. J. & McCabe, D. J., 2002. Species co-occurrence: A meta–analysis of J. M. Diamond's assembly rules model. *Ecology*, 83: 2091–2096.

Grant, P. R., 1999. *Ecology and evolution of Darwin's finches*, 2nd Edition. Princeton University Press, Princeton, USA.

Guimarães, P. R. Jr. & Guimarães, P., 2006. Improving the analyses of nestedness for large sets of matrices. *Environmental Modelling and Software*, 21: 1512–1513.

Hu, G., Feeley, K. J., Wu, J., Xu, G. & Yu, M., 2011. Determinants of plant species richness and patterns of nestedness in fragmented landscapes: evidence from land–bridge islands. *Landscape Ecology*, 26: 1405–1417.

Lomolino, M. V., 1996. Investigating causality of nestedness of insular communities: selective immigrations or extinctions? *Journal of Biogeography*, 23: 699–703.

Longo–Sánchez, M. C. & Blanco, J. F., 2009. Sobre los filtros que determinan la distribución y la abundancia de los macroinvertebrados diádromos y no–diádromos en cada nivel jerárquico del paisaje fluvial en islas. *Actualidades Biológicas*, 31: 179–195.

MacNally, R. & Timewell, C. A. R., 2005. Resource availability controls bird–assemblage composition through interspecific aggression. *The Auk*, 122: 1097–1111.

Méndez, M., 2004. La composición de especies de aves en islas y paisajes fragmentados: un análogo ecológico de las muñecas rusas. *El Draque*, 5: 199–212.

Parada, A. & García–Quintas, A., 2012. Avifauna de los archipiélagos del sur de Ciego de Ávila y Camagüey, Cuba: una revisión taxo–ecológica actualizada. *Mesoamericana*, 16: 35–55.

Parada, A., Socarrás, E., Primelles, J. & Hernández, D., 2012. New bird species and distributional records for Jardines de la Reina archipelago, Cuba, during autumn and spring migrations 2009–10. *Cotinga*, 34: 55–60.

Patterson, B. D., 1990. On the temporal development of nested subset patterns of species composition. *Oikos*, 59: 330–342.

Patterson, B. D. & Atmar, W., 1986. Nested subsets and the structure of insular mammalian faunas and archipelagos. *Biological Journal of the Linnean Society*, 28: 65–82.

Pianka, E. R., 1999. Putting communities together: Ecological assembly rules. Perspectives, advances, retreats. *Tree*, 14: 501–502.

Rohde, K., Worthen, W. B., Heap, M., Hugueny, B. & Guégan, J. F., 1998. Nestedness in assemblages of metazoan ecto–and endoparasites of marine fish. *International Journal for Parasitology*, 28: 543–549.

StatSoft Inc., 2007. *STATISTICA (data analysis software system), version 8.0*. Statsoft, Inc., Tulsa, OK.

Ulrich, W., Almeida–Neto, M. & Gotelli, N. J., 2009. A consumer's guide to nestedness analysis. *Oikos*, 118: 3–17.

11

Drusia (Escutiella) alexantoni n. sp. (Gastropoda, Pulmonata, Parmacellidae), a new terrestrial slug from the Atlantic coast of Morocco

A. Martínez–Ortí & V. Borredà

Martínez–Ortí, A. & Borredà, V., 2013. *Drusia (Escutiella) alexantoni* n. sp. (Gastropoda, Pulmonata, Parmacellidae), a new terrestrial slug from the Atlantic coast of Morocco. *Animal Biodiversity and Conservation*, 36.1: 59–67.

Abstract

Drusia (Escutiella) alexantoni *n. sp. (Gastropoda, Pulmonata, Parmacellidae), a new terrestrial slug from the Atlantic coast of Morocco.*— We describe a new parmacellid, *Drusia (Escutiella) alexantoni* n. sp. from the Moroccan Atlantic coast. The species most closely related to the new taxon are D. *(E.) deshayesii* and D. *(D.) valenciennii*. The new parmacellid differs from D. *(E) deshayesii* mainly by the presence of external spots and bands on both the back and the shield, a reproductive system with uneven atrial appendices of the horn–shaped organ, and a different reticulated pattern of the inner epiphallus. It differs from D. *(D.) valenciennii* mainly for the appearance of the shell and the pattern and disposition of the bumps inside the penis, the presence of an elbow–shape in this organ, and the reticulated appearance of the inner wall of the epiphallus. An updated dichotomous key of the family Parmacellidae is provided.

Key words: Slug, Parmacellidae, *Drusia (Escutiella) alexantoni*, New species, Morocco, North Africa.

Resumen

Drusia (Escutiella) alexantoni *sp. n. (Gastropoda, Pulmonata, Parmacellidae), una nueva babosa del litoral atlántico de Marruecos.*— Se describe un nuevo parmacélido, *Drusia (Escutiella) alexantoni* sp. n., de la costa atlántica marroquí. Las especies más afines al nuevo taxon son D. *(E.) deshayesii* y D. *(D.) valenciennii*. De la primera se diferencia por presentar externamente manchas y bandas sobre el dorso y escudo, un aparato reproductor con apéndices atriales del órgano corniforme bastante desiguales, y por el distinto aspecto del reticulado del interior del epifalo. De D. *(D.) valenciennii* se diferencia principalmente por la forma de su concha, así como por el aspecto y la disposición de los mamelones del interior del pene, la presencia de un marcado acodamiento en este órgano, así como por el aspecto reticulado del interior del epifalo. Se proporciona además una clave dicotómica actualizada de la familia Parmacellidae.

Palabras clave: Babosa, Parmacellidae, *Drusia (Escutiella) alexantoni*, Nueva especie, Marruecos, Norte de África.

Alberto Martínez–Ortí & Vicent Borredà, Museu Valencià d'Història Natural i iVBiotaxa, L'Hort de Feliu–Alginet, P. O. Box 8460, 46018 València, Espanya (Spain) and Dept. de Zoologia, Fac. de Ciències Biològiques, Univ. de València, c/ Dr. Moliner 50, 46100 Burjassot, València, Espanya (Spain).

Corresponding author: A. Martínez–Ortí. E–mail: amorti@uv.es

Introduction

In a recent article (Martínez–Ortí & Borredà, 2012) we revised the systematics of the family Parmacellidae P. Fischer, 1856 and we proposed a new systematic scenario for this family, which would be formed by four genera: *Candaharia* Godwin–Austen, 1888 (2 subgen., 4 spp.), from Central Asia; *Cryptella* Webb et Berthelot, 1833 (7 spp.) from the Canary Islands; *Parmacella* Cuvier, 1804 (2 spp.) from Libya and Egypt; and *Drusia* Gray, 1855 (2 subgen., 4 spp.), with a wide distribution detailed below.

In the previously mentioned paper, the genus *Drusia* Gray, 1855 was divided into two subgenera: *D. (Escutiella)* Martínez–Ortí & Borredà, 2012 and *D. (Drusia)* s. str. The subgenus *D. (Drusia)* includes three species: *D. (D.) valenciennii* (Webb et Van Beneden, 1836), from the South of the Iberian peninsula; *D. (D.) tenerifensis* (Alonso, Ibáñez & Díaz, 1985), from Tenerife and *D. (D.) ibera* (Eichwald, 1841) from the Caucasus–Caspian Sea area. Subgenus *D. (Escutiella)* was described to include only one species: *D. (E.) deshayesii* (Moquin–Tandon, 1848), from Algeria and Northern Morocco.

In January 2011, we carried out a malacological prospection along the Moroccan Atlantic coast, collecting numerous specimens of a parmacelle which we propose as a new species to be included in the subgenus *D. (Escutiella)*.

Results

After a detailed morpho–anatomical study of the collected specimens we observed that they corresponded to a parmacelle closely related to the species *Drusia (Drusia) valenciennii* and *Drusia (Escutiella) deshayesii*, particularly to the latter, but we believe it is a new species, and we propose naming it *Drusia (Escutiella) alexantoni* n. sp.

Family Parmacellidae P. Fisher, 1856

Genus *Drusia* Gray, 1855
Subgenus *D. (Escutiella)* Martínez–Ortí & Borredà, 2012

Drusia (Escutiella) alexantoni n. sp.

Typical locality
Road from Marrakech to Essaouira, 12 km before Essaouira, Taftchet. Essaouira. Morocco (UTM = 29RMQ4388) (January 2, 2011). Collectors: A. Martínez–Ortí, A. López Alabau and A. Pérez Ferrer (MVHN–100111GH01).

Other localities
Road of Essaouira to Agadir–Smimov, Smimov. Essaouira. Morocco (UTM = 29RMQ3274) (January 3, 2011) (Collectors: A. Martínez–Ortí, A. López Alabau and A. Pérez Ferrer) (MVHN–100111GH02; five specimens); Agadir–Ida–Outanane, close to a lake with coots beside the road (February 6, 2009) (Martínez, 2009) (UTM = 29RMP46).

Type material
Formed by 29 specimens. The holotype is deposited at the Museu Valencià d'Història Natural (Valencia, Spain) with the code MVHN–100111GH01a. There are 13 paratypes (in ethanol 70%) with the code MVHN–100111GH01b and four paratypes (in ethanol 96%) with the code MVHN–100111GH01c, all at the same institution. In addition, three paratypes (in ethanol 70%) were deposited at the Museu de Ciències Naturals de Barcelona (Zoologia, MZB) with the code MZB 2012–0728; three paratypes (in ethanol 70%) at the Nationaal Natuurhistorisch Museum–Naturalis of Leiden (The Netherlands) with the code RMNH. MOL.323195; three paratypes (in ethanol 70%) at the Museo Nacional de Ciencias Naturales of Madrid (Spain) with the code MNCN–15.05/60078; and two paratypes (in ethanol 70%) at the Senckenberg Museum of Frankfurt am Main (Germany) with the code SMF 341354.

Etymology
Species dedicated to Alejandro Pérez Ferrer and Antonio López Alabau, co–collectors of the studied specimens and enthusiastic Valencian amateurs of the malacology.

Common name
Slug of Barbary; Babosa de Berbería; Limace de Berberie.

Diagnosis
Parmacelle of great size. Young specimens present an olive brown dorsum with black lines and spots, especially on the shield, while adult specimens are light orange–brown and with lighter lines and spots. Toward the back of the shield, multiple lines or black bands of different thicknesses converge on the protoconch showing individual pattern variation (character less patent in adults). This protoconch is bright greenish, covered in adults and protruding slightly on the body surface of young individuals. Inside the reproductive atrium there is a thick ligula that extends inside the largest of the horn–shaped accessorial appendices. It has a penis with a lateral bulge, giving it an elbow–like shape and it has two thick internal bumps. The interior of the epiphallus has a characteristic reticulated form with thick longitudinal folds that can spread out. Between these folds there are other less patent transverse folds that are almost perpendicular.

External appearance (figs. 1–11): slug of the family Parmacellidae with external features characteristic of this family: large, rough skin, and large, granular shield with the pneumostome in its right posterior portion. Light orange dorsal keel on the caudal part of the animal. Orange dark keel clearly visible in the posterior part of the body, especially in well–developed adult specimens. Very acuminated tail. Foot is of aulacopod type and the sole is light in colour. Caudal gland absent. Adult individuals reach 15 cm in length. Young individuals present a dorsal olive brown background with black lines and spots, especially on the shield; dorsal black bands or lines converge

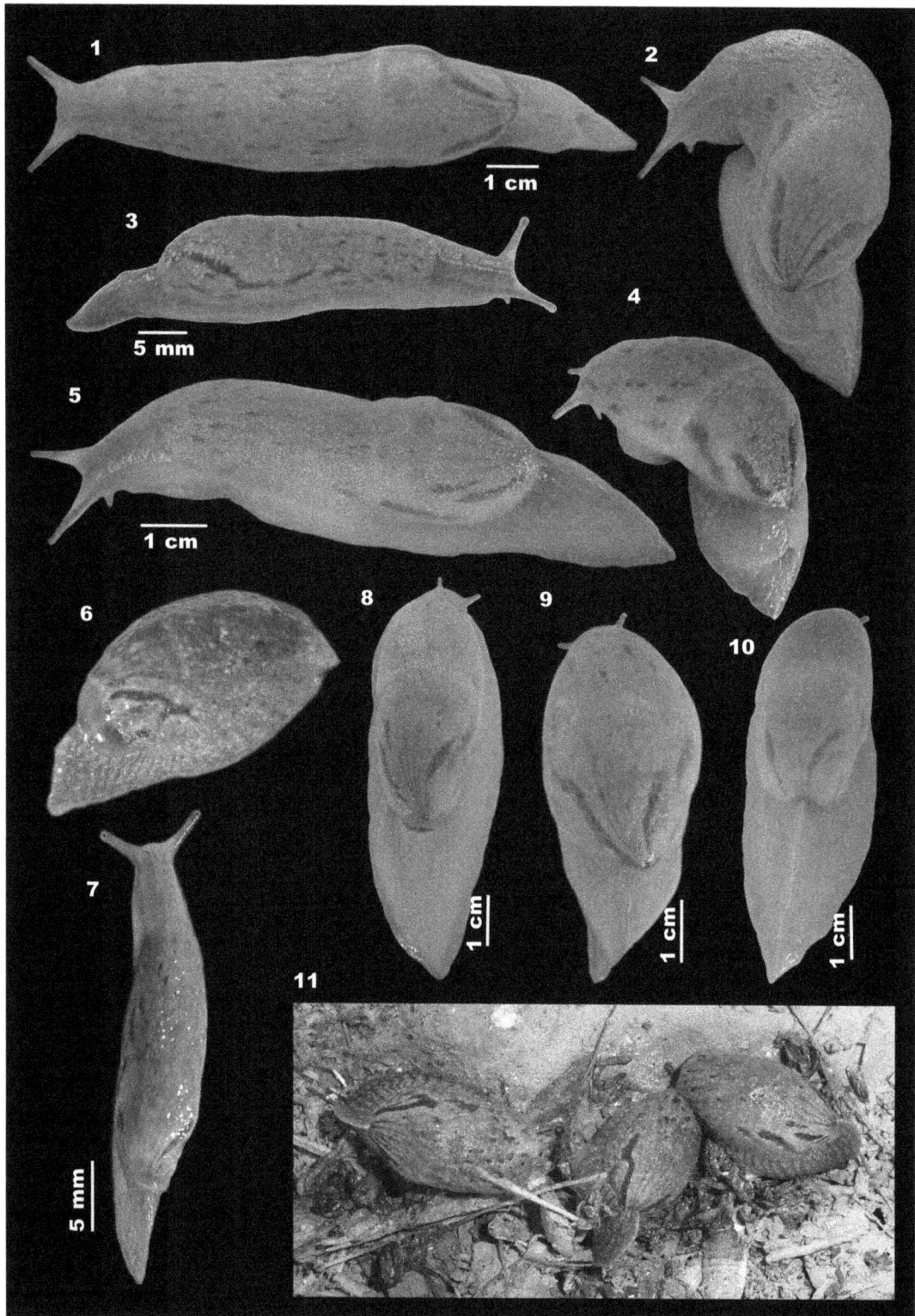

Figs. 1–11. *Drusia (E.) alexantoni* n. sp.: 1–2. Adult holotype; 3. Juvenile holotype; 4–5. Two paratypes; 6–7. Young paratypes showing the protoconch; 8–10. Shield pattern variability of three paratypes; 11. View of a group of several paratypes in the type locality.

Figs. 1–11. Drusia (E.) alexantoni *sp. n.:* 1–2. Holotipo adulto; 3. Holotipo juvenil; 4–5. Dos paratipos; 6–7. Paratipos juveniles mostrando la protoconcha; 8–10. Variabilidad de los dibujos de los escudos de tres paratipos; 11. Vista de un grupo de varios paratipos en la localidad tipo.

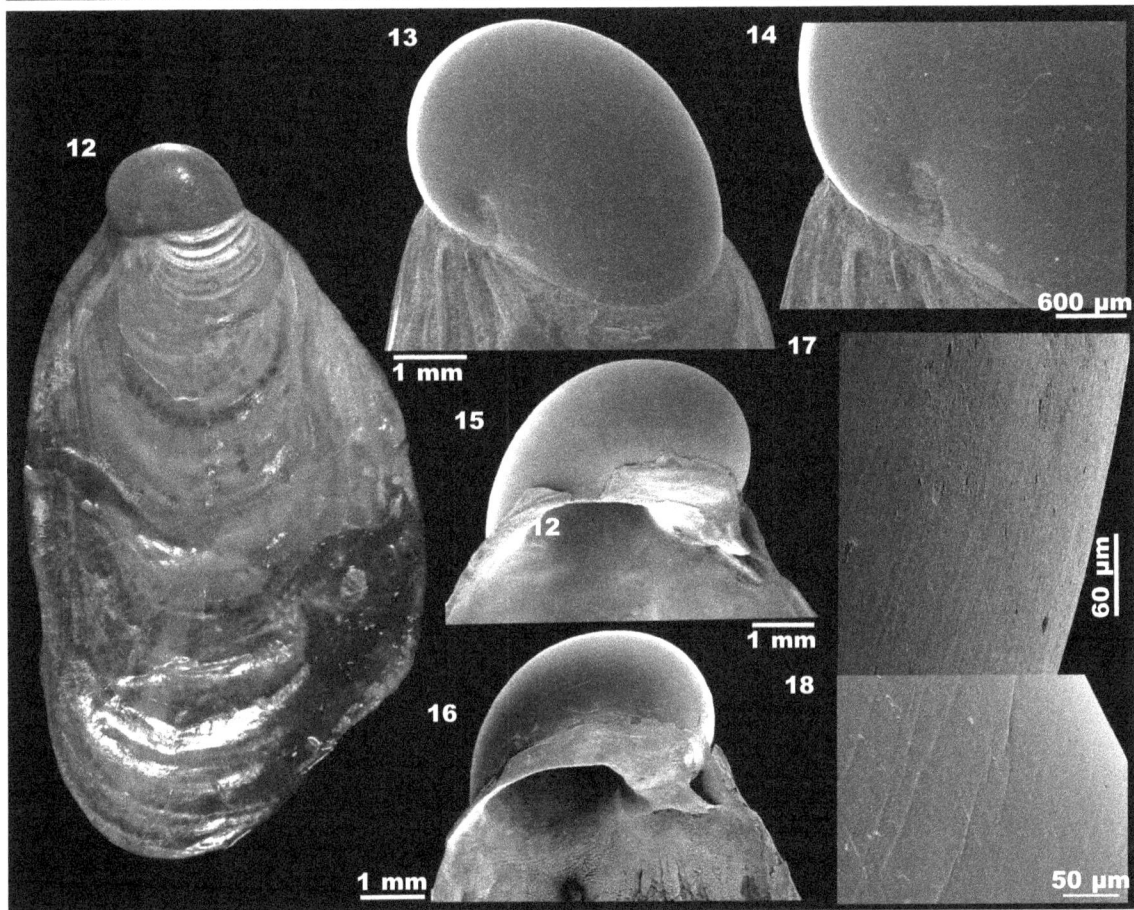

Figs. 12–18. Shell of *Drusia (E.) alexantoni* n. sp.: 12. Paratype (21.5 mm length) (digital photography); 13–14. Protoconch (scanning electron microscope); 14. Detail of the nucleus of the protoconch; 15–16. Posterior region of the shell showing the anchorage denticulation for the muscles; 17–18. Aspect of the surface of the protoconch; 18. Detail of the irregularly reticulated protoconch.

Figs. 12–18. Concha de Drusia (E.) alexantoni *sp. n.: 12. Paratipo (21,5 mm de longitud) (fotografía digital); 13–14. Protoconcha (microscopio electrónico de barrido); 14. Detalle del núcleo de la protoconcha; 15–16. Región posterior de la concha, con denticulación para el anclaje muscular; 17–18. Aspecto de la ornamentación de la superficie de la protoconcha; 18. Detalle del reticulado irregular de la protoconcha.*

toward the shield end, having individual pattern variation. The greenish, bright protoconch is slightly protruded in young individuals and even in sub–adult specimens (figs. 6–7). In well–developed adults, the overall tone of the body is light orange brown, with more visible bands and spots found only on the edge of the shield, while the rest of the dorsum shows a uniform appearance. In general, adult coloration is lighter than in younger animals.

Shell (figs. 6–7, 12–18): the shell is located under the mantle in the posterior part of the shield. It consists of a protoconch, from where a spiral begins, attached to a flat lamina, the limacella (or spatula). The protoconch is greenish, shiny, smooth, and relatively wide. The spiral is clearly visible. The limacella is white, slightly curved and paddle–shaped; it is slightly narrow in comparison

and not strictly flat, being more cupped than in other species of the family. The protoconch protrudes slightly from the posterior end of the mantle in young and sub–adult specimens; it is well–developed and presents a well–marked oval–circular opening (figs. 15–16). In the outer flange an arrowhead–shaped, anchoring tooth is appreciable (figs. 15–16). Although at a glance the protoconch looks smooth and glistening, high magnification reveals a characteristic form, consisting of longitudinal and transverse lines forming an irregular grid in some areas (figs. 17–18). The size of the shell from two of the adult paratypes varies from 12.0 to 14.0 mm in width and from 21.5 to 24.0 mm in length.

Reproductive system (figs. 19–24): hermaphrodite gland partly covered by digestive organs is bilobed and formed by irregular acini. In young specimens it

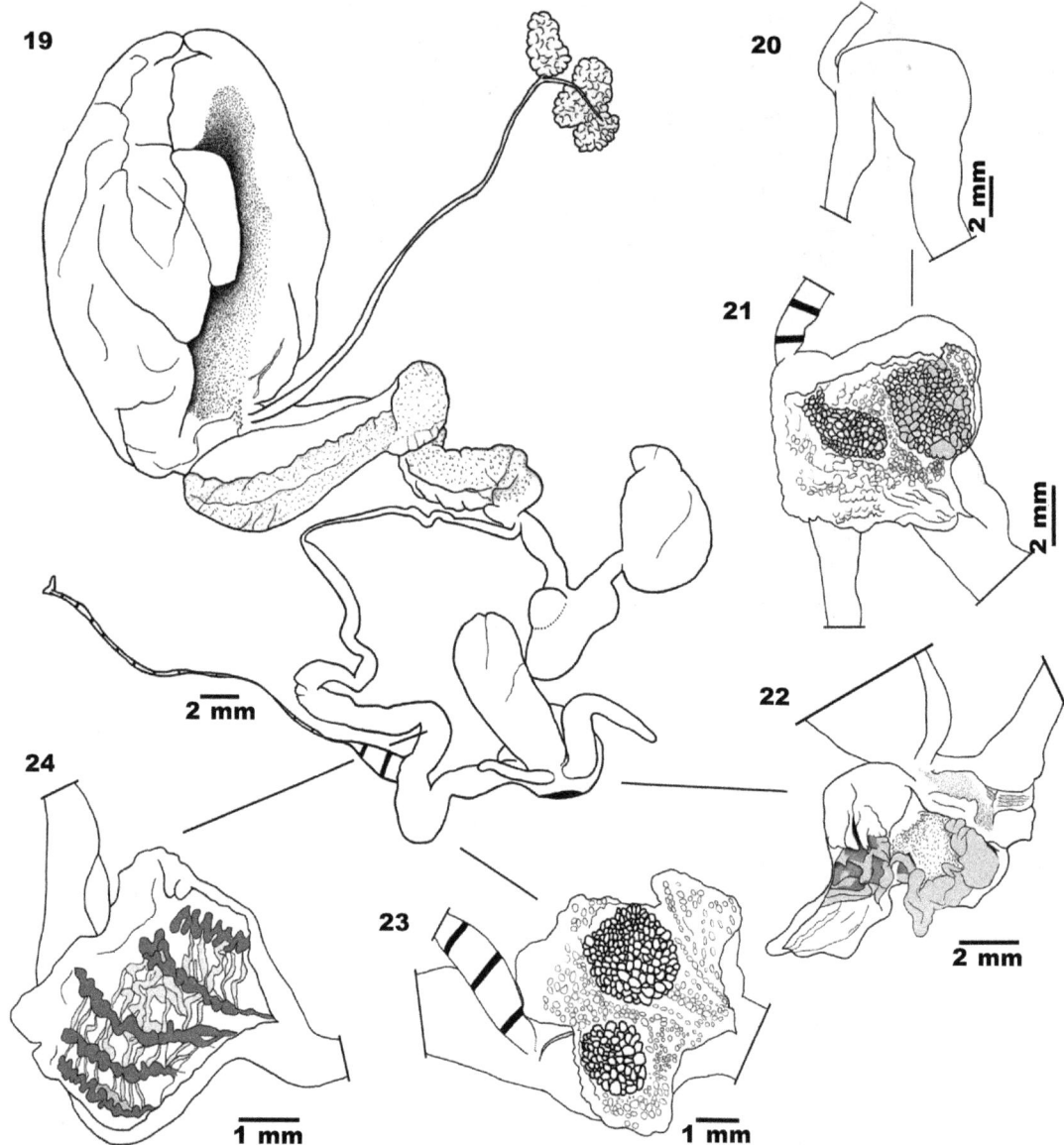

Figs. 19–24. Reproductive system of *Drusia (E.) alexantoni* n. sp.: 19, 22–24. A paratype genitalia: 19. Complete genitalia; 22. Atrium containing the ligula; 23. Detail of the interior of the penis showing the bumps; 24. Form of the inner wall of the epiphallus. 20–21. Penis of another paratype.

Figs. 19–24. Aparato reproductor de Drusia (E.) alexantoni *sp. n.: 19, 22–24. Genitalia de un paratipo: 19. Genitalia completa; 22. Atrio con la ligula; 23. Detalle del interior del pene mostrando los mamelones. 24. Ornamentación de la pared interior del epifalo. 20–21. Pene de otro paratipo.*

is lighter and in adults it is darker in colour, greyish, with the same colour as the hepatopancreas. Hermaphrodite duct long and winding. Very large, triangular, whitish and irregular albumen gland, larger than in *D. (E.) desayesii* and *D. (D.) valenciennii*. Ovispermiduct relatively short, shorter than the albumen gland; distally it separates into feminine and masculine ducts. The masculine duct consists of vas deferens, epiphallus and penis, and together is longer than the ovispermiduct. The vas deferens is flared at its distal part,

turning into the epiphallus, which presents a series of very thick longitudinal folds that can spread out along with other transverse, perpendicular, some of them oblique, less patent folds which give it a reticular appearance interiorly (fig. 24). This reticular appearance is similar to that of *D. (E.) deshayesii*, although this species has both the transverse and longitudinal folds similarly well–marked. The retractor muscle is inserted in the distal part of the epiphallus and it enlarges markedly turning into the penis. The

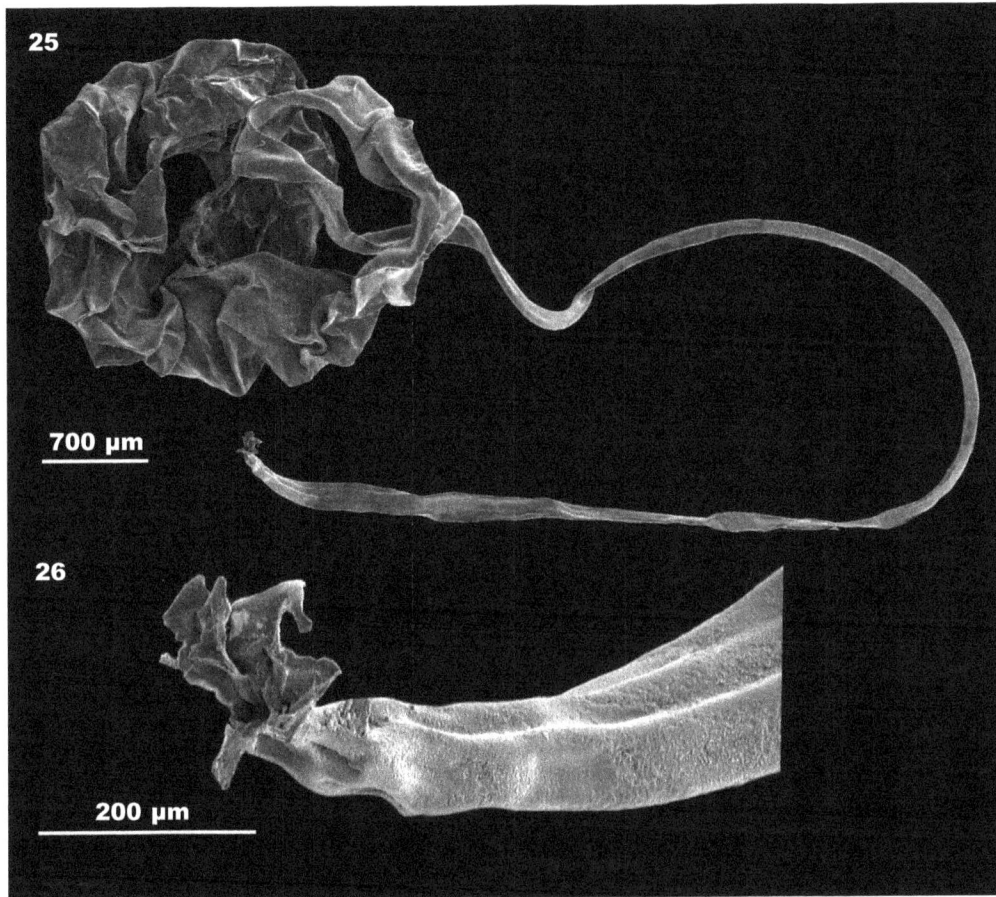

Figs. 25–26. Spermatophore of *Drusia (E.) alexantoni* n. sp.: 25. General view; 26. Anchoring disk detail.

Figs. 25–26. Espermatóforo de Drusia (E.) alexantoni *sp. n.: 25. Vista general; 26. Detalle del disco de anclaje.*

penis has a lateral protrusion close to the retractor muscle, giving it an elbow–like shape. Interiorly, the penis is completely covered with tight papillae. Inside the penis, in its proximal part, there is a bump next to the area of insertion of the muscle retractor (figs. 21, 23). Another larger bump is present in a distal position inside the elbow area. No complete spermatophores have been recovered (figs. 25–26). Inside the bursa copulatrix of four adult paratypes occurred several spermatophores (up to four in one of them), partially digested but quite complete. The spermatophores have the characteristics of the parmacelle morphology, and they are formed by a spiral from which a long filament emerges ending in a star–shaped fixing disk. We did not find entire anchoring disks whose morphology is a character of possible taxonomic value among the partially digested spermatophores, but some of them fairly complete (fig. 26). The female duct begins with a short and cylindrical free oviduct which ends in a widened structure which also converges at the duct of the bursa copulatrix. This widened structure is smooth and ovoid, with a hemispherical bulge in front of the

end of the short bursa duct; the bursa is rather large and has very thin walls, although its size and shape vary greatly depending on the presence and degree of digestion of the spermatophores (fig. 19). The widened area increases its width becoming more glandular in aspect, having a bean–shape; it is the so–called peri-vaginal gland. The vagina is surrounded by this gland and ends in the atrium, which is rather short and has two conspicuous appendices attached, unequal in size and shape (figs. 19, 22). They are the atrial appendices; together they constitute the corniform organ, which has an irregular croissant shape. In the interior of the atrium, as is typical in the genus *Drusia*, there is a highly developed fleshy ligula that expands through the larger corniform organ appendix (Martínez–Ortí & Borredà, 2012) (fig. 22).

Other characters (figs. 27–36): jaw of oxygnathous type and crescent–shaped (figs. 27–29), similar to that of *D. (E.) deshayesii*. In addition, it has a serra-ted edge, visible as tiny teeth at high magnification (fig. 29). The radulae of two examined paratypes

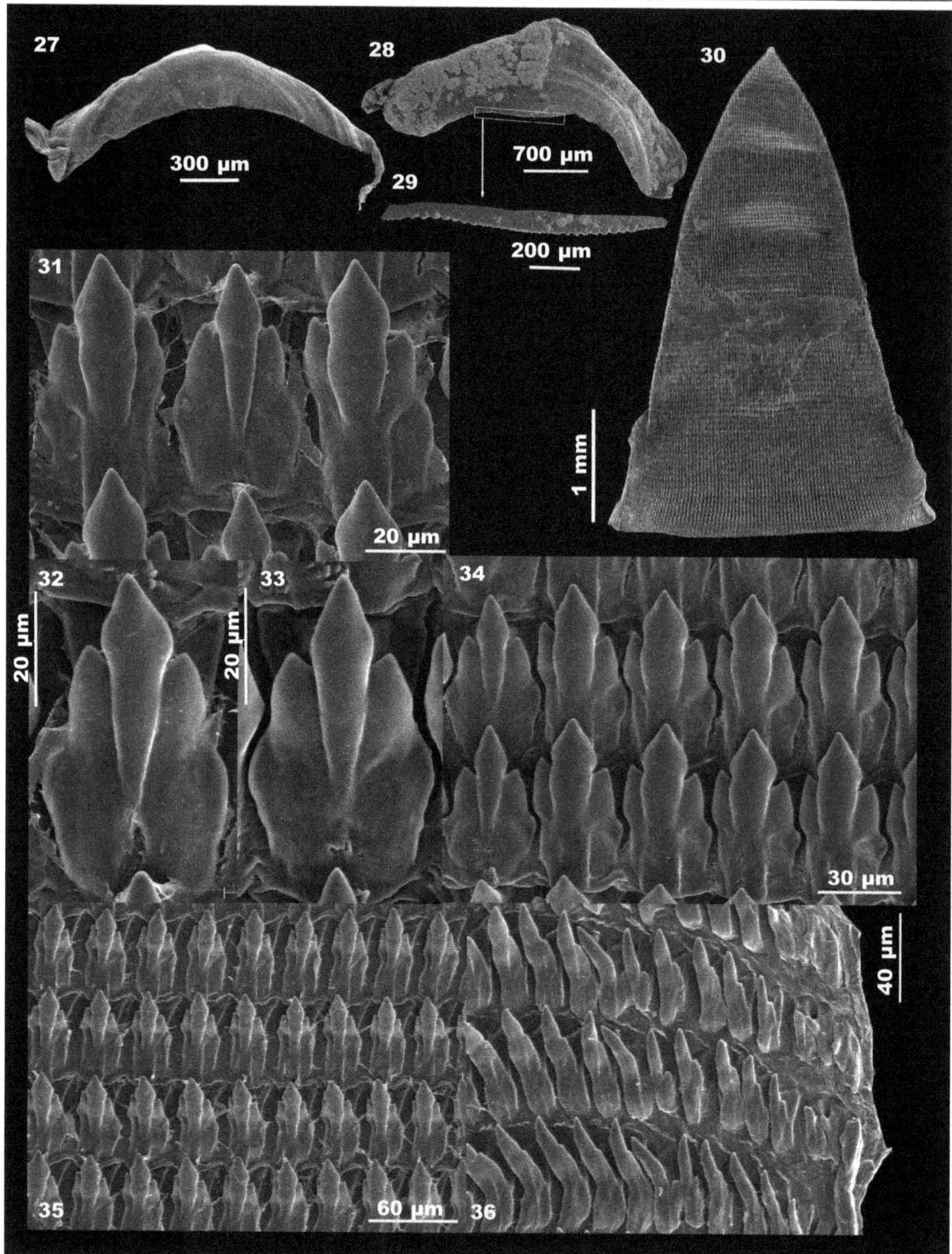

Figs. 27–36. Jaw and radula of *Drusia (E.) alexantoni* n. sp.: 27–29. Jaw: 27. Paratype; 28–29. Other paratype; 29. Detail of the serrated edge. 30–35. Radula; 30. General view of the radula; 31. Central tooth and first lateral teeth; 32–33. Central tooth; 34. Lateral teeth next to the central tooth; 35. Transition from the lateral teeth toward the edge of the radula. 36. Last lateral teeth.

Figs. 27–36. Mandíbula y rádula de Drusia (E.) alexantoni *sp. n.: 27–29. Mandíbula: 27. Paratipo; 28–29. Otro paratipo; 29. Detalle del borde aserrado. 30–35. Rádula; 30. Vista general de la rádula; 31. Diente central y primeros laterales; 32–33. Diente central; 34. Dientes laterales próximos al diente central; 35. Transición de los dientes laterales hacia el borde de la rádula. 36. Últimos dientes laterales.*

New key for the determination of the family Parmacellidae P. Fischer, 1856.

Nueva clave para la determinación de la familia Parmacellidae P. Fischer, 1856.

1	Vagina surrounded by a perivaginal gland not thickened and provided with a long finger–shape caecum	*Candaharia* (Central Asia)
	Vagina with a swollen perivaginal gland, well–developed and bean–shaped. No caecum.	2
2	Genital atrium without appendices. Bursa copulatrix without thickening	*Cryptella* (Canary Islands)
	Genital atrium with two appendices, or at least one. Duct of the bursa with a thickening where the spermatophores are attached	3
3	Atrial appendices of similar size. Elongated and well–developed distal part of the atrium from the insertion of appendices to the genital pore. Without intraatrial stimulators, only fleshy folds, with small ridges on its wall	*Parmacella* (Libya, Egypt) 4
	Atrial appendices of different size. Short distal part of the atrium. One or more intraatrial large and fleshy stimulator folds	*Drusia* 5
4	Ornamented protoconch with small parallel spiral grooves. Very long epiphallus with two bends	*P. festae*
	Smooth protoconch. Epiphallus shorter and with a single curvature	*P. olivieri*
5	Adults presenting dorsum with a shield that has dark stains and/or bands. Smooth penis without extrusion. Interior of the epiphallus not reticulated. Protoconch amber coloured and limacella in form of broad paddle	*D. (Drusia)* s. str. 6
	Adults with dorsum and shield of uniform reddish–brown colour, or only with small lines at the end of the shield. Epiphallus internally reticulated. Penis with side extrusion, sometimes elbow–shaped. Greenish protoconch and a little wide limacella in the form of elongated paddle. Morocco and Algeria	*D. (Escutiella)* 8
6	Shell with a spatula (limacella) shaped shovel, very wide. Animals of large size (70–95 mm in ethanol). Anchoring disk of the spermatophore curved like an umbrella. Tenerife, Canary Islands	*D. (D.) tenerifensis*
	Shell with spatula oval, much more narrow. Specimens of smaller size. Anchoring disk of the spermatophore almost flat	7
7	Wide spatula (limacella) of the shell (long/wide < 1.60). Stimulator fold in the Interior of the atrium thin and not very developed. Georgia, Kazakhstan, and other countries in the E of the Caspian Sea	*D. (D.) ibera*
	Spatula much narrower (long/wide > 1.85). Atrial appendices of very different sizes, sometimes only one. Stimulator fold of the atrium unique, pleated and very thick, occupying almost all of the intraatrial space. South of the Iberian peninsula, Spain and Portugal	*D. (D.) valenciennii*
8	Juvenile with dorsum and shield with black bands and spots, which tend to disappear in adults. Atrial appendices of the corniform organ quite unequal. Interior of the epiphallus with thick reticulate. Huge albumen gland. Atlantic coast of Morocco, Essaouira to Agadir	*D. (E.) alexantoni* n. sp.
	Dorsum and shield, both juveniles and adults, of reddish–brown uniform colour, no bands or spots. Slender reticulate inside the epiphallus. Only slightly unequal atrial appendices. Northern Morocco and Algeria	*D. (E.) deshayesii*

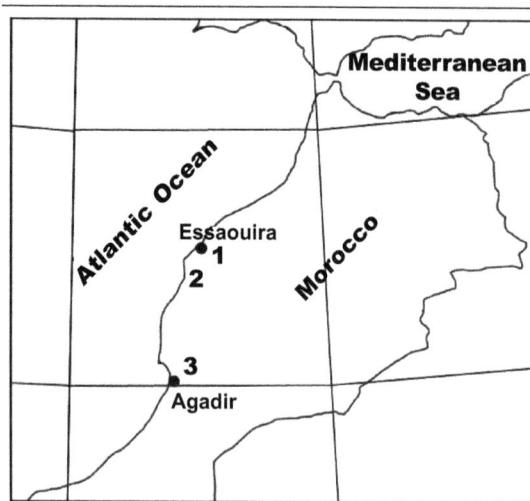

Fig. 37. Map of geographical distribution of *Drusia (E.) alexantoni* n. sp.

Fig. 37. Mapa de distribución geográfica de Drusia (E.) alexantoni *sp. n.*

consist of 100 and 116 rows and both measure 4.65 mm length and 3.0 mm wide. Its radular formula is: 51 + C + 51. Teeth are generally similar to *D. (E.) deshayesii* (figs. 30–36) (Martínez–Ortí & Borredà, 2012). The central tooth presents a deep cut in the shape of an isosceles triangle at the base of the mesocone and reaching the vertex and lower end of this triangle (figs. 31–33). The ectocones also present wing–shape expansions. Other teeth present at the base of the external ectocone with additional wing–shape expansions directed outwards (Martínez–Ortí & Borredà, 2012).

Geographical distribution and habitat

D. (E.) alexantoni n. sp. has been found on the Moroccan Atlantic coast, from Essaouira to Agadir (fig. 37), in crops of argan (*Argania spinosa* (L.) Skeels). One of the authors (Martínez–Ortí) collected all the specimens living in colonies underneath the stones and small walls between these crops along with the Papillionaceae plant *Ononix natrix* L. which is possibly part of their diet. It has also been cited in lacustrine riparian environments (Martínez, 2009).

Discussion

This new species undoubtedly belongs to the genus *Drusia* and we decided to include it in the subgenus *D. (Escutiella)* due to the appearance of its shell and other features. Besides, it is very similar to *D. (E.) deshayesii*

due to the following reproductive characters: i) penis with a lateral protrusion, ii) inside the penis there are two thick and solid bumps and iii) reticulated epiphallus inside with thick longitudinal folds.

It differs from *D. (E.) deshayesii* by i) a reproductive system with uneven atrial appendices of the horn–shaped organ, ii) lateral protrusion that gives it an elbow–like shape that is not present in *D. (E.) deshayesii*, iii) the arrangement and number of bumps inside the penis, only two of them in *D. (E.) alexantoni* n. sp and up to four in *D. (D.) desayeshii*, iv) the reticulated appearance of the inner wall of the epiphallus is different, with the longitudinal folds being much larger in *D. (E.) alexantoni* n. sp. and v) the very large albumen gland of the new species.

These reproductive characters are taxonomically more relevant than the external appearance which in juveniles, with spots and bands, could be confused with the subgenus *D. (Drusia)* s. str. and with the species *Drusia (D.) valenciennii*. Equally, the two appendices of the corniform organ are very unequal in the new species, which makes it more like *D. (D.) valenciennii*. However, due to the set of characters mentioned and described above, it seems much more related to *D. (E.) deshayesii* and we have included it in the subgenus *D. (Escutiella)*. The radula maximum dimensions of *D. (E.) alexantoni* n. sp. are 4.65 x 3.00 mm, being slightly smaller than in *D. deshayesii* (6.75 x 3.95 mm) and *D. (D.) valenciennii* (7.00 x 4.00 mm). In addition, the radular formula of *D. (E.) alexantoni* n. sp. (51 + C + 51) is clearly different from *D. deshayesii* (70 + C + 70) and *D. valenciennii* (65 + C+ 65) (Martínez–Ortí & Borredà, 2012).

Martínez–Ortí & Borredà (2012) provide a dichotomous key to identify the species in the family Parmacellidae but due to the discovery of *D. (E.) alexantoni* n. sp. and the new morpho–anatomical features provided it requires slight modifications (see above).

Acknowledgements

We thank Dr. Fouad Achemchem from l' Université Ibn Zohr Agadir (Morocco) for his help during the sampling. We also thank the staff at the *Sección de Microscopía Electrónica* of the S. C. S. I. E., Universitat de València for their help using the SEM Hitachi S–4100.

References

Martínez, F., 2009. Parmacella sp. http://www.bio-diversidadvirtual.org/insectarium/Parmacella+sp-img62285.html (accessed July 10, 2012)

Martínez–Ortí, A. & Borredà, V., 2012. New systematics of Parmacellidae P. Fischer 1856 (Gastropoda, Pulmonata), with the recovery of the genus–name *Drusia* Gray 1855 and the description of *Escutiella* subgen. nov. *Journal of Conchology*, 41(1): 1–18.

12

Red squirrels from south–east Iberia: low genetic diversity at the southernmost species distribution limit

J. M. Lucas, P. Prieto & J. Galián

Lucas, J. M., Prieto, P. & Galián, J., 205. Red squirrels from south–east Iberia: low genetic diversity at the southernmost species distribution limit. *Animal Biodiversity and Conservation*, 38.1: 129–138.

Abstract

Red squirrels from southeast Iberia: low genetic diversity at the southernmost species distribution limit.— South-east Iberia is the southernmost limit of this species in Europe. Squirrels in the region mainly inhabit coniferous forests of *Pinus*. In this study, we analyzed the pattern of mitochondrial genetic variation of southern Iberian red squirrels. Fragments of two mitochondrial genes, a 350–base pair of the displacement loop (D–loop) and a 359–bp of the cytochrome b (Cytb), were sequenced using samples collected from 88 road–kill squirrels. The genetic variation was low, possibly explained by a recent bottleneck due to historical over–exploitation of forest resources. Habitat loss and fragmentation caused by deforestation and geographic isolation may explain the strong genetic subdivision between the study regions. Six new haplotypes for the D–loop and two new haplotypes for the Cytb fragments are described. A Cytb haplotype of south–east Iberia was found to be present in Albania and Japan, suggesting local extinction of this haplotype in intermediate areas. No significant clustering was found for the south–east of Spain or for the other European populations (except Calabria) in the phylogenetic analysis.

Key words: *Sciurus vulgaris*, Mitochondrial DNA, Genetic diversity, Population bottleneck

Resumen

Ardillas rojas del sureste ibérico: baja diversidad genética en el límite austral de la distribución de la especie.— El sureste ibérico es el límite más austral de la distribución de esta especie en Europa, donde las ardillas habitan principalmente en bosques de *Pinus*. En este estudio, se investigó el patrón de variación genética mitocondrial de las ardillas rojas del sureste ibérico. Se secuenciaron fragmentos de dos genes mitocondriales, 350 pares de bases de la región control (D–loop) y 359 pb del citocromo b (Cytb) utilizando muestras obtenidas a partir de 88 ardillas atropelladas. Se encontró una baja variación genética, lo cual podría explicarse por la existencia de un cuello de botella reciente causado por la sobreexplotación histórica de los recursos madereros de la zona. La pérdida y fragmentación del hábitat debidas a la deforestación y al aislamiento geográfico podrían explicar la fuerte subdivisión genética observada entre las regiones del estudio. Se describen seis nuevos haplotipos para el fragmento D–loop y dos para el Cytb. Un haplotipo encontrado en el sureste ibérico para el Cytb se observó también en Albania y Japón, lo que sugiere una extinción local de este haplotipo en áreas intermedias. En los análisis filogenéticos, no se detectó un agrupamiento significativo de las ardillas del sureste ibérico, ni de ninguna otra población europea (excepto en Calabria).

Palabras clave: *Sciurus vulgaris*, ADN mitocondrial, Diversidad genética, Cuello de botella poblacional

J. M. Lucas, & J. Galián, Depto. de Zoología y Antropología Física, Fac. de Veterinaria, Univ. de Murcia, Campus de Espinardo, 30100 Murcia, España (Spain).– P. Prieto, Parque Natural de Cazorla, Segura y las Villas, c/ Martinez Falero 11, 23470 Cazorla, Jaén, España (Spain).

Corresponding author: José Manuel Lucas, e–mail: lucas@um.es

Introduction

The red squirrel (*Sciurus vulgaris* Linnaeus, 1758) is widely distributed from Iberia in the west across the Palaearctic to the island of Hokkaido (Japan), and from the UK, Ireland, Scandinavia and Siberia to the Mediterranean (Corbet, 1978; Lee & Fukuda, 1999; Lurz et al., 2005). In the Iberian Peninsula, this native sciurid is continuously distributed from Girona to Galicia and the Northern Iberian Mountain Range, the Northern Plateau and the Central Mountain Range, and southwards to Valencia. It is discontinuously distributed from Cataluña to Andalucía, and widely spread in the Baetic Mountain Ranges, including Murcia, Albacete and Alicante (Valverde, 1967; Purroy, 2014). As the result of recent reintroductions, the species can also be found in central and north Portugal (Mathias & Gurnell, 1998; Ferreira et al., 2001; Ferreira & Guerreiro, 2002) (fig. 1A). Most Iberian squirrels occupy pure pine forest: *Pinus halepensis* in the lower altitudes, *P. pinaster* and *P. nigra* in middle levels, and *P. mugo* in the higher locations (Valverde, 1967). In south–east Iberia, the most common species of pine is *P. halepensis (*Aleppo pine), however, even at relatively medium/high altitudes. This is especially evident in the region of Murcia where red squirrels are found in urban parks and adjacent copses, in small to large villages, and even in cities where Aleppo pine can be found. In these localities, they have even been seen feeding on date palms (pers. obs.).

The species is extremely variable in color. Considerable regional variation is superimposed on a striking polymorphism and equally striking seasonal differences (Corbet, 1978). Many studies of the morphological diversity of Spanish squirrels have been made in the past century, especially in the early nineteen hundreds (Cabrera, 1905; Miller, 1907, 1909, 1912), which led to an intense taxonomical discussion. More recently, the first researcher to provide new material morphological variation was Valverde (1967). He assigned his samples to four previously described subspecies (*S. v. alpinus* Desmarest, 1822, *S. v. numantius* Miller, 1907, *S. v. infuscatus* Cabrera, 1905 and *S. v. segurae* Miller, 1912) and suggested the existence of a new subspecies, which he named *S. v. hoffmanni* Valverde, 1967 from Sierra Espuña (southeast Spain). However, subsequent authors considered that only two subspecies are present in Iberia: *S. v. fuscoater* Altum, 1876 and *S. v. infuscatus* (Corbet, 1978; Lurz et al., 2005; Sidorowicz, 1971) (fig. 1B).

Valverde (1967) emphasized the importance of the *hoffmanni* subspecies because of its ecological and morphological features. These squirrels represent the southeastern limit of the Iberian distribution of the species in the xerothermic forest–margin of the Iberian Peninsula, where it lives in pure Aleppo pine forest. Moreover, *S. v. hoffmanni* would be the largest of the European red squirrels, with the palest fur. Thus, this form should represent the ecological limit and the most extreme phenotype of Iberian squirrels (Valverde, 1967). According to the author, *S. v. hoffmanni* is restricted to the Regional Park of Sierra Espuña, but currently the *hoffmanni* phenotype can be easily observed in the Regional Park of Carrascoy–El Valle further south of this region, separated from Espuña by the Guadalentín River. The Regional Park of Sierra Espuña is about 80 km east of the Natural Park of Sierra de Cazorla, Segura y Las Villas and they are connected by a northwestern green corridor (Special Protected Area of Sierra de Burete, Lavia y Cambrón, and the Northwestern Mountains of Murcia). The Natural Park of Sierra de Cazorla, Segura y Las Villas is the largest protected area in Spain, 214,300 ha, and it was designated by UNESCO as a Biosphere Reserve in 1983. There are good–sized red squirrel populations in this area, and they are still under taxonomic discussion (*S. v. baeticus* Cabrera, 1905 = *S. v. segurae* = *S. v. infuscatus*).

Beyond the taxonomical discussion, no recent studies of the ecological characteristics of squirrels from southeast Iberia have been published. Only one study has investigated the genetics of some Iberian populations (Lucas & Galián, 2009), and it found extremely low genetic variation in the population of the Regional Park of Sierra Espuña.

We investigated whether the low genetic diversity found in Sierra Espuña can be considered a pattern in Southeast Iberia or whether it is a peculiarity of this population. In order to study the relationships between the southeastern Iberian squirrels and the other European populations, we compared our results with those in the literature. To achieve these objectives two mitochondrial gene fragments (D–Loop and Cytb) were analyzed using samples from road–kill animals.

Material and methods

Sample collection

In southeastern Spain, most natural areas are crossed by roads and frequented by a large number of visitors. As found in other European populations (Shuttleworth, 2010), road–kill squirrels are frequent in the study area both in natural and suburban environments.

The study area was divided into five regions according to geographical and ecological barriers or distance between samples clusters (fig. 2). Samples from CSV and ESP were collected from the reported distribution of the *segurae* and *hoffmanni* subspecies. All of the samples comprised approximately 2 mm² of muscle tissue and were preserved in absolute ethanol, then stored at –20°C until DNA purification.

DNA extraction and sequencing

Total genomic DNA was extracted from tissue samples using a Qiagen DNAeasy Tissue Kit, according to the manufacturer's protocol. A total of 754–bp were amplified from two gene regions of the mitochondrial DNA. A 395–bp fragment of the D–loop was amplified in 12.5–µl reactions, following the protocol described by Hale et al., 2004, using 1 µl of tissue DNA and the red squirrel–specific primers, H16359 (Barratt et al., 1999) and RScont6 (Hale et al., 2004). A 359–bp region of the Cytb was amplified using the same pro-

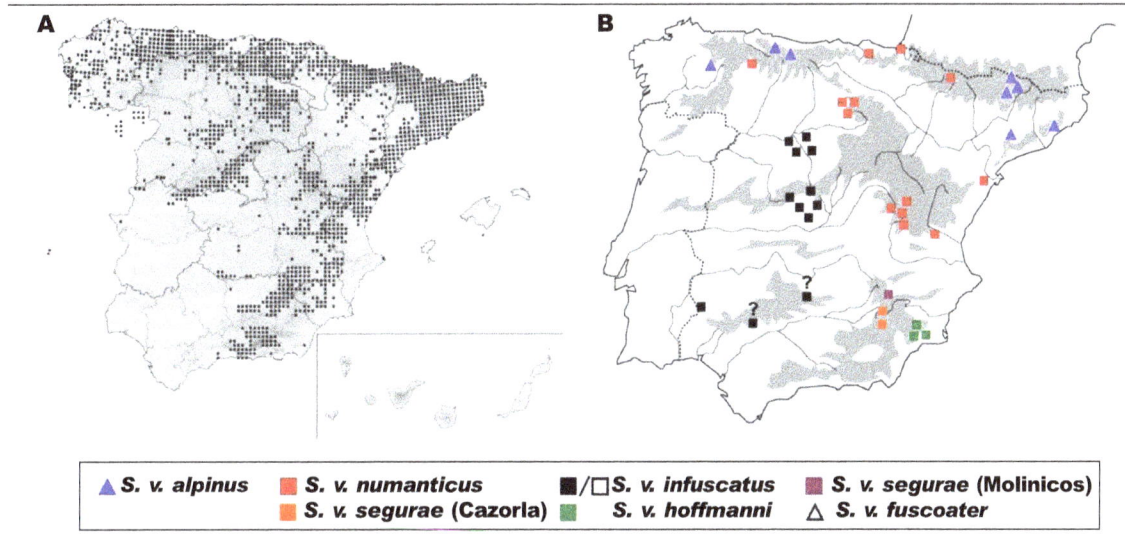

Fig. 1. Map of the species distribution in the Iberian peninsula (A) modified from Palomo & Gisbert (2002). Geographic distribution of red squirrel subspecies (B), obtained from Valverde (1967) and Mathias & Gurnell (1998). Square shape (□) and triangle shape (△) refer to the subspecies *infuscatus* and *fuscoater*, respectively, as synonymised in more recent studies (Sidorowicz, 1971; Corbet, 1978; Lurz et al., 2005).

Fig. 1. Mapa de distribución de la especie en la península ibérica (A), modificado de Palomo & Gisbert (2002). Distribución geográfica de las subspecies de ardilla roja (B), a partir de la información de Valverde (1967) y Mathias & Gurnell (1998). Los cuadrados (□) y los triángulos (△) hacen referencia a las subspecies infuscatus *y* fuscoater *respectivamente, sinonimizadas en trabajos más recientes (Sidorowicz, 1971; Corbet, 1978; Lurz et al., 2005).*

tocol, except that we used the primers SV14226F and SV14647R from Grill et al. (2009). Negative (sterile water) and positive (known squirrel DNA) controls were always used and the products were visualized on 2% agarose gels alongside a 100–bp size standard to determine the success of the amplification. The PCR products were sequenced in both forward and reverse directions for each sample by Macrogen Inc., Korea.

Data analysis

Consensus sequences for each individual were obtained by aligning the forward and reverse complementary sequences of each gene (D–loop and Cytb) with Geneious 4.8.3. D–loop sequences were aligned in MUSCLE (Edgar, 2004) and Cytb sequences with ClustalW algorithm (Larkin et al., 2007). The haplotypes were identified with TCS 1.21 (Clement et al., 2000) and compared with those available in the GenBank using BLAST (Altschul et al., 1990). The relative frequencies of the Cytb and D–loop haplotypes were calculated with Arlequin 3.1.2.3 (Excoffier & Lischer, 2010). Haplotype diversity was calculated separately for each gene. Due to the low diversity found in Cytb sequences, both genes were combined to investigate the nucleotide diversity. The molecular diversity indices were determined using DnaSP 5 (Librado & Rozas, 2009).

The pairwise genetic distances between regions, which were measured as F_{ST}, were calculated from a distance matrix of D–loop haplotypes based on the Tamura–Nei model (Tamura & Nei, 1993) in Arlequin 3.1.2.3 (Excoffier & Lischer, 2010).

The genealogical relationships between the D–loop haplotypes of southeast Iberia were assessed by constructing a median–joining network in NETWORK 4.6.1 (Bandelt et al.,1999). Haplotype networks including sequences from the GenBank were also calculated for both genes.

Phylogenetic analyses were conducted in MEGA6 (Tamura et al., 2013) using the maximum likelihood (ML) method with the nearest neighbour interchange algorithm. Nucleotide sequences of red squirrels from other European populations (Hale et al., 2004; Grill et al.,2009; Doziéres et al., 2012) and of the Japan squirrel *Sciurus lis* (Oshida & Masuda, 2000) were downloaded from GenBank and aligned with our data set. These sequences showed a 100% overlap with the sequences we analysed. The model of nucleotide substitution that best fitted the data set was determined with MEGA6 (Tamura et al., 2013). The stability of the ML tree topologies were tested using 1,000 bootstrap replicates.

Results

A total of 88 samples from the five regions were genotyped successfully. Twenty of the samples from ESP were used in previous work (Lucas & Galián,

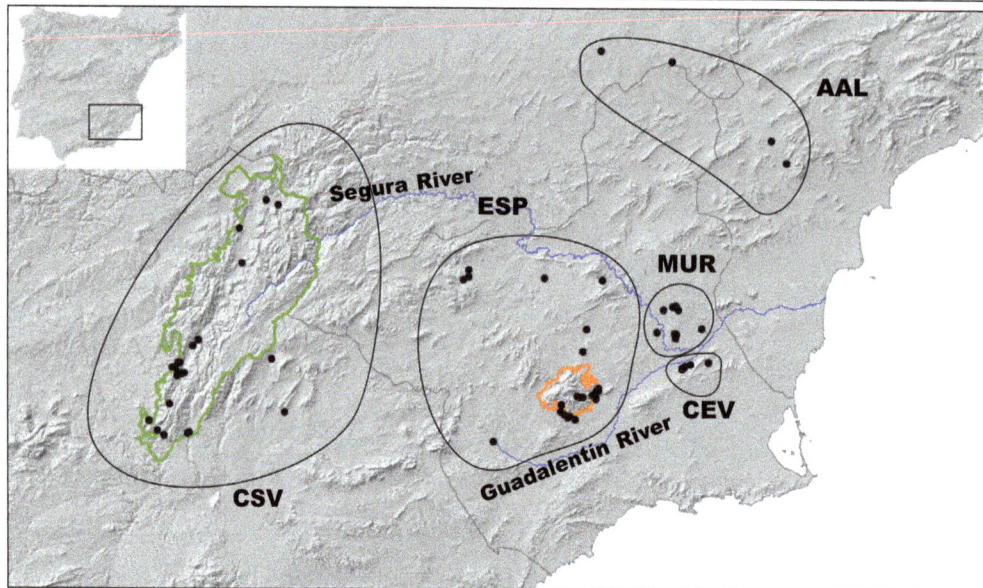

Fig. 2. Map of the study area. The black dots represent *Sciurus vulgaris* specimens. The green line marks the area of the Natural Park of Sierra de Cazorla, Segura y Las Villas, and the orange line delimits the area of the Natural Park of Sierra Espuña. The five regions in the study area are bounded by black lines: CSV. Natural Park of Sierra de Cazorla, Segura y Las Villas and surroundings; ESP. Regional Park of Sierra Espuña and surroundings; MUR. Copses and periurban parks near the city of Murcia; CEV. Regional Park of Carrascoy–El Valle; AAL. Albacete and Alicante.

Fig. 2. Mapa del área de estudio. Los puntos negros representan los individuos de Sciurus vulgaris. La línea verde indica el límite del Parque Natural de Sierra de Cazorla, Segura y Las Villas, y la línea naranja delimita el área del Parque Regional de Sierra Espuña. Las líneas negras definen las cinco regiones en las que se divide el área de estudio: CSV. Parque Natural de Sierra de Cazorla, Segura y Las Villas y al-rededores; ESP. Parque Regional de Sierra Espuña y alrededores; MUR. Bosquetes y parques periurbanos próximos a la ciudad de Murcia; CEV. Parque Regional de Carrasco y–El Valle; AAL. Albacete y Alicante.

2009). Fragments of the D–loop and Cytb (395–bp and 359–bp respectively) were obtained for each sample. As we found that a tRNA was present within the nucleotide spans of the D–loop fragment, they were trimmed to 350–bp to adjust the sequence length to the target gene. Aligned sequence data were submitted to the GenBank database with accession numbers KJ146734–KJ146742. We found a total of six D–loop haplotypes which have never been reported, and a total of three Cytb haplotypes, two of which were also found to be exclusive to the south east of Spain (SvCb2 and SvCb3). SvCb1 was identical to haplotypes previously found in Albania (Grill et al., 2009) and Japan (Oshida et al., 2009).

Three of the six haplotypes identified for the D–loop were found in CSV and two were present throughout the whole study area. One of the three Cytb haplotypes was exclusive to CSV but the others were present in more than one region (table 1).

The concatenated alignment was 709–bp long and contained eight variable positions. These sequences were collapsed into seven haplotypes. The nucleotide (π) diversity of the concatenated sequence was zero in CEV, low in the ESP region, intermediate in MUR

and AAL, and higher in CSV (table 2). The haplotype diversity (Hd) of the two genes varied in the same way when treated separately, although it was lower in the case of the Cytb. Genetic differentiation between regions was high in almost all cases (table 3).

In the haplotype network (fig. 3), three haplotypes were placed as external nodes, two belonging to CSV (one of them unique to this region) and one exclusive to AAL. The two most common haplotypes (SvCR1 and SvCR2) were both placed as internal nodes, as was haplotype SvCR4. This haplotype was exclusive to CSV and located in the center of the network, also being connected to SvCR3 (exclusive to AAL).

Haplotype SvCb1 was placed in the center of the Cytb network (data not shown). The SvCb2 and SvCb3 haplotypes were directly connected to this and both differed in two nucleotide positions. Haplotype networks using sequences from the GenBank (data not shown) did not show any grouping by geographic region. In the Cytb network, the only haplotype that showed a clear differentiation was that found in Calabria by Grill et al. (2009).

A phylogenetic analysis was conducted for the D–loop haplotypes, including haplotypes from Hale

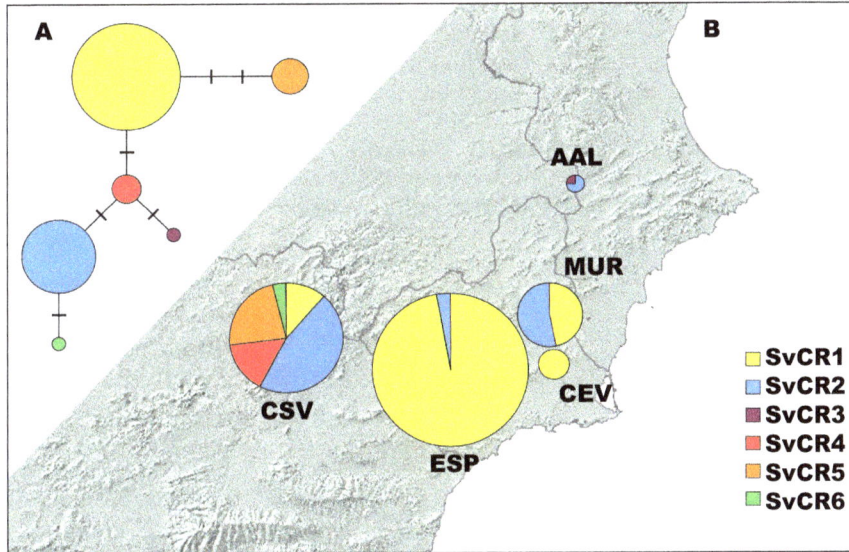

Fig. 3. Median–joining network of the six new D–loop haplotypes (A) and their spatial distribution (B): A. The circles (nodes) in the network represent the haplotypes and the areas of the circles are proportional to the number of samples for each haplotype. The perpendicular short black lines represent mutations; B. Each pie in the distribution map represents the proportion of haplotypes in each region and the size of the pie is proportional to the number of individuals.

Fig. 3. Red haplotípica (basada en el algoritmo de unión de medianas (median–joining) para los seis nuevos haplotipos del fragmento D–loop (A) y distribución espacial de los mismos (B): A. En la red haplotípica, los círculos (nodos) representan los haplotipos y las áreas son proporcionales al número de muestras de cada haplotipo; B. En el mapa de distribución, cada gráfica representa la proporción de haplotipos en cada región y su tamaño es proporcional al número de muestras.

et al. (2004), Grill et al. (2009) and Doziéres et al. (2012). A 252–bp alignment was generated. The tree with the highest log likelihood (–762.4389) was obtained in the maximum likelihood analysis of the D–loop haplotypes (fig. 4A). This phylogenetic tree was conducted under the Hasegawa–Kishino–Yano (HKY85) model (Hasegawa et al., 1985) with rate heterogeneity among sites (gamma distribution shape

Table 1. Haplotype frequencies in the five regions and the overall study area. (For abbreviations see figure 2; SE Spain refers to the overall study samples.)

Tabla 1. Frecuencias haplotípicas en las cinco regiones y en toda el área de estudio. (Para las abreviaturas, véase la figura 2; SE Spain se refiere al total de muestras.)

Haplotype	ESP	MUR	CEV	AAL	CSV	SE Spain
SvCb1	0.972	0.400	1	0.250	0.654	0.761
SvCb2	0.023	0.600	0	0.750	0.192	0.193
SvCb3	0	0	0	0	0.154	0.045
SvCR1	0.972	0.467	1	0	0.115	0.591
SvCR2	0.028	0.533	0	0.750	0.462	0.273
SvCR3	0	0	0	0.250	0	0.011
SvCR4	0	0	0	0	0.154	0.045
SvCR5	0	0	0	0	0.231	0.068
SvCR6	0	0	0	0	0.038	0.011

Table 2. Summary of the diversity indices: N. Number of sequences/individuals; π Nucleotide diversity with standard deviation; h. Number of haplotypes; Hd. Haplotype diversity with standard deviation. (For other abbreviations see figure 2; SE Spain refers to the overall study samples.)

Tabla 2. Resumen de los índices de diversidad: N. Número de secuencias/individuos; π. Diversidad nucleotídica con desviación estándar; h. Número de haplotipos; Hd. Diversidad haplotípica con desviación estándar. (Para las otras abreviaturas, véase la figura 2; SE Spain se refiere al total de muestras.)

	N	π	h_{D-loop}	h_{Cytb}	$h_{Combined}$	Hd_{D-loop}	Hd_{Cytb}	$Hd_{Combined}$
CSV	26	0.00339 ± 0.00030	5	3	6	0.723 ± 0.064	0.532 ± 0.092	0.831 ± 0.032
ESP	36	0.00024 ± 0.00022	2	2	2	0.056 ± 0.052	0.056 ± 0.052	0.056 ± 0.052
MUR	15	0.00226 ± 0.00022	2	2	2	0.533 ± 0.052	0.533 ± 0.052	0.533 ± 0.052
CEV	7	0.00000	1	1	1	0.000	0.000	0.000
AAL	4	0.00212 ± 0.00112	2	2	2	0.500 ± 0.265	0.500 ± 0.265	0.500 ± 0.265
SE Spain	88	0.00222 ± 0.00020	6	3	7	0.576 ± 0.044	0.385 ± 0.055	0.607 ± 0.050

parameter of 0.17). No significant clustering of the haplotypes was found for the southeast of Spain or for the rest of the European populations.

A second analysis was performed for the combined data set, that included nine haplotypes from other European populations (Grill et al., 2009). A 611–bp alignment was generated. The maximum likelihood tree of the combined mtDNA sequences (log likelihood of −1,306.7513) was inferred based on the Tamura 3–parameter model (Tamura, 1992) with invariant sites (fig. 4B). The phylogeny showed a clear differentiation for the Calabrian lineage but not for the rest of the sample. The same result was observed by analyzing the Cytb haplotypes (data not shown). Sequences of *S. lis* were always rooted in the phylogenetic trees.

Discussion

Capture and manipulation of living red squirrels may imply a high risk for their health, such as heart attack or dorsal spin fracture (Josep Piqué, pers. comm.).

Collecting tissue samples from road–kill squirrels avoids such risk and has been proven a suitable source of quality DNA for molecular studies (Lucas & Galián, 2009; Doziéres et al., 2012). However, this kind of sampling does not allow the development of a sampling plan where regions are equally represented. In southeast Spain, this disadvantage can be partially compensated for by the abundance of road–kill animals in rural and suburban areas.

In this study, we found a level of genetic diversity similar to that reported for Spain by Hale et al. (2004) and Grill et al. (2009). However, the extremely low genetic diversity of ESP, described by Lucas & Galián (2009), is the most striking result in this study. This contrasts sharply with the relatively high genetic variation found in CSV, despite its ecological connectivity with ESP.

Anthropogenic effects such as farming or direct human exploitation have decreased the distribution ranges and population sizes of many species in the Iberian peninsula (Gómez & Lunt, 2007). In southeastern Spain, the area occupied by ESP and CSV suffered

Table 3. F_{ST} values between pairs of regions (below diagonal) and *P*–values computed based on 1,000 permutations (upper diagonal): *$P < 0.05$, **$P < 0.001$. (For abbreviations see figure 2.)

*Tabla 3. Valores de F_{ST} entre pares de regiones (diagonal inferior) y valores de P calculados a partir de 1.000 permutaciones (diagonal superior): * P < 0,05; ** P < 0,001. (Para las abreviaturas, véase la figura 2.)*

	ESP	MUR	CEV	AAL	CSV
ESP	–	–	0.99902 ± 0.0002	–	–
MUR	0.56208**	–	–	0.24805 ± 0.0161	0.19629 ± 0.0111
CEV	−0.07417	0.39655*	–	–	–
AAL	0.89285**	0.13125	0.82554*	–	0.23828 ± 0.0161
CSV	0.45766**	0.02575	0.30762*	0.05825	–

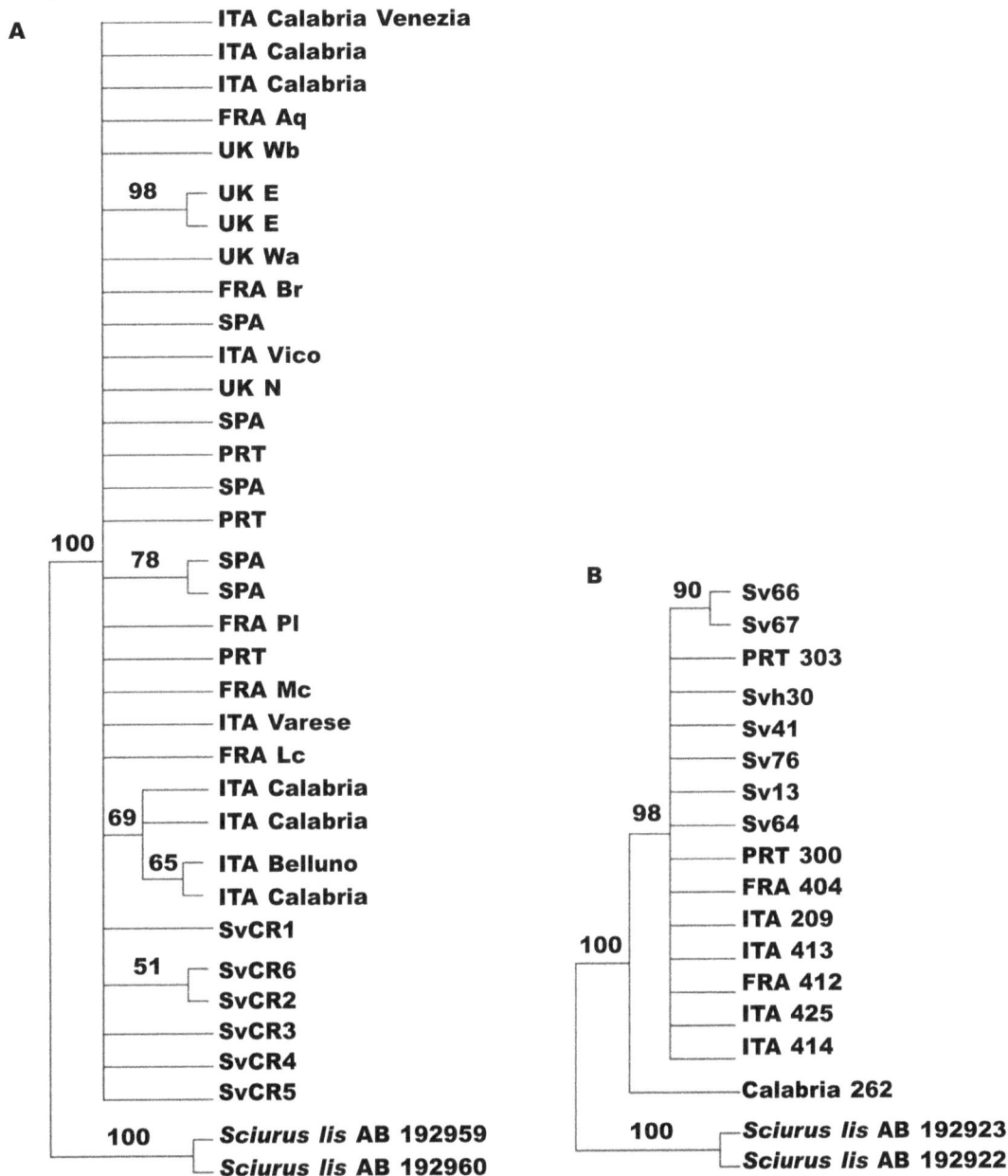

Fig. 4. Condensed maximum–likelihood trees of the D–loop fragment (A) and the combined D–loop and Cyb sequences (B). Branches with less than 50% of bootstrap (1,000 replicates) are collapsed in both trees. The ISO 3166 code is used to designate the country of each sample taken from the literature: A. Taxon labels refer to the D–loop haplotypes from this study (SvCR#) and from other European populations (Hale et al., 2004; Grill et al., 2009); all the French sequences are obtained from Doziéres et al. (2012); B. Labels indicate the sample ID of individuals with different combined haplotypes (Sv##, Svh##) and the specimen numbers from Grill et al. (2009). GenBank accession numbers of the outgroups are indicated in the trees.

Fig. 4. Árboles condensados de máxima verosimilitud para el fragmento del D–loop (A) y para las secuencias concatenadas del D–loop y el Cytb (B). Las ramas presentes en menos del 50% de las 1.000 réplicas obtenidas por muestreo con reemplazo (bootstrap) se han condensado en ambos árboles. Se usa el código ISO 3166 para designar el país de procedencia de cada una de las muestras tomadas de la bibliografía: A. Los nombres de los taxones hacen referencia a los haplotipos del D–loop de este estudio (SvCR#) y a aquellos procedentes de otras poblaciones europeas (Hale et al., 2004; Grill et al., 2009); todas las muestras recogidas en Francia se han obtenido de Doziéres et al. (2012); B. Los nombres de los taxones indican el código de muestra de individuos con distintos haplotipos de secuencias concatenadas (Sv##, Svh##) y el número del espécimen en Grill et al. (2009). Se indican los números de acceso al GenBank de los grupos externos en ambos árboles.

strong deforestation caused by over–exploitation of forest resources in the 18th and 19th centuries (Valverde, 1967; Araque, 2013). As of he second half of the 19th century, reforestation works have been carried out (Codorniu, 1900; González–Pellejero & Álvarez, 2004), helping to preserve red squirrel populations in this area (Valverde, 1967) to date.

As expected given the previous scenario, the proportion of suitable habitats in the landscape decreased critically, increasing the degree of isolation with increasing habitat fragmentation. This situation may have led to a temporary decline in the local squirrel population, which reduced gene flow (Merriam & Wegner, 1992; Andrén & Delin, 1994; Wauters et al., 1994; Amos & Harwood, 1998; Wauters et al., 2010). Therefore, the low genetic variation found in southeast Iberia may be the result of a severe bottleneck, similar to that reported by Trizio et al. (2005) for Alpine squirrels. However, whereas Trizio et al. (2005) found high haplotype diversity but low nucleotide diversity, we found low genetic variation at both levels. This situation contrasts strongly with the high genetic variation found by Gallego & Galián (2008) for the other Pine–specific species *Tomicus destruens* in the Regional Park of Sierra Espuña.

As in other European populations (Hale et al., 2004; Finnegan et al., 2008; Doziéres et al., 2014), we found substantial genetic subdivision between regions (table 3). Habitat loss and fragmentation due to anthropogenic effects and geographical barriers may explain these results. For CEV, where SvCR1 was the only haplotype found, the high haplotype fixation may be explained by the geographical isolation caused by the Guadalentín River or by introduction of animals from other sources such as the Sierra Espuña Regional Park. The strong fixation found in CEV and AAL might also be due to low sample size, which can lead to an overestimation of the F_{ST} values.

Valverde (1967) emphasized the differentiation of the *hoffmanni* subspecies in Sierra Espuña and its differentiation from the populations of Sierra de Cazorla, Segura y Las Villas (*S. v. segurae*) and the rest of the Iberian Peninsula. This classification was achieved using morphological traits and fur colour. Nevertheless, we found no pattern of genetic variation to support this subspecific classification.

Since the internal nodes of haplotype networks are considered as ancestral and the external nodes as more recent status (Castelloe & Templeton, 1994; Templeton, 1998), and since a reduction in population size results in an accelerated increase in genetic distance in the early generations (Chakraborty & Nei, 1974; Nei, 1976; Takezaki & Nei, 1996), our results may be explained by a scenario where widely distributed ancestral haplotypes became extinct due to a bottleneck events. Thus, haplotype SvCR4 occupying the central node of the D–loop network, but in a very low frequency, is a candidate to be considered an ancestral widely distributed haplotype that became extinct in all areas but CSV, especially in ESP which is the region with the largest sample size.

The finding of a Cytb haplotype (SvCb1) that was previously described in Albania and Japan but not in other Eurasian population suggests an ancestral wide distribution of this haplotype, followed by local extinction in intermediate areas.

Iberia and Italy have been reported as potential glacial refuges for the red squirrel (Hale et al., 2004; Finnegan et al., 2008; Grill et al., 2009; Doziéres et al., 2012) and our results confirm that Iberian samples do not show the expected high levels of genetic diversity (Hewitt, 1996; Taberlet et al., 1998). This finding would be supported by a paper by Doziéres et al. (2012) that suggested a postglacial recolonization of Europe from Asia or from the Balkans or, alternatively, a series of recent bottlenecks that reduced the genetic diversity in the Iberian and Italian populations. The finding of haplotype SvCb1 in Iberia, the Balkans and Japan favours the hypothesis of the Iberian Peninsula acting as a glacial refuge. Besides, the low genetic variation found may be explained by the recent bottleneck in these populations.

In contrast with the report by Grill et al. (2009) and Doziéres et al. (2012), we found no significant clustering for the squirrels of Calabria in the phylogenetic analysis of the D–loop haplotypes (fig. 4A). However, these individuals were clearly differentiated in the remaining the phylogenetic trees (fig. 4B). Nevertheless, the results of the phylogenetic analysis are largely dependent on the sequence length (number of informative sites) and the number of individuals analysed. Thus, this could be an explanation of the lack of clustering found in this work for the Calabrian squirrels (fig. 4A). None of the squirrels in Spain were separated in these analyses, suggesting that Iberian squirrels have not been isolated from the rest of the European populations, as found by Doziéres et al. (2012) for French squirrels. Nonetheless, Grill et al. (2009) emphasised the clear separation of the Iberian squirrels, based on the analysis of eight microsatellite loci. We noticed that the squirrels from ESP did not form a monophyletic clade in the philogenetic analyses, in contrast with what we found in previous work (Lucas & Galián, 2009). The inclusion of samples from nearby populations (CSV, CEV, ALL and MUR) shows that, in fact, the population of Sierra Espuña is very close to other Iberian populations.

A more extensive study should be carried out to understand the phylogenetic and demographic relationships between the Iberian populations, not only at a mitochondrial level, but also at a nuclear level. The recent development of next–generation sequencing methods offers a wide potential for obtaining complete genomes, allowing more accurate research into the evolutionary relationships at an intraspecific level (McCormack et al., 2013).

Acknowledgments

We wish to thank the following people who helped us by collecting road–kill squirrel samples: Antonio Ortuño, Ángel Albert, José Manuel López, Carlos González, Jorge Sánchez, Ana Miñano (C. R. F. El Valle), Cristina López, Javier García, Lidia Lorca, José Manuel Vidal, Mario León, Irene Muñoz, Carmelo Andújar, Paula Arribas, José Serrano, José Galián,

Rosa María Ros, and Isabel Sánchez Guiu. We also thank the environmental officers of the Natural Park of Sierra de Cazorla, Segura y Las Villas and the Regional Park of Sierra Espuña for collecting samples. And Obdulia S. Sanchez–Domingo and Ana I. Asensio for technical assistance, and thank Prof. José Serrano for useful comments on the manuscript.

References

Altschul, S. F., Gish, W., Miller, W., Myers, E. W. & Lipman, D. J., 1990. Basic local alignment search tool. *Journal of Molecular Biology*, 215: 403–410.

Amos, W. & Harwood J., 1998. Factors affecting levels of genetic diversity in natural populations. *Philosophical transactions of the Royal Society of London. Series B, Biological Sciences*, 353: 177–186.

Andrén, H. & Delin, A., 1994. Habitat Selection in the Eurasian Red Squirrel, *Sciurus vulgaris*, in Relation to Forest Fragmentation. *Oikos*, 70: 43–48.

Araque E., 2013. Evolución de los paisajes forestales del Arco Prebético. El caso de las Sierras de Segura y Cazorla. *Revista de Estudios Regionales*, 96: 321–344.

Bandelt, H. J., Forster, P. & Rohl, A., 1999. Median–joining networks for inferring intraspecific phylogenies. *Molecular Bbiology and Evolution*, 16: 37–48.

Barratt, E. M., Gurnell, J., Malarky, G., Deaville, R. & Bruford, M. W., 1999. Genetic structure of fragmented populations of red squirrel (*Sciurus vulgaris*) in the UK. *Molecular Ecology*, 8: S55–63.

Cabrera, A., 1905. Las ardillas de España. *Boletin de la Real Sociedad Española de Historia Natural*, 5: 225–231.

Castelloe, J. & Templeton, A. R., 1994. Root probabilities for intraspecific gene trees under neutral coalescent theory. *Molecular Phylogenetics and Evolution*, 3: 102–113.

Chakraborty, R. & Nei, M., 1974. Dynamics of gene differentiation between incompletely isolated populations of unequal sizes. *Theoretical Population Biology*, 5: 460–469.

Clement, M., Posada, D. & Crandall, K. A., 2000. TCS: a computer program to estimate gene genealogies. *Molecular Ecology*, 9: 1657–1659.

Codorniu, R., 1900. *Apuntes relativos a la repoblación forestal de la Sierra de Espuña presentados al Congreso Agrícola de Murcia*. Tipográfica de Las Provincias de Levante, Murcia.

Corbet, G., 1978. *The Mammals of the Palaearctic Region: A Taxonomic Review*. British Museum (Natural History), Cornell University Press, London.

Dozières, A., Chapuis, J.–L., Thibault, S. & Baudry, E., 2012. Genetic Structure of the French Red Squirrel Populations: Implication for Conservation. PLoS ONE, 7: e47607. Doi:10.1371/journal.pone.0047607

Edgar, R. C., 2004. MUSCLE: multiple sequence alignment with high accuracy and high throughput. *Nucleic Acids Research*, 32: 1792–1797.

Excoffier, L. & Lischer, H. E., 2010. Arlequin suite ver 3.5: a new series of programs to perform population genetics analyses under Linux and Windows.

Molecular Ecology Resources, 10: 564–567.

Ferreira, A. F., Guerreiro, M., Álvares, F. & Petrucci–Fonseca, F., 2001. Distribución y aspectos ecológicos de *Sciurus vulgaris* en Portugal. *Galemys*, 13: 155–170.

Ferreira, A. & Guerreiro, M., 2002. *Estudo da dinámica populacinal do esquilo–comun (*Sciurus vulgaris*) no Parque Florestal de Monsanto*. Informe Técnico, Parque ecológico de Monsanto, Lisboa.

Finnegan, L. A., Edwards, C. J. & Rochford, J. M., 2008. Origin of, and conservation units in, the Irish red squirrel (*Sciurus vulgaris*) population. *Conservation Genetics*, 9: 1099–1109.

Gallego, D. & Galián, J., 2008. Hierarchical structure of mitochondrial lineages of Tomicus destruens (Coleoptera, Scolytidae) related to environmental variables. *Journal of Zoological Systematics and Evolutionary Research*, 46: 331–339.

Gómez, A. & Lunt, D. H., 2007. Refugia within refugia: Patterns of phylogeographic concordance in the Iberian Peninsula. In: Phylogeography of southern European refugia: 155–188 (S. Weiss & N Ferrand Eds.). Springer, Dordrecht.

González–Pellejero, R. & Álvarez, A., 2004. El Mapa Forestal de España, una obra secular (1868–1966) concluida por Luis Ceballos. *Ería*, 64–65: 285–318.

Grill, A., Amori, G., Aloise, G., Lisi, I., Tosi, G., Wauters, L. A. & Randi, E., 2009. Molecular phylogeography of European *Sciurus vulgaris*: refuge within refugia? *Molecular Ecology*, 18: 2687–2699.

Hale, M., Lurz, P. W. & Wolff, K., 2004. Patterns of genetic diversity in the red squirrel (*Sciurus vulgaris* L.): Footprints of biogeographic history and artificial introductions. *Conservation Genetics*, 5: 167–179.

Hasegawa, M., Kishino, H., & Yano, T., 1985. Dating the human–ape split by a molecular clock of mitochondrial DNA. *Journal of Molecular Evolution*, 22: 160–174.

Hewitt, G. M., 1996. Some genetic consequences of ice ages,and their role in divergence and speciation. *Biological Journal of the Linnean Society*, 58: 247–276.

Larkin, M. A., Blackshields, G., Brown, N. P., Chenna, R., McGettigan, P. A., McWilliam, H., Valentin, F., Wallace, I. M., Wilm, A., Lopez, R. J., Thompson, D., Gibson, T. J. & Higgins, D. G., 2007. Clustal W and Clustal X version 2.0. *Bioinformatics*, 23: 2947–2948.

Lee, T. H. & Fukuda, H., 1999. The distribution and habitat use of the Eurasian red squirrel (*Sciurus vulgaris* L.) during summer, in Nopporo Forest Park, Hokkaido. *Mammal Study*, 24: 7–15.

Librado, P. & Rozas, J., 2009. DnaSP v5: a software for comprehensive analysis of DNA polymorphism data. *Bioinformatics*, 25: 1451–1452.

Lucas, J. M. & Galián, J., 2009. Análisis molecular de *Sciurus vulgaris hoffmanni* Valverde, 1967 (Rodentia: Sciuridae) e implicaciones para su conservación. *Anales de Biología*, 31: 81–91.

Lurz, P. W. W., Gurnell, J. & Magris, L., 2005. *Sciurus vulgaris*. *Mammalian Species*, 769: 1–10.

Mathias, M. d. L. & Gurnell, J., 1998. Status and conservation of the red squirrel (*Sciurus vulgaris*) in Portugal. *Hystrix*, 10: 13–19.

McCormack, J. E., Hird, S. M., Zellmer, A. J., Carstens, B. C. & Brumfield, R. T., 2013. Applications of next–generation sequencing to phylogeography and phylogenetics. *Molecular Phylogenetics and Evolution,* 66: 526–538.

Merriam, G. & Wegner, J., 1992. Local Extinctions, Habitat Fragmentation, and Ecotones. In: *Landscape Boundaries:* 150. (A. Hansen & F. di Castri, Eds.). Springer, New York.

Miller, G. S., 1907. LX– Four new European squirrels. *Annals and Magazine of Natural History,* 20: 426–430.

– 1909. LIV–Twelve new European mammals. *Annals and Magazine of Natural History,* 3: 415–422.

– 1912. *Catalogue of the Mammals Western Europe: Europe exclusive of Russia,* British Museum.

Nei, M., 1976. Mathematical models of speciation and genetic distance. In: *Population genetics and ecology:* 723. (S. Karlin & E. Nevo, Eds.). Academic Press, New York.

Oshida T. & Masuda, R., 2000. Phylogeny and zoogeography of six squirrel species of the genus *Sciurus* (Mammalia, Rodentia), inferred from cytochrome *b* gene sequences. *Zoological Science,* 17: 405–409.

Oshida, T., Arslan, A. & Noda, M., 2009. Phylogenetic relationships among the Old World *Sciurus* squirrels. *Folia Zoologica,* 58: 14–25.

Palomo, L. J. & Gisbert, J., 2002. *Atlas de los Mamíferos Terrestres de España.* DGCN–SECEM–SECEMU, Madrid.

Purroy, F. J., 2014. Ardilla roja – *Sciurus vulgaris.* In: *Enciclopedia Virtual de los Vertebrados Españoles* (Salvador, A. & Luque–Larena, J. J., Eds.). Museo Nacional de Ciencias Naturales, Madrid. http://www.vertebradosibericos.org/

Shuttleworth, C. M., 2001. Traffic related mortality in a red squirrel (*Sciurus vulgaris*) population receiving supplemental feeding. *Urban Ecosystems,* 5: 109–118.

Sidorowicz, J., 1971. Problems of subspecific taxonomy of squirrel (*Sciurus vulgaris* L.) in Palaearctic. *Zoologischer Anzeiger,* 187: 123–142.

Taberlet, P., Fumagalli, L., Wust–Saucy, A.–G. & Cosson, J.–F., 1998. Comparative phylogeography and postglacial colonization routes in Europe. *Molecular Ecology,* 7: 453–464.

Takezaki, N. & Nei, M., 1996. Genetic distances and reconstruction of phylogenetic trees from microsatellite DNA. *Genetics,* 144: 389–399.

Tamura, K., 1992. Estimation of the number of nucleotide substitutions when there are strong transition–transversion and G + C–content biases. *Molecular Biology and Evolution,* 9: 678–687.

Tamura, K. & Nei, M., 1993. Estimation of the number of nucleotide substitutions in the control region of mitochondrial DNA in humans and chimpanzees. *Molecular Biology and Evolution,* 10: 512–526.

Tamura, K., Stecher, G., Peterson, D., Filipski, A. & Kumar, S., 2013. MEGA6: Molecular Evolutionary Genetics Analysis Version 6.0. *Molecular Biology and Evolution,* 30: 2725–2729.

Templeton, A. R., 1998. Nested clade analyses of phylogeographic data: testing hypotheses about gene flow and population history. *Molecular Ecology,* 7: 381–397.

Trizio, I., Crestanello, B., Galbusera, P., Wauters, L. A., Tosi, G., Matthysen, E. & Hauffe, H. C., 2005. Geographical distance and physical barriers shape the genetic structure of Eurasian red squirrels (*Sciurus vulgaris*) in the Italian Alps. *Molecular Ecology,* 14: 469–481.

Valverde, J. A., 1967. Notas sobre vertebrados. III. Nueva ardilla del SE español y consideraciones sobre las subespecies peninsulares. *Boletín de la Real Sociedad Española de Historia Natural (Biol.),* 65: 225–248.

Wauters, L. A., Hutchinson, Y., Parkin, D. T. & Dhondt, A. A., 1994. The effects of habitat fragmentation on demography and on the loss of genetic variation in the red squirrel. *Proceedings. Biological Sciences /The Royal Society,* 255: 107–111.

Wauters, L. A., Verbeylen, G., Preatoni, D., Martinoli, A. & Matthysen, E., 2010. Dispersal and habitat cuing of Eurasian red squirrels in fragmented habitats. *Population Ecology,* 52: 527–536.

13

Hatchling sex ratio, body weight and nest parameters for *Chelonia mydas* nesting on Sugözü beaches (Turkey)

Ç. Kılıç & O. Candan

Kılıç, Ç. & Candan, O., 2014. Hatchling sex ratio, body weight and nest parameters for *Chelonia mydas* nesting on Sugözü beaches (Turkey). *Animal Biodiversity and Conservation*, 37.2: 177–182.

Abstract

Hatchling sex ratio, body weight and nest parameters for Chelonia mydas *nesting on Sugözü beaches (Turkey).*— We investigated the relationship between nest parameters, hatchling body mass, and sex ratio of green turtle, *Chelonia mydas,* embryos and hatchlings at the temperate nesting rookery of Sugözü Beach (Adana–Turkey). Mean nest temperature and distance from the sea were correlated, while mean nest temperature and incubation period were inversely related. There was no apparent relationship between incubation period and hatchling mass. Hatchling and embryo sex ratios, determined by histological examination, showed a 70.5% and 93.5% female bias, respectively. There was no correlation between sex and body weight of hatchlings, but mean hatchling bodyweight of females (16.8 g) was slightly higher than that of males (16.2 g).

Key words: *Chelonia mydas*, Hatchling mass, Nest parameters, Sex ratio, Sugözü beaches

Resumen

La proporción de sexos y el peso corporal de las crías y los parámetros del nido en la población de Chelonia mydas *que anida en las playas de Sugözü (Turquía).*— Estudiamos la relación entre los parámetros del nido y el peso corporal y la proporción de sexos de los embriones y las crías de tortuga verde, *Chelonia mydas,* en la colonia de cría templada de la playa de Sugözü (Adana, Turquía). Se observó que la temperatura media del nido y la distancia al mar estaban correlacionadas, mientras que la temperatura media del nido y el período de incubación estaban inversamente relacionados. No hubo relación aparente entre el período de incubación y el peso de las crías. La proporción de sexos en las crías y los embriones, que se determinó mediante examen histológico, mostró una desviación en favor del sexo femenino del 70,5 y el 93,5%, respectivamente. No hubo correlación entre el sexo y el peso corporal de las crías, pero el promedio del peso corporal de las hembras (16,8 g) fue ligeramente superior al de los machos (16,2 g).

Palabras clave: *Chelonia mydas*, Peso de las crías, Parámetros del nido, Proporción de sexos, Playas de Sugözü

Çağla Kılıç & Onur Candan, Biology Section, Dept. of Biology, Fac. of Arts and Science, Ordu Univ., 52200 Cumhuriyet Campuss, Ordu (Turkey).

Corresponding author: Çağla Kılıç. E–mail: kiliccagl@gmail.com

Introduction

The sex of a sea turtle depends on incubation temperature (Bull, 1980). The temperature at which hatchling sex ratios are 1:1 is known as the pivotal temperature; when incubation temperature is higher than the pivotal temperature, hatchlings become female, and when it is lower than the incubation temperature, hatchlings become male (Yntema & Mrosovsky, 1980).

Valenzuela et al. (2004), using landmark–based geometric morphometric methods, found that morphological methods, while producing high accuracy when compared with 'true sex' in *Chrysemys picta* (98%) and *Podocnemis expansa* (90%), had a discriminant function analysis cross–validation rate of only 85% when used to determine sex in hatchling turtles. Besides carapace length and width, weights of hatchlings, temperature profiles, nest parameters, and histological examination (Glen et al., 2003; Ischer et al., 2009; Sönmez et al., 2011) have been considered to determine the sex of hatchling turtles in the wild. A non–invasive, accurate methodology relating duration of incubation to hatchling sex ratio has also been developed for artificially incubated eggs (Mrosovsky et al., 2002; Zbinden et al., 2007; Katselidis et al., 2012). Although various methods may be used to determine hatchling sex ratios, their accuracy is variable (Ceriani & Wyneken, 2008) and only histological examination ensures accurate sex determination (Mrosovsky & Benabib, 1990; Mrosovsky & Godfrey, 1995).

We studied the relationship between nest parameters of *Chelonia mydas* L., 1758 on Sugözü beaches during the 2012 nesting season. Temperatures in eight *Chelonia mydas* nests were determined using data loggers, sex of hatchlings was determined by histological examination, and the relationship between nest temperatures, nest parameters, hatchling weights, and sexes was evaluated statistically.

Material and methods

Study site

Chelonia mydas were studied at the nesting site on Sugözü beaches (36° 48.677' N – 35° 51.068' E, 36° 52.795' N – 35° 56.017' E), located in the Adana province within the borders of Yumurtalık and Ceyhan counties. The site consists of four subsections (from west to east): Akkum (1.4 km), Sugözü (1.0 km), Botaş (0.6 km), and Hollanda (0.4 km), and consists of a total of 3.4 km suitable beach habitat. In 2012, the nesting season was between June and September (fig. 1).

Nest parameters

Gemini Data Loggers (Tinyalk Temperature Range H–30ºC/+50ºC Part No: TK–0040, UK) used to determine nest temperatures were placed in the middle of nests during nesting and programmed to provide hourly readings. Daily nest temperatures were determined by averaging 24 hourly measurements each day. Middle third and whole incubation temperatures were calculated.

Five parameters, incubation duration (time between date of egg laying and first day of hatchling emergence), clutch size (total number of eggs in the nest), hatching success (ratio of empty eggshells to total number of eggs [%]), nest depth (distance [cm] from surface of the sand to the bottom of the nest), and nest distance (distance [m] from high tide line to nest) from the sea, were used.

Weighing hatchlings

Dead hatchlings (n = 88) were collected in sterile plastic bags, and washed, dried, and weighed using a Densa JW precision scale (600 g ± 0.01 g).

Histological determination of sex

After removing the genitourinary complex from hatchlings (n = 88) and late embryos (n = 62), the complex was preserved in 4% buffered paraformaldehyde. Samples to undergo histological examination were then embedded in paraffin blocks, further prepared at a thickness of 5 μm with a Thermo–Shandon microtome and stained with Haematoxylon and Eosin (H&E), and sealed with Entellan. Histological examination was carried out with a Nikon E 100 light microscope using both 10X and 40X objectives following the Yntema & Mrosovsky (1980) criteria for distinguishing ovaries and testicles.

Data analyses

The Pearson product–moment correlation coefficient (r) and simple linear regression coefficient of determination (r^2) were calculated and used to assess statistical significance among variables. The relationship between middle third temperature and sex ratio was evaluated by the Mann–Whitney (U) test. All statistical analyses were executed using an SPSS packaged program (SPSS Inc., Released 2006, Version 15.0. Chicago, USA).

Results

Nest temperatures and parameters

Table 1 presents nest temperatures and other parameters. It shows temperatures for the whole period and middle third were similar. There was no relationship between nest temperature and nest depth (n = 8, r^2 = 0.38, P > 0.05) and nest temperature with clutch size (n = 8, r^2 = 0.11, P > 0.05). The relationship between nest temperature and distance from the sea (n = 8, r^2 = 0.85, Pearson's r = 0.927, P = 0.001) was positive and the relationship between nest temperature and incubation duration was negative (n = 8, r^2 = 0.71, Pearson's r = –0.846, P < 0.01). We found a statistically significant difference between middle third temperature and hatchling sex ratio (U = 12.500, P < 0.01).

Hatchling weights and nest parameters

Weights and nest parameters were recorded for 88 hatchlings gathered from 31 nests throughout the nesting

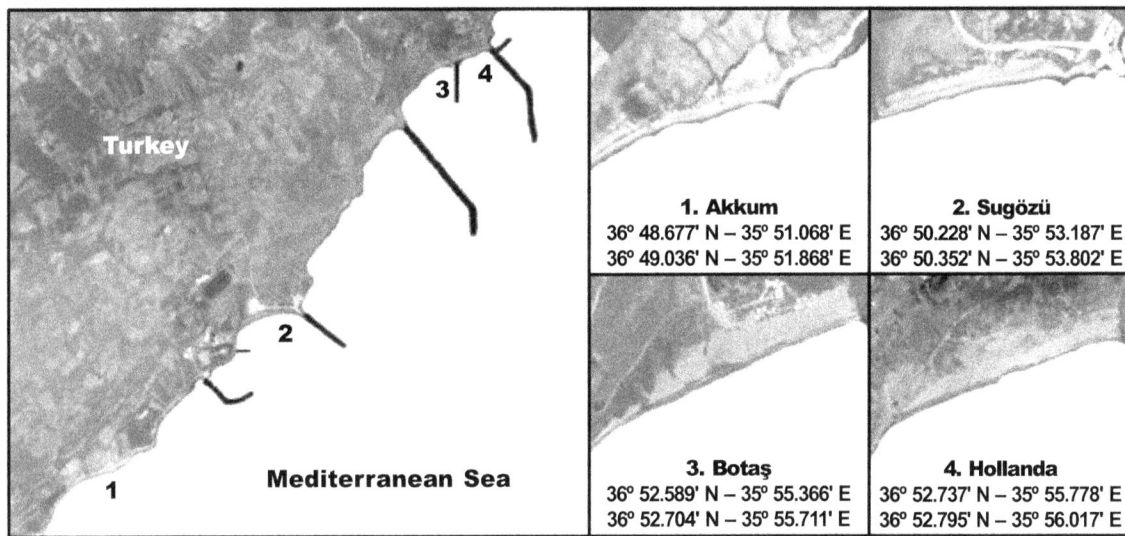

1. Akkum 36° 48.677' N – 35° 51.068' E 36° 49.036' N – 35° 51.868' E	**2. Sugözü** 36° 50.228' N – 35° 53.187' E 36° 50.352' N – 35° 53.802' E
3. Botaş 36° 52.589' N – 35° 55.366' E 36° 52.704' N – 35° 55.711' E	**4. Hollanda** 36° 52.737' N – 35° 55.778' E 36° 52.795' N – 35° 56.017' E

Fig. 1. Study site.

Fig. 1. Lugar del estudio.

season. No relationship was found between hatchling weight and duration of incubation ($n = 31$, $r^2 = 0.01$, $P > 0.05$), nest depth ($n = 31$, $r^2 = 0.01$, $P > 0.05$), distance from the sea ($n = 31$, $r^2 = 0.02$, $P > 0.05$), and hatching success ($n = 31$, $r^2 = 0.02$, $P > 0.05$). When relationships within nest parameters were assessed, only clutch size and nest depth appeared related ($n = 31$, $r^2 = 0.24$, $P < 0.01$).

Determination of sex ratios

Histological examination of 29 gonad samples taken from nests for which temperature data were available showed that female hatchling development occurred in 86% (table 1). When all gonad samples (150) were examined, 120 (80%) were determined to be female. When embryos and hatchlings were

Table 1. Nest temperatures and nest parameters ($n = 8$).

Tabla 1. Temperaturas y parámetros del nido (n = 8).

Nest number	Mean nest temperature		Incubation duration (day)	Clutch size (number)	Nest depth (cm)	Distance from sea (m)
	Whole period (°C)	Middle third (°C)				
1	30.5	30.4	54	103	75	31.4
2	32.6	32.5	48	134	86	35.5
3	32.1	32	45	103	84	40.4
4	31.1	31.3	49	73	82	36.5
5	31.4	31.3	44	46	62	36
6	28.1	27.9	61	146	102	12.9
7	29.6	29.7	57	142	90	19.8
8	29.3	29.2	51	88	99	24.9
Mean	30.6	30.5	51	104	85	29.7

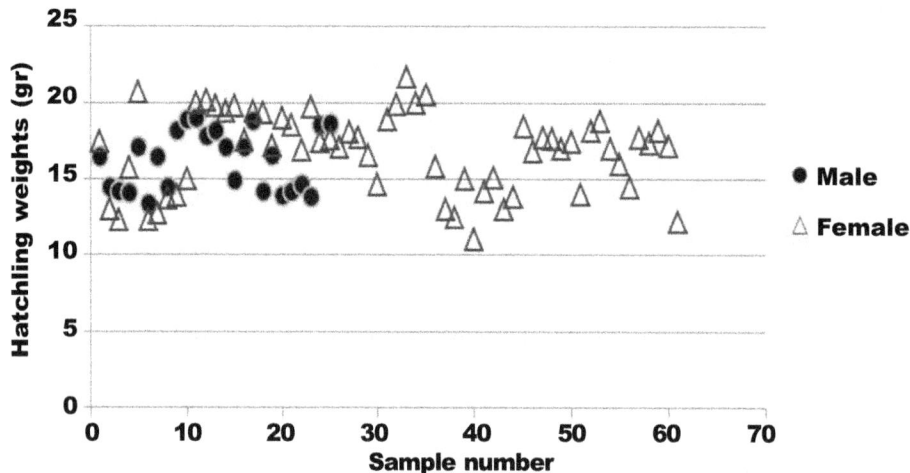

Fig. 2. Female and male hatchling weights.

Fig. 2. Pesos de las crías macho y hembra.

considered separately, 93.5% (58 of 62) of embryos were female, and 70.5% (62 of 88) of hatchlings were female.

Sex and hatchling weights

There was no significant difference in weight between male (mean = 16.2 g) and female (mean = 16.8 g) hatchlings (fig. 2).

Discussion

Most sex ratio studies performed on hatchling sea turtles (Öz et al., 2004; Candan 2006; Kaska et al., 2006; Katselidis et al., 2012; Uçar et al. 2012) along the Mediterranean coastline of Turkey concern nest and sand temperatures (Baran & Kasparek, 1989; Canbolat, 1991; Ergene et al., 2009), marine predation (Türkecan & Yerli, 2007), and morphometric diversification (Türkecan et al., 2008).Studies about sex ratio, body weight, and nest parameters among hatchlings are scarce (Sönmez et al., 2011).

The duration of incubation is reduced by increasing nest temperatures (Yntema & Mrosovsky, 1980; Godley et al., 2001; Wood et al., 2014). Our findings also support an inverse relationship between temperature and incubation duration. We found a positive relationship between nest temperature and nest distance from the sea. Candan (2006) found a similar positive relationship on the same beach. The greater the distance a nest is from the sea, the higher the nest temperature (Uçar et al., 2012).

Incubation duration and hatchling weight were not directly affected by nest temperature, supporting findings by Ischer et al. (2009) and Booth et al. (2013) who reported that *Chelonia mydas* hatchling weights

are not directly affected by nest temperatures Average weights of female and male hatchlings were similar, and we found no significant relationship between hatchling sex and weight. A similar lack of significance was found in a study at Samandağ Beach (Sönmez, 2011), and in a laboratory study of *Chelonia mydas* hatchlings in Oman's Ras Al–Hadd Reserve (Mahmoud et al., 2005).

The sex ration of hatchlings in the Eastern Mediterranean is heavily skewed in favor of females (80%) in studies of natural temperature regimes (Kaska et al., 1998; Broderick et al., 2000; Casale et al., 2000). Katselidis et al. (2012) suggested that a sex ratio heavily biased in favor of females could be obtained from analysis of data from 1, 2 or 3 nesting seasons. When longer time frames, such as 20 or 150 years, are considered, results are likely to produce a 1:1 sex ratio. In our study, middle third (30.6°C) and whole incubation (30.5°C) temperatures were (table 1) greater than the pivotal temperature (28.9°C) defined by Kaska et al. (1998). When all sample gonads taken from nesting pre–hatchlings were examined histologically, the hatchling sex ratio was greater than 80% female, 20% male. Our findings are comparable to those of Kaska et al. (1998) and Elmas (2008), in which female percentages were 78.8% and 82.8% respectively, but they are lower than those of Broderick et al. (2000), at 96% female. While our results were significantly higher than those of Candan (2010), when considered separately, our embryo and hatchling sex ratios were significantly variable. This difference should be considered in other estimates of sex ratios.

Sex ratio varies between seasons, populations, and nesting sites (Merchant–Larios, 1999), and additional studies on sex ratio estimation are required to understand population dynamics in marine turtles. Although we distinguished between hatchling sex, weight, and nest parameters in our study, the relationship between nest temperature and nest distance from the sea with duration of incubation is strong. It is encouraging that our results can be used to estimate sex ratio. Perhaps more precise results would be obtained by performing additional studies using additional samples over several nesting seasons.

References

Baran, I. & Kasparek, M., 1989. *Marine Turtles Turkey, Status survey 1988 and recommendations for conservation and management*. WWF, Verlag, Heidelberg.

Booth, D. T., Feeney, R. & Shibata, Y., 2013. Nest and maternal origin can influence morphology and locomotor performance of hatchling green turtles (*Chelonia mydas*) incubated in field nests. *Marine Biology*, 160: 127–137.

Broderick, A. C., Godley, B. J., Reece, S. & Downie, J. R., 2000. Incubation periods and sex ratios of green turtles: highly female biased hatchling production in the eastern Mediterranean. *Marine Ecology Progress Series*, 202: 273–281.

Bull, J. J., 1980. Sex Determination in Reptiles. *Quarterly Review of Biology*, 55: 3–21.

Canbolat, A. F., 1991. The investigation on the population of the loggerhead sea turtle, *Caretta caretta* (Linnaeus 1758) nesting in Dalyan Beach, Mugla. Turkey. Doğa. *Turkish Journal of Zoology*, 15: 255–274.

Candan, O., 2006. The sex–temperature relation on hatchlings of green sea turtles (*Chelonia mydas*) nesting in hollanda beach (Ceyhan–Adana). M. Sc. Thesis, Hacettepe University.

– 2010. Sex–temperature relation and histological investigation of sex on hatchlings of green sea turtle (*Chelonia mydas*) nesting in sugözü (Ceyhan–Adana) and Kazanlı (Kazanlı–Mersin) beaches. Ph. D. Thesis, Hacettepe University.

Casale, P., Gerosa, G. & Yerli, S. V., 2000. Female–biased Primary Sex Ratio of the Green Turtle, *Chelonia mydas*, Estimated Through Sand Temperatures at Akyatan, Turkey. *Zoology in the Middle East*, 20: 33–42.

Ceriani, S. A. & Wyneken, J., 2008. Comparative morphology and sex identification of the reproductive system in formalin–preserved sea turtle specimens. *Zoology*, 111: 179–187.

Elmas, M., 2008. Early gonadal development and sexual differentiation in green turtle, *Chelonia mydas*. M. Sc. Thesis, Mustafa Kemal University.

Ergene, S., Aymak, C., Uçar, A. H. & Kaçar, Y., 2009. The research on the population of *Chelonia mydas* and *Caretta caretta* nesting on Kazanlı Beach (Mersin) in 2005 nesting season. *Ege University Journal of Fisheries and Aquatic Sciences*, 26: 187–196.

Glen, F., Broderick, A. C., Godley, B. J. & Hays, G. C., 2003. Incubation environment affects phenotype of naturally incubated green turtle hatchlings. *Journal of the Marine Biological Association of the United Kingdom*, 83: 1183–1186.

Godley, B. J., Broderick, A. C., Downie, J. R., Glen, F., Houghton, J. D., Kirkwood, I., Reece, S. & Hays, G. C., 2001. Thermal conditions in nests of loggerhead turtles: further evidence suggesting female skewed sex ratios of hatchling production in the Mediterranean. *Journal of Experimental Marine Biology and Ecology*, 263: 45–63.

Ischer, T., Ireland, K. & Booth, D. T., 2009. Locomotion performance of green turtle hatchlings from the Heron Island Rookery, Great Barrier Reef. *Marine Biology*, 156: 1399–1409.

Kaska, Y., Downie, R., Tippett, R. & Furness, R. W., 1998. Natural temperature regimes for loggerhead and green turtle nests in the eastern Mediterranean. *Canadian Journal of Zoology*, 76: 723–729.

Kaska, Y., Ilgaz, C., Ozdemir, A., Baskale, E., Turkozan, O., Baran, I. & Stachowitsch, M., 2006. Sex ratio estimations of loggerhead sea turtle hatchlings by histological examination and nest temperatures at Fethiye beach, Turkey. *Naturwissenschaften*, 93: 338–343.

Katselidis, K. A., Schofield, G., Stamou, G., Dimopoulos, P. & Pantis, J., 2012. Females first? Past, present and future variability in offspring sex ratio at a temperate sea turtle breeding area. *Animal Conservation*, 15: 508–518.

Mahmoud, I. Y., AlKindi, A. Y., Ba–Omar, T. A., Al–Siyabi, S., Al–Bahry, S. N., Elshafie, A. Q. & Bakheit, C. S., 2005. Emergence pattern of the Green Turtle, *Chelonia mydas*, hatchlings under laboratory and natural conditions. *Zoology in the Middle East*, 35: 19–27.

Merchant–Larios, H., 1999. Determining hatchling sex. In: *Research and Management Techniques for the Conservation of Sea Turtles*: 130–136 (K. L. Eckert, K. A. Bjorndal, F. A. Abrer–Grobois & M. Donnelly, Eds.). IUCN/SSC Marine Turtle Specialist Group Publication, Pennsylvania, USA.

Mrosovsky, N. & Benabib, M., 1990. An Assessment of 2 Methods of Sexing Hatchling Sea–Turtles. *Copeia*, 2: 589–591.

Mrosovsky, N. & Godfrey, M. H., 1995. Manipulating sex ratios: turtle speed ahead. *Chelonian Conservation and Biology*, 1: 238–240.

Mrosovsky, N., Kamel, S., Rees, A. F. & Margaritoulis, D., 2002. Pivotal temperature for loggerhead turtles (*Caretta caretta*) from Kyparissia Bay, Greece. *Canadian Journal of Zoology*, 80: 2118–2124.

Öz, M., Erdoğan, A., Kaska, Y., Düşen, S., Aslan, A., Sert, H., Yavuz, M. & Tunç, M. R., 2004. Nest temperatures and sex–ratio estimates of loggerhead turtles at Patara beach on the South western coast of Turkey. *Canadian Journal of Zoology*, 82: 94–101.

Sönmez, B., Turan, C. & Yalçın–Özdilek, Ş., 2011. The effect of relocation on themorphology of Green Turtle, *Chelonia mydas* (Linnaeus, 1758), hatchlings on Samandağ beach, Turkey. *Zoology in the Middle East*, 52: 29–38.

Türkecan, O., Turkozan, O., Oruc, A., Mangit, F., Demirayak, F. & Yerli, S. V., 2008. A Preliminary Study on The Morphometric Variation of *Chelonia mydas* in Three Different Beaches of Turkey. *NOAA Technical Memorandum NMFS SEFSC*, 569: 1–251.

Türkecan, O. & Yerli, S. V., 2007. Marine Predation on Loggerhead Hatchlings at Beymelek Beach, Turkey. *Israel Journal of Ecology and Evolution*, 53: 167–171.

Uçar, A. H., Kaska, Y., Ergene, S., Aymak, C., Kaçar, Y., Kaska, A. & Ili, P., 2012. Sex Ratio Estimation of the Most Eastern Main Loggerhead Sea Turtle Nesting Site: Anamur Beach, Mersin, Turkey. *Israel Journal of Ecology and Evolution*, 58: 87–100.

Valenzuela, N., Adams, D. C., Bowden, R. M. & Gauger, A. C., 2004. Geometric morphometric sex estimation for hatchling turtles: A powerful alternative for detecting subtle sexual shape dimorphism. *Copeia*, 2004: 735–742.

Wood, A., Booth, D. T. & Limpus, C. J., 2014. Sun exposure, nest temperature and loggerhead turtle hatchlings: Implications for beach shading management strategiesat sea turtle rookeries. *Journal of Experimental Marine Biology and Ecology*, 451: 105–114.

Yntema, C. L. & Mrosovsky, N., 1980. Sexual–Differentiation in Hatchling Loggerheads (*Caretta Caretta*) Incubated at Different Controlled Temperatures. *Herpetologica*, 361: 33–36.

Zbinden, J. A., Davy, C., Margaritoulis, D. & Arlettaz, R., 2007. Large spatial variation and female bias in estimated loggerhead sea turtle hatchling sex ratio of Zakynthos (Greece). *Endangered Species Research*, 3: 305–312.

14

Trophy hunting, size, rarity and willingness to pay: inter–specific analyses of trophy prices require reliable specific data

M. Sarasa

Sarasa, M., 2013. Trophy hunting, size, rarity and willingness to pay: inter–specific analyses of trophy prices require reliable specific data. *Animal Biodiversity and Conservation*, 36.2: 165–175.

Abstract

Trophy hunting, size, rarity and willingness to pay: inter–specific analyses of trophy prices require reliable specific data.— Awareness of the importance of the wildlife trade and human perception in animal conservation is growing. Recent studies carried out on a continental and world scale have analysed the associations between trophy score, rarity and prices. As a large range of ungulates are legally hunted throughout the world and numerous ungulate taxa are threatened, the relationship between rarity and trophy prices has been studied in several species. This article briefly reviews verifiable data on species and trophy prices and compares findings with data used in recent articles. The findings show that several elements of intra–specific data were inadequately addressed and that the trophy prices considered were not necessarily representative of real trophy prices. Furthermore, the body mass used for numerous taxa did not fit current knowledge of species, and several subspecies and rarity indexes that were considered disagreed with recognized subspecies or with the real conservation status of taxa. Thus, caution should be taken when considering some reported results. To improve our understanding of the associations between wildlife trade and wildlife conservation, further studies should take into account reliable specific data, such as that from government agencies, rather than publicity data.

Key words: Ungulate, Recreational hunting, Trophy price, Game management, Wildlife trade.

Resumen

La caza de trofeos, el tamaño, la rareza y la disposición a pagar: los análisis interespecíficos de los precios de los trofeos requieren datos específicos fiables.— Cada día hay más conciencia de la importancia que tienen el comercio de fauna silvestre y la percepción de los animales por parte del hombre en la conservación de los mismos. En determinados estudios llevados a cabo recientemente a escala continental o mundial se han analizado las relaciones existentes entre la puntuación de los trofeos, la rareza y los precios. Numerosas especies de ungulados se cazan legalmente en todo el mundo y varias de ellas son especies amenazadas. Por este motivo, se ha estudiado la relación existente entre la rareza y los precios de los trofeos en varias especies. En el presente artículo se examinan brevemente los datos verificables relativos a las especies y los precios de los trofeos, y se comparan con los datos utilizados en algunos artículos recientes. Los resultados ponen de manifiesto que varios elementos de los datos intraespecíficos se trataron inadecuadamente y que los precios de los trofeos analizados no eran necesariamente representativos de los precios reales. Asimismo, el peso corporal utilizado para muchos ungulados no se ajustaba a los valores documentados para estas especies y varias de las subespecies así como algunos indicadores de rareza analizados no se correspondían con las subespecies reconocidas o con su estado real de conservación. Por consiguiente, los resultados documentados deberían ser considerados con cautela. Para comprender mejor las relaciones existentes entre el comercio y la conservación de la fauna silvestre, los futuros estudios deberían tener en cuenta información específica fiable, por ejemplo de organismos gubernamentales, en vez de información publicitaria.

Palabras clave: Ungulados, Caza recreativa, Precio de trofeo, Gestión cinegética, Comercio de fauna silvestre.

M. Sarasa, Grupo Biología de las Especies Cinegéticas y Plagas (RNM–118), Sevilla, España (Spain).

Current address: Mathieu Sarasa, Fédération Nationale des Chasseurs, 13, Rue du Général Leclerc, F–92136 Issy les Moulineaux Cedex (France).

E–mail: mathieusar@hotmail.com; msarasa@chasseurdefrance.com

Introduction

Public awareness of the wildlife trade is increasing and understanding the need for animal conservation is growing (Johnson et al., 2010; Sarasa et al., 2012a). Human perception of species modulates wildlife conservation, and wildlife conservation policies affect human perception of species. Both international and local perception of wildlife may affect environmental policy and management practices (Pusey et al., 2007; Li et al., 2010). However, the perceived rarity of species and even policy and legal frameworks that compile the conservation status of species and that regulate the trading of wild animals —such as the Convention on International Trade in Endangered Species of Wild Fauna and Flora (CITES) and the International Union for the Conservation of Nature (IUCN)— may themselves increase trading activity because of a 'Limited edition' effect on wildlife trade (Barnes, 1996; Slone et al., 1997; Raymakers, 2002; Stuart et al., 2006). The 'limited edition' effect might be defined as an increase in desire for goods because they are limited in number or supply. The 'limited edition' effect has been a key tool in marketing management for decades (Mazis et al., 1973; West, 1975; Worchel et al., 1975; Balachander & Stock, 2008) and in wildlife trade in recent years it has been called the anthropogenic Allee effect (Courchamp et al., 2006). The 'limited edition' concept carries a sense of immediacy and exclusivity of goods which will only be available for a short time and/or in limited numbers. The concept affects the perceived rarity; it favours stiff prices and benefits and it stimulates impulsive purchases and collector behaviours (Mazis et al., 1973; West, 1975; Worchel et al., 1975; Balachander & Stock, 2008). Exploited rare goods or species, might become even rarer and thus more valuable, sucking them into a vortex toward the extinction of populations or species. This phenomenon might affect, for instance, insects, bird eggs, hunting trophies, and even live animals (Slone et al., 1997; Kiff, 2005; Courchamp et al., 2006; Stuart et al., 2006). The 'limited edition' effect in wildlife trade and conservation has received increasing interest over the last decade. As previously observed in collected insects (Slone et al., 1997), Johnson et al. (2010) highlighted the relationships between trophy score, rarity and prices of 159 taxa hunted in Africa. Palazy et al. (2012) later carried out a world scale analysis of these associations in trophy ungulates. Their compiled data were presented in an Appendix file that provides the opportunity to verify the reliability of such analysis. In this article I briefly reviewed verifiable data on species and trophy prices to compare these to the data detailed in the Appendix file of Palazy et al. (2012). The operational sections of this review are mainly focused on the data set of Palazy et al. (2012), although other studies based on undetailed publicity data from commercial hunting companies (Courchamp et al., 2006; Johnson et al., 2010; Palazy et al., 2011; Prescott et al., 2012) are probably affected by this issue to some extent.

Misrepresented data

Palazy et al. (2012) attempted to cover a larger range of taxa than previous studies, but several elements of intra–specific data noted thereafter were not adequately addressed. To compare trophy prices between species (see table 1 for detailed presentation of trophy price indexes), Palazy et al. (2012) used annual trophy prices from hunting companies, assuming that governments fix trophy fees and that trophy prices from hunting companies are representative of trophy fees and of the perceived value of trophies. However, this is not necessarily the case because, as detailed in table 1, trophy price is calculated using different formulas, and trophy fee is just one factor of the factors taken into account.

Auction hammer prices, complementary prices and profits of hunting companies cause substantial differences between trophy fees and trophy prices. Moreover, in several countries, for instance in Spain, local hunters, national hunters and international hunters may use different formulas to calculate trophy prices. For instance, Palazy et al. (2012) used a trophy price of USD 7,800 for both subspecies of the Iberian ibex *Capra pyrenaica*, Schinz 1838. Nevertheless, *C. p. victoriae*, Cabrera 1911, generally has longer and thicker horns than *C. p. hispanica*, Schimper 1848 (Granados et al., 2001) and so is often more appreciated by hunters and more expensive. Hunting permits for trophies are usually increased for auction (starting price in 2008: USD 6,635 at Riaño for *C. p. victoriae*; USD 3,650 in Andalucía for *C. p. hispanica*) and the perceived value of trophies, that is, their final sale price, consists of the hammer price plus a complementary price depending on the trophy score (Diario de León.es, 2008c; Junta de Andalucía, 2008) (tables 1, 2). The volatility of demand is hence a major factor in trophy prices. At Riaño, a record hammer price reached USD 39,870 in 2012, leading to a final price of USD 89,625 because of the score–based complementary price that reached USD 49,755 (Diario de León.es, 2012). According to table A1 of Palazy et al. (2012), only trophies of Markhor *Capra falconeri*, Wagner 1839, and of rhinoceros species would be more expensive than this trophy of Iberian ibex. However, this suggests trophy prices are misrepresented in their data set. The Iberian ibex is not an isolated case and table 2 highlights that numerous other species are also affected by this issue. Mismatches between trophy prices used by Palazy et al. (2012) and true trophy prices were also recorded within and between other species (table 2). In 23 taxa with verifiable data, only two presented absolute mismatches lower than 10% (mean; min; max: 35%; –140%; 92%). For instance, the prices of Iberian ungulates were over– or under–estimated, and the reported price differences between species from Europe, Asia or Africa mismatch the true differences between trophy fees reported by several authors (table 2). Thus, the prices used by Palazy et al. (2012) —probably distorted by call prices and by exaggerated prices of hunting companies (table 1)— are not representative of real trophy prices. Festa–Bianchet (2012) already suggested that

Table 1. Definitions of trophy price indexes.

Tabla 1. Definiciones de los índices de precios de los trofeos.

Trophy fees (TF)	Fixed amount of money paid to the management institution (*e.g.* governments or parks) for hunting one trophy individual.
Starting price for auction (SPA)	Initial amount of money expected by a management institution that sells by auction trophy hunting individuals.
Auction hammer price (AHP)	Final amount of money proposed by a purchaser to a management institution that sells by auction trophy hunting individuals. Auction hammer price is higher than the starting price for auction when the demand is greater that the supply.
Trophy score (TS)	Numerical value attributed to trophies according to measurements and appreciations (*e.g.* length, thickness, complexity, preferred shape, deformities, etc.) based on referenced hunters' aesthetic preferences.
Complementary price depending on the trophy score (CP)	Additional cost that can be fixed on the basis on the hunting trophy score of the individual hunted in the field. This additional cost is particularly common when the exact trophy value of individuals is estimable with difficulty from a distance or to apply penalties to mismatches between agreed hunting permits and observed hunting events.
Profit of hunting companies (PHC)	Difference between the fees paid to management institutions by hunting companies and the fees paid by hunters to the latter for hunting trophy individuals.
Call prices of hunting companies (CPHC)	Marketing tool of companies that can present underestimated trophy prices in their advertising to attract potential customers. Once obtained, the customer's interest and confidence allows commercial profits to be derived from overblown complementary costs.
Exaggerated prices of hunting companies (EPHC)	Commercial tool that can be used by companies to increase their commercial profits by using overblown price of their hunting permits. EPHC are largely due to scarcity of information about sales of hunting permits and on management institutions' trophy fees.
Trophy price (TP) or final sale price	Amount of money paid by hunters for hunting one trophy individual. It reveals the perceived trophy values. When hunters directly pay management institutions that apply fixed trophy prices: TP = TF or TP = TF + CP When hunters directly pay management institutions that sell by auction trophy hunting individuals: TP = AHP or TP = AHP + CP When hunters pay for trophy hunting through hunting companies: TP = TF + PHC or TP = TF + CP + PHC or TP = AHP +PHC or TP = AHP + CP + PHC

marketing may have a stronger effect than rarity on the cost of a hunt with hunting operators, although the two concepts are sometimes linked to each other. Price mismatches could be a serious concern in Palazy et al.'s analyses and interpretations, particularly taking into account that hunting companies just represent a variable, and often a minority part, of the total trophy hunting activity (Sharp & Wollscheid, 2009). This critical reappraisal was possible in Palazy et al.'s study because they presented a detailed data set.

Table 2. Mismatches between available specific data and data used in Palazy et al. (2012) (*): [1] Starting price for auction. [2] Auction hammer price. [3] Fee paid to the management institution for one individual. [4] The distinction between West Siberian Moose *A. a. pfizenmayeri* and East Siberian moose *A. a. buturlini* has not been widely accepted; body mass presented for *A. a. pfizenmayeri* (Rodgers, 2001). [5] Alashan wapiti *C. e. alashanicus* and Gansu deer *C. e. kansuensis* are considered as synonyms by Dolan (1988) and Groves (2006) recommended that the recognition of these taxa should be left for further studies; body mass presented for *C. e. kansuensis*. [6] The valid name of this species is *Damaliscus pygargus*, not *Damaliscus dorcas*; the two well–differentiated subspecies are the Bontebok *D. p. pygargus* and the Blesbok *D. p. phillipsi* (Lloyd & David, 2008). [7] Unclear taxonomic position; *S. c. brachyceros* would include planiceros (Van Hooft et al., 2002). [8] It is still unclear whether *C. caucasica* and *C. cylindricornis* are two separate species or if they are a single species with geographically dependent variability (Weinberg, 2008); Mid–Caucasian tur is considered a potential hybrid of *C. caucasica* and *C. cylindricornis* (Kopaliani & Gurielidze, 2009). [9] The taxonomy of *Capra sibirica* subspecies is not yet resolved and *C. s. hemalayanus* is not a recognized subspecies (Reading & Shank, 2008). [10] Two subspecies are recognized: Defassa Waterbuck *K. e. defassa* and Ellipsen Waterbuck *K. e. ellipsiprymnus*; *K. e. crawshayi* is included in *K. e. ellipsiprymnus* and *K. e. unctuosus* is included *K. e. defassa* (IUCN SSC Antelope Specialist Group, 2008; Lorenzen et al., 2006). [11] Argali *O. a. ammon* and *darwini* could be considered a single ESU or subspecies (Tserenbataa et al., 2004). [12] Three subspecies are recognized and *T. s. bea* is included in *T. s. strepsiceros* (Kingdon, 1997; Nersting & Arctander, 2001). [13] Three subspecies are recognized and *T. s. burlacei* is *T. s. cottoni* (Nersting & Arctander, 2001). [14] The two last records of the table A1 in Palazy et al. (2012) referred to *T. s. strepsiceros* and can not be considered rigorously as different tax.

Tabla 2. Diferencias entre los datos específicos disponibles y los datos utilizados en Palazy et al. (2012) (*): [1] Precio de salida para la subasta. [2] Precio de remate de la subasta. [3] Tasa pagada a la institución encargada de la gestión por un individuo. [4] La distinción entre el alce de Yakutia A. a. pfizenmayeri y el alce de Kamchatka A. a. buturlini aún no se ha aceptado ampliamente; peso corporal presentado para A. a. pfizenmayeri (Rodgers, 2001). [5] El uapití de Alashan C. e. alashanicus y el ciervo Gansu C. e. kansuensis se consideran sinónimos en Dolan (1988) y Groves (2006) recomendó que el reconocimiento de estos taxones se dejara para estudios posteriores; peso corporal presentado para C. e. kansuensis. [6] El nombre válido de esta especie es Damaliscus pygargus, no Damaliscus dorcas; las dos subspecies bien diferenciadas son el bontebok D. p. pygargus y el blesbok D. p. phillipsi (Lloyd & David, 2008). [7] Posición taxonómica poco clara; S. c. brachyceros incluiría a los búfalos del grupo S. c. planiceros (Van Hooft et al., 2002). [8] Aún no está claro si C. caucasica y C. cylindricornis son dos especies distintas o una sola con variabilidad geográfica (Weinberg, 2008); el tur del Cáucaso central se considera un posible híbrido de C. caucasica y C. cylindricornis (Kopaliani & Gurielidze, 2009). [9] La taxonomía de las subspecies de Capra sibirica aún no se ha resuelto y C. s. hemalayanus no es una subspecie reconocida (Reading & Shank, 2008). [10] Se reconocen dos subspecies: el antílope defasa K. e. defassa y el antílope acuático de Ellipsen K. e. ellipsiprymnus; K. e. crawshayi se incluye en K. e ellipsiprymnus y K. e. unctuosus se incluye en K. e. defassa (Grupo de especialistas sobre el antílope de la Comisión de Supervivencia de Especies de la Unión Internacional para la Conservación de la Naturaleza, 2008; Lorenzen et al., 2006). [11] Los muflones de Argal O. a. ammon y O. a darwini podrían considerarse una única UES o subspecie (Tserenbataa et al., 2004). [12] Se reconocen tres subspecies y T. s. bea se incluye en T. s. strepsiceros (Kingdon, 1997; Nersting & Arctander, 2001). [13] Se reconocen tres subspecies y T. s. burlacei se incluye en T. s. cottoni (Nersting & Arctander, 2001). [14] Los dos últimos registros de la tabla 1 en Palazy et al. (2012) hacen referencia a T. s. strepsiceros y no pueden considerarse rigurosamente como taxones distintos.

Trophy price
Species

Subspecies (in table A1) (*)	Record (in table A1) (*)	Trophy price (in USD) (*)	Trophy price index (in USD)	Reference
Capra pyrenaica				
C. p. hispanica	34th	7,800	3,650[1]	(Junta de Andalucía, 2008)
C. p. victoriae	35th	7,800	6,635[1]	(Diario de León.es, 2008c)
C. p. victoriae	35th	7,800	17,120–20,300[2]	(Diario de León.es, 2008b)
Rupicapra pyrenaica				
R. p. parva	168th	4,900	2,654[1]	(Diario de León.es, 2008c)
R. p. parva	168th	4,900	4,378–4,436[2]	(Diario de León.es, 2008a)
R. p. pyrenaica	169th	4,900	4,237–4,767[2]	(Heraldo.es, 2008)

Table 2. (Cont.)

Species

Subspecies (in table A1) (*)	Record (in table A1) (*)	Trophy price (in USD) (*)	Trophy price index (in USD)	Reference
Cervus elaphus				
C. e. hispanicus	58th	4,500	1,990[1]	(Diario de León.es, 2008c)
C. e. hispanicus	58th	4,500	3,509[2]	(Diario de León.es, 2008a)
Capreolus capreolus	40th	2,216	1,327[1]	(Diario de León.es, 2008c)
	40th	2,216	1,768[2]	(Diario de León.es, 2008a)
Capra falconeri				
C. f. falconeri	32th	70,000		
C. f. jerdoni			20,000–35,000[3]	(Frisina & Tareen, 2009)
Ovis vignei				
O. v. cycloceros	144th	8,000	6,500–11,000[3]	(Frisina & Tareen, 2009)
Diceros bicornis	77th	150,000	195,000–210,000[3]	(Davies et al., 2009)
Syncerus caffer				
S. c. caffer	176th	7408	600[3]	(Lamprey & Mugisha, 2009)
Potamochoerus porcus	148th	632	150[3]	(Lamprey & Mugisha, 2009)
Taurotragus oryx				
T. o. pattersonianus	182th	2125	500[3]	(Lamprey & Mugisha, 2009)
Hippopotamus amphibious	87th	2328	500[3]	(Lamprey & Mugisha, 2009)
Aepyceros melampus				
A. m. rendilis	3rd	663	250[3]	(Lamprey & Mugisha, 2009)
Ourebia ourebia	119th	645	150[3]	(Lamprey & Mugisha, 2009)
Redunca redunca				
R. r. wardi	163th	605	250[3]	(Lamprey & Mugisha, 2009)
Damaliscus lunatus				
D. l. jimela	73th	910	350[3]	(Lamprey & Mugisha, 2009)
Phacochoerus aethiopicus	146th	454	250[3]	(Lamprey & Mugisha, 2009)
Kobus ellipsiprymnus				
K. e. defassa	94th	676	500[3]	(Lamprey & Mugisha, 2009)

Body mass
Species

Subspecies (in table A1) (*)	Record (in table A1) (*)	Male body mass (*)	Male body mass	Reference
Alces alces				
A. a. alces	9th	400	375–475	
A. a. cameloides	12th	453.5	250–350	(Bishop, 1988; Haigh et al., 1980;
A. a. buturlini [4]	11th	453.5	340–6,554	Rodgers, 2001; Wallin et al., 1996)
A. a. andersoni	10th	453.5	350–570	
A. a. gigas	13th	453.5	400–700	

Table 2. (Cont.)

Species

Subspecies (in table A1) (*)	Record (in table A1) (*)	Male body mass (*)	Male body mass	Reference
Capra pyrenaica				
C. p. hispanica	34th	72.5	50.4–65	(Couturier, 1962; Granados et al., 2001)
C. p. victoriae	35th	72.5	61.9–90	
Ceratotherium simum				
C. s. cottoni	53th	2800	100–1,600	(Groves et al., 2010)
C. s. simum	54th	2800	2,000–2,400	
Cervus elaphus				
C. e. alashanicus [5]	56th	180.5	2405	http://www.scirecordbook.org/gansu–deer/
C. e. hippelaphus	57th	180.5	160	(Geist & Bayer, 1988)
C. e. hispanicus	58th	180.5	80–160	(Carranza, 2011)
C. e. kansuensis[5]	59th	180.5	2,405	http://www.scirecordbook.org/gansu–deer/
C. e. nelsoni	60th	180.5	350	(Geist & Bayer, 1988)
C. e. sibiricus	61th	180.5	300	http://www.scirecordbook.org/altai–wapiti/
C. e. songaricus	62th	180.5	300	(Gao et al., 2011)
Cervus nippon	63th	47.6	30–120	(Groves, 2006)
Damaliscus dorcas [6]				
D. d. dorcas [6]	71th	68	46.5	(Hayward et al., 2006)
D. d. phillipsi [6]	72th	68	52.5	
Syncerus caffer				
S. c. aequinoctialis	174th	522	500–590	(Hayward et al., 2006; Solounias et al., 1994)
S. c. brachyceros	175th	522	400–500	http://www.scirecordbook.org/nile–buffalo/
S. c. caffer	176th	522	432–754	http://www.scirecordbook.org/central–
S. c. nanus	177th	522	265	african–savanna–buffalo/
S. c. planiceros [7]	178th	522	320–410	http://www.scirecordbook.org/west– african–savanna–buffalo

Conservation status

Species

Subspecies (in table A1) (*)	Record (in table A1) (*)	Conservation status (*)	Proposed conservation status	Reference
Capra pyrenaica				
C. p. hispanica	34th	Least concern	Least concern	(Acevedo & Cassinello, 2009; Pérez et al., 2002)
C. p. victoriae	35th	Least concern	Vulnerable	
Naemorhedus goral	112th	Near threatened	Endangered	(CITES, 2011)
Ovis ammon				
O. a. hodgsoni	127th	Near threatened	Endangered	(CITES, 2011)
O. a. polli	130th	Near threatened	Vulnerable	(Schaller & Kang, 2008)

Table 2. (Cont.)

Other unclear or unrecognized taxonomic distinctions

Species	Subspecies in table A1 (*)	Record in table A1 (*)
Capra cylindricornis [8]		31th
Capra caucasica [8]	*C. c. caucasica* [8]	29th
	C. c. dinniki [8]	30th
Capra sibirica	*C. s. hemalayanus* [9]	38th
Kobus ellipsiprymnus	*K. e. crawshayi* [10]	93th
	K. e. defassa [10]	94th
	K. e. ellipsiprymnus [10]	95th
	K. e. unctuosus [10]	96th
Ovis ammon	*O. a. ammon* [11]	123th
	O. a. darwini [11]	126th
Tragelaphus strepsiceros	*T. s. bea* [12]	198th
	T. s. burlacei [13]	199th
	T. s. chora	200th
	T. s. strepsiceros [14]	201th
	T. s. strepsiceros [14]	202th

However, other studies that used trophy prices from hunting companies (Courchamp et al., 2006; Johnson et al., 2010; Palazy et al., 2011) were exposed to this concern as well. The variability of trophy price setting systems between countries and taxa is a key factor of this question. Thus, to promote the reliability of results, a detailed presentation of trophy prices and of setting systems should be required for each record and taken into account. Further studies should consider trophy prices from governmental agencies rather than publicity data to improve the accuracy of row data, of results and of biological inferences.

By separating sub–species, Palazy et al. (2012) were able to address a large range of taxa. Beyond unclear and unrecognized taxonomic distinctions (table 2), the 'subspecies' of 34 species (103 units in all) were considered as having a single mean body mass (BM) per species, but the subspecies of other taxa were considered to have different BM. Differences in BM were probably considered when found. However, in the Iberian ibex, *C. p. hispanica* is known to be smaller than *C. p. victoriae* (Couturier, 1962; Fandos & Vigal, 1988; Granados et al., 1997; Granados et al., 2001). Similarly, among other species, Alaska moose *Alces alces gigas,* Miller 1899, is the heaviest subspecies of moose (Flerov, 1952; Peterson, 1955; Bishop, 1988) and subspecies of *Cervus elaphus,* Linnaeus 1758, differ in size (Lowe & Gardiner, 1989; Haigh & Hudson, 1993; Novak, 1999), but this was not taken

into account (table 2). The data of Palazy et al. (2012) on BM are not representative of the 202 ungulate units used and as such their results would have been artificially smoothed. Of the 24 taxa with verifiable data, only five presented absolute mismatches lower than 10% (mean; min; max: –4%; –94%; 49%). These observed errors in BM cannot be due to consistent methodology because, as highlighted in the several examples (table 2), data on subspecies body mass are already available in the scientific literature in common data bases on the Internet (http://wokinfo.com; http://scholar.google.com; etc.). Thus, further studies could detail the references for BM to encourage the use of reliable data.

Hunting institution databases record the trophies that have been harvested over long periods of time (see, for instance, Monteith et al., 2013). However, trophy scores are estimated on the basis of the global biometry of the horns and the aesthetic preferences of hunters to compare trophies within a species. They do not therefore accurately take into account confounding factors such as the age of the animal. Trophy scores do not adequately reflect real horn size, at least in wild sheep (König & Hoefs, 1984), just as classical measurements of animal weapons are not necessarily representative of true horn growth (Sarasa et al., 2012b). Wild sheep represents 11% of the sample in Palazy et al. (2012) and most of their sample consists of horned ungulates. This is also a major concern of the biological inferences derived from interspecific analysis of trophy score data. Courchamp et al. (2006) assumed that trophy scores allow interspecific comparisons but the reliability of this factor as a proxy for interspecific comparisons of trophy size is an issue that is still somewhat open. Confusion between trophy score and trophy size might not have affected the analyses *per se*. However, this questionable proxy favours confusion between two different concepts and it leads to an overblown perception of the biological reliability of trophy scores and of the inferences derived from these studies. Thus, the limitations of trophy scores as a proxy of trophy size should have been commented and mentions of trophy size should be avoided when referring to trophy score. Other studies that used trophy scores (Johnson et al., 2010; Palazy et al., 2011) were exposed to this concern as well.

Palazy et al. (2012) used IUCN categories as a proxy of rarity. IUCN statuses are interesting proxies of the conservation status for a focal species at the temporal scale of decades. The long temporal scale of IUCN status is due to the definition of threatened status. It is often based on the restricted size of populations and/or on population reductions over the last 10 years or three generations. As a result, IUCN status is a relative index of conservation status for each species, but not an absolute conservation index. IUCN categories may be intrinsically vague and are a problem for those trying to classify species (Regan et al., 2000). Thus, IUCN statuses are not an accurate source of information for inter–specific analyses. Assuming that IUCN statuses might be considered as accurate proxies of the human perception of the rarity of species, other problems remain. Palazy et al. (2012) considered subspecies to increase their sample size while conservation status of most subspecies is not detailed in the IUCN red list. Mismatches were also observed. Several mismatches in rarity values probably resulted from partial and incomplete updating of IUCN pages and of the overblown sample size in Palazy et al.'s study. For instance, both subspecies of Iberian ibex were considered to be of 'Least Concern' by Palazy et al. (2012). However, *C. p. victoriae* is classified as 'Vulnerable' since it only inhabits a few, small areas, while *C. p. hispanica* is of 'Least Concern' where its viability depends on ongoing conservation programmes (Pérez et al., 2002; Acevedo & Cassinello, 2009). *Ovis ammon polii,* Blyth 1841, is considered to be 'Near Threatened' by Palazy et al. (2012), but a status of 'Vulnerable' seems more accurate (Schaller & Kang, 2008). Thus, to remedy IUCN red list updating limits, specific scientific literature should be assessed in detail and researchers specialized in focal species should be contacted to avoid using incomplete and unrepresentative data. Specialists of focal species have updated knowledge of the conservation status and of the actual perceived rarity of species; well–informed dwellers/hunters sometimes forestall potential changes in conservation status by policy and legal frameworks (Rivalan et al., 2007). In Palazy et al. (2012), 81 subspecies of 25 species were considered to have a single conservation status per species and

the proxies of rarity are not necessarily representative of the real conservation status of the 202 considered records (table 1).

Conclusion

A critical question in inter–specific studies is that unreliable data should be discarded as much as possible because it produces unreliable results. Study designs should be adjusted to ensure the best resolution in sampling while preserving the reliability of the data. In Palazy et al.'s study, taking into account that many data represent species but not subspecies, analysis of the 112 considered species rather than 202 questionable taxa may have been less overblown. The results of Palazy et al. (2012) are potentially interesting because they converge with those of Johnson et al. (2010). Nevertheless, as in other articles on the subject, in Palazy et al. (2012) several elements of intra–specific data were not properly addressed in at least 25–35% of the sample [25% if we only take into account mismatches in trophy prices, body mass, IUCN status and unclear or unrecognized taxonomic distinctions; 35% if we also take into account that classical measurements of weapons and trophy scores misrepresent trophy size (König & Hoefs, 1984; Sarasa et al., 2012b)]. Thus, caution should be taken when considering the reported results. Moreover, while tourist/ foreign hunters spend far more per head than non–tourist/local hunters, international trophy hunting with commercial hunting operators is associated with a global total in spending that is much less than that of stay–at–home hunters (Sharp & Wollscheid, 2009). The economics of commercial hunting operators is a minor part of the total economics of hunting (Sharp & Wollscheid, 2009). This should be also taken into account to avoid overestimating the scientific importance of analysis of publicity data from commercial hunting companies compared to the total economics of hunting. To some extent, other studies that were based on undetailed publicity data from commercial hunting companies (Courchamp et al., 2006; Johnson et al., 2010; Palazy et al., 2011; Prescott et al., 2012) are probably affected by this issue. Biological data and economic data should be adequately addressed in future studies. These should prefer representative lists of trophy prices from governmental agencies rather than publicity data; systematics, body mass, and rarity indexes should be properly represented when data are already available; trophy size should be properly characterized, avoiding aesthetic and incomplete proxies such as trophy score and horn length alone. A rigorous compilation of row data is required so that high quality studies may support the understanding of wildlife trade and the conservation of threatened species.

Acknowledgements

Thanks to Agnès Sarasa and Michael Lockwood for the English revision. This study received no specific financial support.

References

Acevedo, P. & Cassinello, J., 2009. Biology, ecology and status of Iberian ibex *Capra pyrenaica*: a critical review and research prospectus. *Mammal Review,* 39: 17–32.

Balachander, S. & Stock, A., 2008. Limited Edition products: when and when not to offer them. *Marketing Science,* 28: 336–355.

Barnes, J. I., 1996. Changes in the economic use value of elephant in Botswana: the effect of international prohibition. *Ecological Economics,* 18: 215–230.

Bishop, R. H., 1988. The moose in Alaska. In: *Audubon Wildlife Report 1988/1989*: 495–512 (W. J. Chandler & L. Labate, Eds.). Academic Press, St. Louis, MO.

Carranza, J., 2011. Ciervo – *Cervus elaphus*. In: *Enciclopedia Virtual de los Vertebrados Españoles* (A. Salvador & J. Cassinello, Eds.). Museo Nacional de Ciencias Naturales, Madrid. http://www.vertebradosibericos.org/

CITES, 2011. *Appendices I, II and III of the Convention on International Trade in Endangered Species of Wild Fauna and Flora* http://www.cites.org/eng/app/appendices.php.

Courchamp, F., Angulo, E., Rivalan, P., Hall, R. J., Signoret, L., Bull, L. & Meinard, Y., 2006. Rarity value and species extinction: the anthropogenic Allee effect. *PLoS Biology,* 4(12): e415. DOI: 10.1371/journal.pbio.0040415.

Couturier, M., 1962. *Le bouquetin des Alpes (Capra aegagrus ibex ibex* L.). Arthaud, Grenoble.

Davies, R., Hamman, K. & Magome, H., 2009. Does recreational hunting conflict with photo–tourism? In: *Recreational hunting, conservation and rural livelihoods: science and practice*: 233–251 (B. Dickson, J. Hutton, W. M. Adams, Eds). Wiley–Blackwell, Oxford.

Diario de León.es, 2008a. *La caza mayor de 92 piezas en Ancares reportará 120.000 euros a sus pedanías.* http://www.diariodeleon.es/noticias/bierzo/la–caza–mayor–de–92–piezas–en–ancares–reportara–120–000–euros–a–sus–pedanias_375855.html. (Accesed on 28 March 2012).

– 2008b. *La subasta de caza de la reserva alcanza la cifra de 353.600 euros.* http://www.diariodeleon.es/noticias/comarcas/la–subasta–de–caza–de–reserva–alcanza–cifra–de–353–600–euros_374737.html. (Accesed on 28 March 2012).

– 2008c. *Riaño subastará el domingo 137 piezas de caza y 22 batidas de jabalíes.* http://www.diariodeleon.es/noticias/montanaoriental/riano–subastara–domingo–137–piezas–de–caza–y–22–batidas–de–jabalies_374325.html. (Accesed on 28 March 2012).

– 2012. *Un magnate ruso paga 67.500 euros por cazar una cabra montés en Picos.* http://www.diariodeleon.es/noticias/provincia/un–magnate–ruso–paga–67–500–euros–por–cazar–una–cabra–montes–en–picos_678273.html. (Accesed on 28 March 2012).

Dolan, J. M., 1988. A deer of many lands: a guide to the subspecies of the red deer *Cervus elaphus* L. *Zoonooz,* LXII: 4–34.

Fandos, P. & Vigal, C. R., 1988. Body weight and horn length in relation to age of the Spanish wild goat. *Acta Theriologica,* 33: 339–344.

Festa–Bianchet, M., 2012. Rarity, willingness to pay and conservation. *Animal Conservation,* 15: 12–13.

Flerov, K. K., 1952. *Musk deer and deer.* Fauna of U.S.S.R., Mammals, 1(2). U.S.S.R. Academy of Sciences, Moscow.

Frisina, M. R. & Tareen, S. N. A., 2009. Exploitation prevents extinction: case study of endangered Himalayan sheep and goats. In: *Recreational hunting, conservation and rural livelihoods: science and practice*: 141–156 (B. Dickson, J. Hutton, W. M. Adams, Eds.). Wiley–Blackwell, Oxford.

Gao, Q. H., Wang, H. E., Zeng ,W. B., Wei, H. J., Han, C. M., Du, H. Z., Zhang, Z. G. & Li, X. M., 2011. Embryo transfer and sex determination following superovulated hinds inseminated with frozen–thawed sex–sorted Y sperm or unsorted semen in Wapiti (*Cervus elaphus songaricus*). *Animal Reproduction Science,* 126: 245–250.

Geist, V. & Bayer, M., 1988. Sexual dimorphism in the Cervidae and its relation to habitat. *Journal of Zoology,* 214: 45–53.

Granados, J. E., Pérez, J. M., Soriguer, R. C., Fandos, P. & Ruiz Martínez, I., 1997. On the biometry of the Spanish ibex, *Capra pyrenaica*, from Sierra Nevada (southern Spain). *Folia Zoologica,* 46: 9–14.

Granados, J. E., Pérez, J. M., Márquez, F. J., Serrano, E., Soriguer, R. C. & Fandos, P., 2001. La cabra montés (*Capra pyrenaica*, Schinz 1838). *Galemys,* 13: 3–37.

Groves, C., 2006. The genus *Cervus* in eastern Eurasia. *European Journal of Wildlife Research,* 52: 14–22.

Groves, C. P., Fernando, P. & Robovský, J., 2010. The Sixth Rhino: A Taxonomic Re–Assessment of the Critically Endangered Northern White Rhinoceros. *PLoS One,* 5(4): e9703. DOI:9710.1371/journal.pone.0009703.

Haigh, J. C., Stewart, R. R. & Mitton, W., 1980. Relations among linear measurements and weight for moose (*Alces alces*). *Alces,* 16: 1–10.

Haigh, J. D. & Hudson, R. J., 1993. *Farming wapiti and red deer.* Mosby–Year Book Inc., St. Louis, MO.

Hayward, M. W., Henschel, P., O'Brien, J., Hofmeyr, M., Balme, G. & Kerley, G. I. H., 2006. Prey preferences of the leopard (*Panthera pardus*). *Journal of Zoology,* 270: 298–313.

Heraldo.es, 2008. *Las subastas de caza de sarrios dejan más de 120.000 euros en Los Valles y Benasque.* http://www.heraldo.es/noticias/las_subastas_caza_sarrios_dejan_mas_120_000_euros_los_valles_benasque.html?p=1020943041. (Accesed on 28 March 2012).

IUCN SSC Antelope Specialist Group, 2008. *Kobus ellipsiprymnus.* In: *IUCN 2011 IUCN Red List of Threatened Species* Version 20112 www.iuc.redlist.org.

Johnson, P. J., Kansky, R., Loveridge, A. J. & Macdonald, D. W., 2010. Size, rarity and charisma: valuing African wildlife trophies. *PLoS One,* 5(9):

e12866. DOI:12810.11371/journal.pone.0012866.

Junta de Andalucía, 2008. Subasta de permisos de recechos y aguardos para la temporada cinegética 2008–2009. http://www.juntadeandalucia.es/medioambiente/web/1_consejeria_de_medio_ambiente/dg_gestion_medio_natural/instituto_andaluz_de_la_caza_y_la_pesca_continental/oferta_publica_de_caza/segunda_subasta/pliego_recechos_segunda_subasta.pdf. (Accesed on 28 March 2012).

Kiff, L. F., 2005. History, present status, and future prospects of avian eggshell collections in North America. *The Auk,* 122: 994–999.

Kingdon, J., 1997. *The Kingdon field guide to African mammals.* Academic Press, London.

König, R. & Hoefs, M., 1984. Volume and density of horns of Dall rams. *Biennial Symposium of Northern Sheep and Goat Council,* 4: 295–304.

Kopaliani, N. & Gurielidze, Z., 2009. Status of Turs in Georgia and Conservation Action Plan. In: *Status and Protection of Globally Threatened Species in the Caucasus:* 61–68 (N. Zazanashvili & D. Mallon, Eds.). CEPF, WWF, Contour Ltd., Tbilisi.

Lamprey, R. H. & Mugisha, A., 2009. The re–introduction of recreational hunting in Uganda. In: *Recreational hunting, conservation and rural livelihoods: science and practice:* 212–232 (B. Dickson, J. Hutton, W. M. Adams, Eds). Wiley–Blackwell, Oxford.

Li, S., Wang, D., Gu, X. & McShea, W. J., 2010. Beyond pandas, the need for a standardized monitoring protocol for large mammals in Chinese nature reserves. *Biodiversity and Conservation,* 19: 3195–3206.

Lorenzen, E. D., Simonsen, B. T., Kat, P. W., Arctander, P. & Siegismund, H. R., 2006. Hybridization between subspecies of waterbuck (*Kobus ellipsiprymnus*) in zones of overlap with limited introgression. *Molecular Ecology,* 15: 3787–3799.

Lowe, V. P. W. & Gardiner, A. S., 1989. Are the New and Old World wapitis (*Cervus canadensis*) conspecific with red deer (*Cervus elaphus*)? *Journal of Zoology,* 218: 51–58.

Lloyd, P. & David, J., 2008. *Damaliscus pygargus.* In: *IUCN 2011 IUCN Red List of Threatened Species* Version 20112 wwwiucnredlistorg.

Mazis, M. B., Settle, R. B. & Leslie, D. C., 1973. Elimination of phosphate detergents and psychological reactance. *Journal of Marketing Research,* 10: 390–395.

Monteith, K. L., Long, R. A., Bleich, V. C., Heffelfinger, J. R., Krausman, P. R. & Bowyer, R. R., 2013. Effects of harvest, culture, and climate on trenes in size of horn–like structures in trophy ungulates. *Wildlife Monographs,* 183: 1–26.

Nersting, L. G. & Arctander, P., 2001. Phylogeography and conservation of impala and greater kudu. *Molecular Ecology,* 10: 711–719.

Novak, R. M., 1999. *Walker's Mammals of the World,* Vol. 2. 6th edition. Johns Hopkins Univ. Press, Baltimore.

Palazy, L., Bonenfant, C., Gaillard, J.–M. & Courchamp, F., 2011. Cat Dilemma: Too Protected To Escape Trophy Hunting? *PLoS ONE* 6(7), e22424. DOI: 22410.21371/journal.pone.0022424.

Palazy, L., Bonenfant, C., Gaillard, J. M. & Courchamp, F., 2012. Rarity, trophy hunting and ungulates. *Animal Conservation,* 15: 4–11.

Pérez, J. M., Granados, J. E., Soriguer, R. C., Fandos, P., Márquez, F. J. & Crampe, J. P., 2002. Distribution, status and conservation problems of the Spanish Ibex, *Capra pyrenaica* (Mammalia: Artiodactyla). *Mammal Review,* 32: 26–39.

Peterson, R. L. 1955. *North American moose.* University of Toronto Press, Toronto.

Prescott, G. W., Johnson, P. J., Loveridge, A. J. & Macdonald, D. W., 2012. Does change in IUCN status affect demand for African bovid trophies. *Animal Conservation,* 15: 248–252.

Pusey, A. E., Pintea, L., Wilson, M. L., Kamenya, S. & Goodall, J., 2007. The contribution of long term research at Gombe National Park to chimpanzee conservation. *Conservation Biology,* 21: 623–634.

Raymakers, C., 2002. International trade in sturgeon and paddlefish species—the effect of CITES listing. *International Review of Hydrobiology,* 87: 525–537.

Reading, R. & Shank, C., 2008. Capra sibirica. In: IUCN 2011 IUCN Red List of Threatened Species Version 20112 www.iucnredlist.org.

Regan, H. M., Colyvan, M. & Burgman, M. A., 2000. A proposal for fuzzy International Union for the Conservation of Nature (IUCN) categories and criteria. *Biological Conservation,* 92: 101–108.

Rivalan, P., Delmas, V., Angulo, E., Bull, L. S., Hall, R. J., Courchamp, F., Rosser, A. M. & Leader–Williams, N., 2007. Can bans stimulate wildlife trade? *Nature* 447: 529–530.

Rodgers, A., 2001. *Moose.* Voyageur Press, Stillwater.

Sarasa, M., Alasaad, S. & Pérez, J. M., 2012a. Common names of species, the curious case of *Capra pyrenaica* and the concomitant steps towards the 'wild–to–domestic' transformation of a flagship species and its vernacular names. *Biodiversity and Conservation,* 21: 1–12.

Sarasa, M., Soriguer, R. C., Granados, J.–E., Casajus, N. & Pérez, J. M., 2012b. Mismeasure of secondary sexual traits: an example with horn growth in the Iberian ibex. *Journal of Zoology,* 288: 170–176.

Schaller, G. B. & Kang, A., 2008. Status of Marco Polo sheep *Ovis ammon polii* in China and adjacent countries: conservation of a Vulnerable subspecies. *Oryx,* 42: 100–106.

Sharp, R. & Wollscheid, K.–U., 2009. An overview of recreational hunting in North America, Europe and Australia. In: *Recreational hunting, conservation and rural livelihoods: science and practice:* 25–38 (B. Dickson, J. Hutton & W. M. Adams, Eds.). Wiley–Blackwell, Oxford.

Slone, T. H., Orsak, L. J. & Malver, O., 1997. A comparison of price, rarity and cost of butterfly specimens: Implications for the insect trade and for habitat conservation. *Ecological economics,* 21: 77–85.

Solounias, N., Fortelius, M. & Freeman, P., 1994. Molar wear rates in ruminants: a new approach. *Annales Zoologici Fennici,* 31: 219–227.

Stuart, B. L., Rhodin, A. G. J., Grismer, L. L. & Hansel,

T., 2006. Scientific description can imperil species. *Science,* 312: 1137–1137.

Tserenbataa, T., Ramey, R. R., Ryder, O. A., Quinn, T. W. & Reading, R. P., 2004. A population genetic comparison of argali sheep (*Ovis ammon*) in Mongolia using the ND5 gene of mitochondrial DNA; implications for conservation. *Molecular Ecology,* 13: 1333–1339.

Van Hooft, W. F., Groen, A. F. & Prins, H. H. T., 2002. Phylogeography of the African buffalo based on mitochondrial and Y–chromosomal loci: Pleistocene origin and population expansion of the Cape buffalo subspecies. *Molecular Ecology,* 11: 267–279.

Wallin, K., Cederlund, G. & Pehrson, Å., 1996. Predicting body mass from chest circumference in moose *Alces alces. Wildlife Biology,* 2: 53–58.

Weinberg, P., 2008. *Capra cylindricornis.* In: *IUCN 2011 IUCN Red List of Threatened Species* Version 2012 (www.iucnredlist.org).

West, S. G., 1975. Increasing the attractiveness of college cafeteria food: A reactance theory perspective. *Journal of Applied Psychology,* 60: 656–658.

Worchel, S., Lee, J. & Adewole, A., 1975. Effects of supply and demand on ratings of object value. *Journal of Personality and Social Psychology,* 32: 906–914.

15

Interaction of landscape variables on the potential geographical distribution of parrots in the Yucatan Peninsula, Mexico

A. H. Plasencia–Vázquez, G. Escalona–Segura & L. G. Esparza–Olguín

Plasencia–Vázquez, A. H., Escalona–Segura, G. & Esparza–Olguín, L. G., 2014. Interaction of landscape variables on the potential geographical distribution of parrots in the Yucatan Peninsula, Mexico. *Animal Biodiversity and Conservation*, 37.2: 191–203.

Abstract

Interaction of landscape variables on the potential geographical distribution of parrots in the Yucatan Peninsula, Mexico.— The loss, degradation, and fragmentation of forested areas are endangering parrot populations. In this study, we determined the influence of fragmentation in relation to vegetation cover, land use, and spatial configuration of fragments on the potential geographical distribution patterns of parrots in the Yucatan Peninsula, Mexico. We used the potential geographical distribution for eight parrot species, considering the recently published maps obtained with the maximum entropy algorithm, and we incorporated the probability distribution for each species. We calculated 71 metrics/variables that evaluate forest fragmentation, spatial configuration of fragments, the ratio occupied by vegetation, and the land use in 100 plots of approximately 29 km², randomly distributed within the presence and absence areas predicted for each species. We also considered the relationship between environmental variables and the distribution probability of species. We used a partial least squares regression to explore patterns between the variables used and the potential distribution models. None of the environmental variables analyzed alone determined the presence/absence or the probability distribution of parrots in the Peninsula. We found that for the eight species, either due to the presence/absence or the probability distribution, the most important explanatory variables were the interaction among three variables, particularly the interactions among the total forest area, the total edge, and the tropical semi–evergreen medium–height forest. Habitat fragmentation influenced the potential geographical distribution of these species in terms of the characteristics of other environmental factors that are expressed together with the geographical division, such as the different vegetation cover ratio and land uses in deforested areas.

Key words: Deforestation, Fragmentation, Land use, Parrots, Vegetation cover

Resumen

La interacción de las variables del paisaje sobre la distribución geográfica potencial de los loros en la península de Yucatán, México.— La pérdida, degradación y fragmentación de las zonas boscosas están poniendo en peligro a las poblaciones de loros. En este estudio se determinó la influencia de la fragmentación en relación con la cobertura vegetal, los usos del suelo y la configuración espacial de los fragmentos, sobre los modelos de distribución geográfica potencial de los loros en la península de Yucatán, México. Se utilizó la distribución geográfica potencial de ocho especies de loros, teniendo en cuenta los mapas publicados recientemente y obtenidos con el algoritmo de máxima entropía, y se incorporó el mapa de probabilidad de distribución de cada especie. Se calcularon 71 parámetros y variables que evalúan la fragmentación forestal, la configuración espacial de los fragmentos, la proporción ocupada por vegetación y los usos del suelo en 100 parcelas de aproximadamente 29 km² distribuidas al azar dentro de las zonas de presencia y ausencia predichas para cada especie. Además, se tuvo en cuenta la relación entre las variables ambientales y la probabilidad de distribución de las especies. Se empleó una regresión de mínimos cuadrados parciales para analizar la relación existente entre las variables empleadas y los modelos de distribución potencial. Ninguna de las variables ambientales analizadas determina por sí sola la presencia, la ausencia ni la probabilidad de distribución de los loros en la península. Se observó que para las ocho especies, ya sea debido a la presencia y la ausencia o a la probabilidad de distribución, las variables explicativas más importantes son la interacción entre tres variables, en especial la interacción entre la superficie forestal total, la longitud total de los perímetros de los fragmentos y la cantidad de bosque tropical subperennifolio de altura mediana. La fragmentación del hábitat influye sobre la distribución geográfica potencial de estas especies en combinación con otros factores ambientales asociados a la misma, como son la proporción de las diferentes coberturas vegetales y los usos del suelo que se desarrollan en las áreas deforestadas.

Palabras clave: Deforestación, Fragmentación, Uso del suelo, Loros, Cobertura vegetal

A. H. Plasencia–Vázquez, G. Escalona–Segura & L. G. Esparza–Olguín, El Colegio de la Frontera Sur (ECOSUR), Libramiento Carretero Campeche km 1.5, Av. Rancho, Polígono 2–A, Parque Industrial de Lerma, C.P. 24500, San Francisco de Campeche, Campeche, México.

Corresponding author: A. H. Plasencia–Vázquez. E–mail: alexpla79@gmail.com

Introduction

Forest loss and fragmentation are significant factors that contribute to extinction of species worldwide (Hanski et al., 2013). Different species respond differently to these anthropogenic perturbations, particularly those that are affected by the size of the remaining fragment and its connectivity with other fragments or the main forest mass (Donovan & Lamberson, 2001). A reduction in the average size of forest fragments will affect bird populations if these fragments are too small to satisfy species specific requirements (Bregman et al., 2014). It has been shown that the effect of fragment size depends on the characteristics of the surrounding forest–fragment mosaic (Brotons et al., 2002). In addition, in abandoned agricultural areas subject to secondary succession, or in areas where badly planned reforestation has taken place, some species of farmland birds are affected due to both habitat loss and an increase in forest edge density (Rey Benayas et al., 2008; Reino et al., 2009).

Few studies have examined the effects of forest fragmentation on parrots (*e.g.* Evans et al., 2005), one of the most threatened species in Mexico as their habitats face disappearance throughout their range of distribution (Norma Oficial Mexicana, 2010). In Mexico, current research on psittacines has focused on the potential distribution of several species (Monterrubio–Rico et al., 2010; Monterrubio–Rico et al., 2011), the effects of land use changes (Ríos–Muñoz & Navarro–Sigüenza, 2009), habitat loss, and the illegal trafficking of species (Marín–Togo et al., 2012). Eight of the twenty–two species of psittacines in Mexico are present in the Yucatan Peninsula (MacKinnon, 2005), and some of them still have high populations (*e.g.* Macías–Caballero & Iñigo–Elías, 2003; Plasencia–Vázquez & Escalona–Segura, 2014a). However, the loss of forest areas, together with degradation and fragmentation have intensified (Céspedes–Flores & Moreno–Sánchez, 2010) and are is endangering viable parrot populations in the region.

Ecological studies have tended to analyse the independent impact of environmental factors on psittacine distribution. Significantly, it has not been considered whether a series of concomitant environmental factors can modify the impact of forest fragmentation on species. Such factors include a loss of vegetation cover (Waltert et al., 2005) and types of land use (Ríos–Muñoz & Navarro–Sigüenza, 2009) within the habitat mosaic surrounding forest fragments. In addition, different levels of anthropogenic disturbance (Marín–Togo et al., 2012) and fragment degradation (Raman, 2004) can influence the impact of fragmentation on psittacines.

Psittacine distribution patterns may reflect the combined action of all these factors and would be established in association with the spatial configuration of the forest fragments. Thus, the aim of this study was to determine the influence of fragmentation in terms of vegetation cover, land use, and the spatial configuration of forest fragments, on the patterns of potential geographical distribution of parrots in the Yucatan Peninsula, Mexico.

Material and methods

Study area

The study was conducted in the three Mexican states that comprise the Yucatan Peninsula: Campeche, Quintana Roo, and Yucatán (17° 48' – 21° 35' N, 86° 43' – 92° 27' W), in southeastern Mexico (Barrera, 1962). The entire Yucatan Peninsula is a flat–lying karst landscape with few hills and little topographical variation (Lugo–Hubp et al., 1992). It is divided into two biogeographical regions, the first towards the northwest and the second towards the southeast. The vegetation of the Yucatan Peninsula changes gradually along an environmental gradient associated with rainfall distribution patterns that stretch from the semi–arid northwest to the wetter southeast. The changes in vegetation largely reflect rainfall patterns with dry low deciduous scrub and semi–deciduous forest dominating towards the northwest, and semi–wet and tropical semi–evergreen forest predominant towards the southeast (Pennington & Sarukhán, 1998).

Species and models of potential geographical distribution

Eight species of psittacines present in the Yucatan Peninsula were analysed: Olive–throated Parakeet (*Eupsittula nana),* White–fronted Amazon (*Amazona albifrons*), Yellow–lored Amazon (*Amazona xantholora*), Red–lored Amazon (*Amazona autumnalis*), White–crowned Parrot (*Pionus senilis*), Brown–hooded Parrot (*Pyrilia haematotis*), Yellow–headed Amazon (*Amazona oratrix*), and Southern Mealy Amazon (*Amazona farinosa*) (MacKinnon, 2005). Potential geographical distribution models obtained by Plasencia–Vázquez et al. (2014) were used for *A. xantholora* and *A. oratrix,* while potential geographical distribution models obtained by Plasencia–Vázquez & Escalona–Segura (2014b) were used for the remaining six species. The distribution models of the eight species were obtained using the MaxEnt program (Phillips et al., 2006), considered the most appropriate program when few records of species are available (Hernandez et al., 2006). The same methodology was used to obtain the models for all eight species.

Vegetation and land use

A land use and vegetation map from INEGI (National Institute for Statistics and Geography) Serie IV was used to describe the types of land use and vegetation present within the potential geographical distribution areas. These maps cover the Yucatan Peninsula and are the most up–to–date maps available. All vegetation types were taken into account: halophytic vegetation, hammocks, coastal dunes, tule vegetation, popal marshes, mangrove forest, halophytic pasture, induced pasture, tropical semi–evergreen medium–height forest (tree height 15–30 m, TSeMhF), tropical semi–deciduous medium–height forest (tree height 15–30 m, TSdMhF), tropical deciduous medium–height forest (tree height 15–30 m); tropical semi–deciduous low–height forest (tree height 4–15 m), tropical semi–evergreen low–height thorn forest (tree height 4–15 m), tropical deciduous low–height thorn forest (tree height 4–15 m), tropical deciduous low–height forest (tree height 4–15 m), tropical evergreen low–height forest (tree height 4–15 m), tropical semi–evergreen high–height forest (tree height + 30 m), tropical evergreen high–height forest (tree height + 30 m), riparian forest, savanna, oak forest, palm forest, and induced palm grove. The land use variables included: human settlements, urban zones, livestock raising (Liv), irrigated agriculture and rain–fed agriculture. Some types of agriculture could not be classified and were entitled N/C. Ground fresh water sources were also included.

Using the presence/absence map for each species, obtained from the potential geographical distribution models (Plasencia–Vázquez & Escalona–Segura, 2014b; Plasencia–Vázquez et al., 2014), 100 hexagonal plots, each with an approximate area of 29 km², were randomly selected. This area is consistent with the home range of similar psittacine species as presently there is no data on the home range of the studied species (Tamungang et al., 2001; White et al., 2005; Ortiz–Maciel et al., 2010). Plots with 50% or more of their area potentially occupied by the corresponding species were classified as presence and the remaining plots as absence. The numbers of plots classified as absence (A) or presence (P) for each species were the following: E. nana (14A, 86P), A. albifrons (11, 89), A. xantholora (22, 78), A. autumnalis (58, 42), P. haematotis (49, 51), P. senilis (42, 58), A. farinosa (57, 43), and A. oratrix (79, 21). The same procedure was followed for the distribution probability (DP) of each species, calculating the mean DP for each plot.

The same 100 plots that were used in the aforementioned procedures were delineated on the land use and vegetation maps. The area within each plot occupied by the different land uses, water bodies, and vegetation types was then calculated. All these calculations and procedures were carried out using the ArcView 3.2 program (ESRI, 1999).

Forest fragmentation in the Yucatan Peninsula

The land use and vegetation map from Series IV of the National Institute of Statistics and Geography in Mexico (INEGI, 2010) was also used to measure the levels of forest fragmentation within the potential geographical distribution areas of psittacines in the Yucatan Peninsula. Only forested areas (including mangrove forest, hammocks, tropical forest, and oak forest) that are potential habitat for the parrot species were selected; all other vegetation types, ground fresh water sources, and land uses were eliminated. All these forest areas were grouped together in order to obtain one map that represented the total forest mass in the Yucatan Peninsula. The ArcView 3.2 program (ESRI, 1999) was used for the above procedures.

Using the Patch Analyst extension in ArcView 3.2 (ESRI, 1999), a series of indexes were calculated for the plots defined for each species. These indexes were then superimposed onto the forest mass map. Indexes that describe fragmentation over the whole landscape were selected, as suggested by Fahrig (2003). The following variables were calculated: total forest area (TFA), number of patches, mean patch size, median patch size, patch size coefficient of variance, patch size standard deviation, total edge (TE), edge density, mean patch edge (MPE), mean shape index, area–weighted mean shape index, mean perimeter–area ratio (MPAR), mean patch fractal dimension, area–weighted mean patch fractal dimension, Shannon's diversity index, and Shannon's evenness index.

In addition, the distance between fragments (DF) and their spatial configuration were calculated. For distances, the shortest distance between the edge of one fragment to another and the distance between the centroids of all the fragments were calculated for each plot. The sum of the crossed multiples of the matrixes of distances between the fragment edges and their centroids was then calculated to obtain a distances summary variable. Moran's I Index for four neighbours (Moran, 1950) with the ROOKCASE for Excel 97/2000 complement (Sawada, 1999) was calculated using the matrix of distances between fragment edges in each plot in order to obtain a measure of the spatial distribution of fragments in the landscape.

Statistical analysis

To reduce the co–linearity between fragmentation, spatial configuration, land use, and vegetation type variables, Pearson correlations between all variables were calculated and those with coefficient values of $|r| < 0.7$ were selected (Dormann et al., 2012). Based on the natural history, habitat and knowledge obtained through field observations of these species (Fitzpatrick et al., 2013), one of variables that possibly presents a strong relationship with psittacine distribution patterns in the Yucatan Peninsula was selected from the pairs of strongly correlated variables. Land use and resultant vegetation variables that were represented in less than 10% of the 100 plots were eliminated.

A partial least squares regression (PLS) was performed to determine the possible relationship between the matrix of fragmentation, spatial configuration, land use and vegetation variables and the potential geographical distribution of the species of psittacines, based on the distribution probability (DP) and presence/absence (P/A). The PLS regression is a statistical method with great potential for ecological studies, and it is appropriate for analyses that endeavour to explain complex phenomena defined by the combination of a variety of explicative variables (Carrascal et al., 2009). The dependent presence/absence and distribution probabilities, with two and three categories respectively, were used as qualitative variables during the PLS regression analysis. The probability of presence values, obtained for each species in the study plots, were converted into three percentage intervals of equal width (high, medium and low probability).

The automatic mode was used as the stop condition and the third level interactions between the explicative variables were taken into account. The Q^2 index, which measures the total goodness of fit and the predictive quality of the models, was calculated.

Lost data were disregarded and the correlation matrix was considered to determine the type of relationship established between explicative and dependent variables. The variable importance in projection (VIP) was also considered. Regarding the PLS regression, after eliminating several variables that were highly correlated or represented in less than 10% of the plots, the number of explicative variables fluctuated between 13 and 15, depending on the specific species of psittacine. The XLSTAT (2014) complement for Excel was used for the above analyses.

Results

Urban zones, irrigated agriculture, and non–classified agriculture N/C were discarded from the total of six land use variables. Human settlements were only taken into account during the analysis of *A. albifrons* and *E. nana*. The vegetation variables that presented the highest percentages of representation in plots were tropical semi–deciduous medium–height forest (22–31%), tropical semi–evergreen medium–height forest (52–74%), and tropical semi–evergreen low–height thorn forest (35–52%). The mangrove forest vegetation type was used for the analysis with the species *A. albifrons*, *A. oratrix* and *P. senilis*. Tule vegetation was taken into account for the three aforementioned species and *E. nana*.

The number of components that were automatically selected by the PLS regression varied according to species and the P/A and DP data analysis. The following list presents psittacine species and the number of PLS regression components for the P/A and DP data presented in parenthesis: *A. albifrons* (7 P/A and 2 DP), *A. autumnalis* (4 and 6), *A. farinosa* (3 and 3), *E. nana* (4 and 2), *A. oratrix* (5 and 5), *A. xantholora* (4 and 5), *P. haematotis* (3 and 2), and *P. senilis* (4 and 3). The Q² index attained low values for the eight psittacine species, including the cumulated value for the total obtained components (table 1). The DP data presented lower Q² cumulated (Q²cum) values than the P/A data. *A. albifrons* presented the highest Q²cum.value using the P/A data (table 1).

In summary, with respect to the confusion matrix and the reclassification of the observations, the P/A data presented the highest percentage of observations of well–classified species (table 2). The percentages of well–classified observations coincided for both P/A and DP data only in the case of *A. Oratrix*.

For all eight species of parrot, the most important explicative variables for both P/A and DP data were third level interactions (table 3). Using P/A and DP data, the most frequent effect was the interaction between total forest area, total edge, and tropical semi–evergreen medium–height forest. This effect contributed to the models of six species. The most frequent variables in the third level interactions that had the greatest influence on the models were total edge, total forest area and distance between fragments (appendix 1). In the matrixes of the correlation coefficients between the different fragmentation variables and the species P/A and DP, it was possible to distinguish which independent variables had a positive or negative influence (tables 4, 5).

Table 1. Accumulated predictive quality (Q²cum) of the models performed with the presence/absence and distribution probability data of eight parrot species found in the Yucatan Peninsula, Mexico: P/A. Presence/Absence; DP. Distribution probability distribution.

Tabla 1. Calidad predictiva acumulada (Q²cum) de los modelos realizados con los datos de presencia y ausencia y de probabilidad de distribución de las ocho especies de loros presentes en la península de Yucatán, México: P/A. Presencia/Ausencia; Pd. Probabilidad de distribución.

Species	P/A Q²cum	DP Q²cum
Amazona albifrons	0.62	0.14
Amazona autumnalis	0.18	−0.03
Amazona farinosa	0.14	0.05
Eupsittula nana	0.49	0.08
Amazona oratrix	0.43	0.42
Amazona xantholora	0.09	0.14
Pyrilia haematotis	0.22	0.11
Pionus senilis	0.30	0.38

Table 2. Confusion matrix with the total percentage of presence/absence (P/A) observations and well–classified distribution probability (DP) for eight parrot species in the Yucatan Peninsula.

Tabla 2. Matriz de confusión con el porcentaje total de observaciones de presencia y ausencia (P/A) y de probabilidad de distribución (DP) bien clasificada, para las ocho especies de loros de la península de Yucatán.

Species	Correct P/A (%)	Correct DP (%)
A. albifrons	91	56
A. autumnalis	74	70
A. farinosa	55	46
E. nana	92	51
A. oratrix	80	80
A. xantholora	84	65
P. haematotis	65	60
P. senilis	79	77

Table 3. Variables with greater importance during the projection (VIP) for the components automatically generated during the PLS regressions made with the presence/absence and distribution probability data of eight parrot species found in the Yucatan Peninsula: TFA. Total forests area; TE. Total edge; MPAR. Mean perimeter–area ratio; MPE. Mean patch edge; DF. Distance among the fragments; TSeMhF. Tropical semi–evergreen medium–height forest; TSdMhF. Tropical semi–deciduous medium–height forest; Liv. Livestock raising.

Tabla 3. Variables con mayor importancia durante la estimación (VIP) de los componentes generados automáticamente durante las regresiones de mínimos cuadrados parciales realizadas con los datos de presencia y ausencia y de probabilidad de distribución de las ocho especies de loros presentes en la península de Yucatán: TFA. Superficie forestal total; TE. Longitud total de los perímetros de los fragmentos; MPAR. Media de la proporción entre la superficie y el perímetro; MPE. Media del perímetro de fragmento; DF. Distancia entre los fragmentos; TSeMhF. Bosque tropical subperennifolio de altura mediana; TSdMhF. Bosque tropical subcaducifolio de altura mediana; Liv. Ganadería.

Species	VIP Presence/Absence	VIP Distribution of probability
A. albifrons	TFA*TE*TSeMhF	TFA*TE*TSeMhF
A. albifrons	TFA*TE*TSdMhF	TFA*TE*DF
E. nana	TE*MPAR*DF	TE*MPAR*DF
E. nana		TFA*TE*TSeMhF
A. xantholora	TFA*TE*TSeMhF	TFA*TE*TSeMhF
A. xantholora		TE*MPAR*DF
A. autumnalis	TE*Liv*DF	TE*Liv*DF
A. autumnalis	TFA*TE*MPAR	TFA*TE*MPAR
A. farinosa	TE*Liv*DF	TFA*TE*TSdMhF
A. farinosa	TFA*TE*MPAR	TFA*TE*TSeMhF
A. oratrix	TE*MPAR*DF	TE*MPAR*DF
A. oratrix	TFA*TE*TSeMhF	TFA*TE*TSeMhF
P. haematotis	TE*Liv*DF	TE*Liv*DF
P. haematotis	TFA*TE*TSeMhF	TFA*TE*TSeMhF
P. senilis	TFA*MPE*TSeMhF	TFA*TE*MPE
P. senilis	TFA*MPE*TSdMhF	TE*Liv*DF
P. senilis	MPAR*Liv*DF	

The potential presence of *A. albifrons* was favoured in those areas where the landscape is dominated by tropical semi–evergreen medium–height forest and tropical semi–deciduous medium–height forest. The distribution probability of this parrot was lower in areas with small forest fragments that are separated by large distances. *Eupsittula nana* was present mainly in areas with large forest fragments that were close together and which tended to be compact and simple (few irregularities). If, in addition to these characteristics, the fragments were composed of tropical semi–evergreen medium–height forest, then the distribution probability of this species was even higher. *Amazona xantholora* was present in landscapes characterized by large fragments of tropical semi–evergreen medium–height forest, and its distribution probability decreased with an increase in fragment irregularity and distances between forest fragments.

Amazona autumnalis was potentially found in areas where forest fragments tended to be irregular, separated by larger distances, where most of the total landscape area is not forested and agricultural activity is low. *Pyrilia haematotis* was mainly found in areas with similar characteristics to those inhabited by *A. autumnalis*, although the probability of finding this species was higher in areas where tropical semi–evergreen medium–height forest fragments were in close proximity. Regarding *P. senilis*, its potential distribution was favoured in those areas where the landscape was dominated by proximate regular tropical semi–evergreen medium–height forest. This particular species was less abundant in sites with agricultural activity or where tropical semi–deciduous medium–height forest was the dominant vegetation.

A. farinosa preferred landscapes characterized by clustered, regular forest fragments where there was very little or zero agricultural activity. This distribution probability of this species was higher at sites dominated by tropical semi–evergreen medium–height forest, while at sites where tropical semi–deciduous medium–height forest dominated there was a medium probability of occurrence. *Amazona oratrix* was potentially found in areas

Table 4. Coefficient of correlation matrix between fragmentation variables integrated to third level interactions, and the presence of eight species of parrots living in the Yucatan Peninsula: 1. *A. albifrons*; 2. *A. autumnalis*; 3. *A. farinosa*; 4. *E. nana*; 5. *A. oratrix*; 6. *A. xantholora*; 7. *P. haematotis*; 8. *P. senilis*. (For abbreviations of variables see table 3.)

Tabla 4. Matriz con los coeficientes de correlación entre las variables de fragmentación integradas en interacciones de tercer orden y la presencia de las ocho especies de loros presentes en la península de Yucatán. (Para las abreviaturas de las especies ver arriba y para las de las variables ver tabla 3.)

Variables	Presence of species							
	1	2	3	4	5	6	7	8
TFA	0.15	−0.14	0.21	0.26	−0.47	0.36	−0.10	0.12
MPE	–	–	–	–	–	–	–	0.05
TE	0.16	0.17	0.20	0.13	−0.12	0.07	0.20	–
MPAR	–	0.02	−0.06	−0.16	0.24	−0.30	0.13	−0.19
Liv	−0.01	−0.04	−0.14	−0.32	0.16	−0.37	−0.01	−0.19
TSdMhF	0.03	−0.23	0.02	0.13	−0.21	0.02	−0.05	−0.29
TSeMhF	0.08	0.26	0.06	0.00	−0.29	0.33	0.01	0.44
DF	0.04	0.25	0.00	−0.45	0.29	−0.20	0.13	−0.17

where small forest fragments predominated, separated by larger distances. The distribution probability of this psittacine species increased in landscapes characterized by very irregular forest fragments and few areas of tropical semi–deciduous medium–height forest.

Discussion

There is no precise information on the distribution areas of psittacines in the Yucatan Peninsula, and the information in the literature varies considerably (*e.g.* Howell & Webb, 1995; Peterson & Chalif, 1998; Forshaw, 2006). Thus, the use of the potential geographical distribution models developed by Plasencia–Vázquez et al. (2014) and Plasencia–Vázquez & Escalona–Segura (2014b) was crucial. However, these models are only a hypothesis on environments similar to those where these parrot species have been observed and are probably between the limits of the fundamental and occupied niche (Peterson et al., 2011). Consequently, the results on the influence of different environmental factors on potential geographical distribution of psittacines are only an approximation and therefore may represent information that does not entirely equate with reality.

Although the distribution of species is determined by a variety of factors, it is always important to know which variables are responsible for most of the observed variation (Carrascal, 2004). In the Yucatan Peninsula, the eight species of parrot that remain in the region are generally associated with tropical forest and for the majority, the potential presence and distribution probability are associated with the most preserved areas characterized by little anthropogenic alteration, a finding which has been found for species of psittacines in other areas and regions of Mexico (Morales–Pérez, 2005). In general,

most parrot species are associated with forest areas as these provide food resources such as fruit, seeds, leaves, and flowers in addition to arboreal nesting sites (Morales–Pérez, 2005). Therefore, a reduction in forests would imply a loss of trophic resources and tree cavities for nesting that would lead to a decrease in psittacine populations (Rice, 1999; Berovides & Cañizares, 2004). Currently, tropical semi–evergreen medium–height forest covers the largest area within the Yucatan Peninsula and is the vegetation type that exerts most influence on the potential distribution of most psittacines in the region, coinciding with results obtained by Plasencia–Vázquez & Escalona–Segura (2014b).

In areas subject to forest loss and fragmentation, a decrease in reproductive success, due to an increase in egg predation and nest parasitism, has been observed at the edges of forest fragments (Willson et al., 2001). The structural characteristics of forest edges result in an increase in nest predation rates, so reproductive success rates increase towards the forest interior where vegetation is much denser (Hartley & Hunter, 1998). Furthermore, at the forest edge, parrot nests are more easily found and removed by poachers that participate in the illegal trafficking of parrot species. In the forests of the Calakmul Biosphere Reserve, in the state of Campeche, Mexico, larger frugivorous species are more likely to be found in areas with mature and senescent vegetation than in forests comprising younger successional stages of vegetation (Weterings et al., 2008). In general, these mature forests are found in well conserved natural areas or are secondary forests that have not been managed or disturbed over a long period of time, thus resembling mature primary forests (Guariguata et al., 1997). Most areas with younger stages of vegetation have been subject to recent anthropogenic modifications, such as *acahuales* (fallow agricultural land undergoing secondary

Table 5. Coefficient of correlation matrix between fragmentation variables integrated to third level interactions, and distribution probability of eight species of parrots living in the Yucatan Peninsula: DP. Distribution probability (H. High; M. Medium; L. Low). (For abbreviations of species see table 4 and for abbreviations of variables see table 3.)

Tabla 5. Matriz con los coeficientes de correlación entre las variables de fragmentación integradas en interacciones de tercer orden y la probabilidad de distribución de las ocho especies de loros presentes en la península de Yucatán: DP. Probabilidad de distribución (H. Alta; M. Media; L. Baja). (Para las abreviaturas de las especies ver tabla 4 y para las de las variables ver tabla 3.)

| Variables | DP | Species | | | | | | | |
		1	2	3	4	5	6	7	8
TFA	H	−0.06	−0.01	0.15	0.12	−0.37	0.10	0.07	0.17
	M	0.13	−0.15	0.05	0.15	−0.28	0.07	−0.08	0.02
	L	−0.10	0.15	−0.17	−0.25	0.49	−0.13	0.05	−0.14
TE	H	0.06	0.05	0.01	0.02	−0.27	0.10	0.00	−0.08
	M	0.23	0.15	0.19	0.06	0.14	−0.15	0.11	0.01
	L	−0.25	−0.17	−0.18	−0.08	0.06	0.08	−0.11	0.05
MPE	H	−	−	−	−	−	−	−	0.03
	M	−	−	−	−	−	−	−	0.05
	L	−	−	−	−	−	−	−	−0.06
MPAR	H	−	−0.04	−0.01	−0.11	0.09	−0.11	−0.05	−0.09
	M	−	0.03	0.03	0.08	0.17	−0.11	0.02	−0.06
	L	−	0.00	−0.02	0.01	−0.21	0.17	0.00	0.11
Liv	H	−0.11	−0.08	−0.17	−0.10	0.16	−0.17	−0.07	−0.13
	M	−0.19	−0.01	0.06	−0.22	0.07	−0.22	−0.04	−0.18
	L	0.24	0.05	0.08	0.31	−0.17	0.32	0.07	0.24
TSdMhF	H	−0.18	−0.04	−0.19	−0.19	−0.08	−0.10	−0.09	−0.14
	M	0.00	−0.23	0.30	0.01	−0.15	−0.16	−0.18	0.00
	L	0.10	0.24	−0.11	0.15	0.18	0.21	0.21	0.10
TSeMhF	H	0.23	0.09	0.17	0.21	−0.21	0.21	−0.10	0.16
	M	0.05	0.22	−0.12	0.03	−0.25	0.22	0.17	0.10
	L	−0.17	−0.27	−0.03	−0.21	0.35	−0.34	−0.13	−0.19
DF	H	0.25	−0.01	−	−0.09	0.08	−0.09	−0.05	−0.08
	M	−0.08	0.24	−	−0.17	0.30	−0.12	−0.01	−0.09
	L	−0.05	−0.23	−	0.25	−0.31	0.17	0.03	0.13

succession, initially characterized by tall herbaceous species), and have been undergoing regeneration for only a short period of time. Younger early stage vegetation may not provide the conditions required by psittacines for their development; there is a marked absence of large older trees with cavities that are required by the parrots for nesting and large tree species that provide the fruit and seeds that parrot species feed on are absent or have not grown sufficiently. Therefore, in certain areas, the presence of parrot species could be determined by the amount of time that *acahuales* have been regenerating.

This study showed that in the Yucatan Peninsula, the effect of habitat loss and fragmentation depends on the species of parrot and varies according to the sensitivity of each species to these changes, their habitat requirements and biology (Gurrutxaga, 2006). Several of the studied species are capable of exploiting some of the resources present in the landscape matrix surrounding the forest fragments and even degraded environments, such as agroecosystems, can support viable populations of certain psittacine species (Romero–Balderas et al., 2006). The effectivity of the matrix as a habitat depends on the interaction between its structural characteristics and the ecological requirements of the species (Antogiovanni & Metzger, 2005). The lesser the structural contrast between the matrix and the native habitat of the species, the more favourable the conditions for providing suitable habitat (Antogiovanni & Metzger, 2005).

Although all the species analysed in this study are found in areas that have been modified to a greater or lesser extent by different factors, the effect on their potential geographical distribution varies considerably. Widely–distributed species such as *A. albifrons*, *A. xantholora* and *E. nana*, which present extensive areas of distribution within the Yucatan Peninsula, were not expected to be affected by fragmentation and would be found even within agricultural and urban matrixes. However, this study showed that these species display a greater preference for more conserved sites

Many species of parrot occupy agricultural areas simply because they provide large and easily accessible concentrations of food, spending little time and energy on foraging. In some areas, these anthropogenic resources are exploited to such an extent that parrots are considered agricultural pests (Bucher, 1992). Nevertheless, the fact that these species take advantage of these modified areas does not mean they always prefer them to natural areas. Due to the destruction of their natural habitat, reproduction and feeding areas, combined with the illegal capture and trafficking for the pet trade, wild parrot species currently experience relentless human pressure, leading to a reduction in their areas of distribution and forcing them to occupy new habitats which generally lack the appropriate ecological characteristics for their successful development.

For *A. farinosa* and *P. senilis*, the present study showed that most of their potential distribution areas are still found within well–preserved areas, and their distribution probability is greater in less fragmented sites, as observed for most psittacines in Mexico (Morales–Pérez, 2005). Contrary to their habitat characteristics described in the literature (*e.g.* Howell & Webb, 1995; Peterson & Chalif, 1998; Forshaw, 2006), *P. haematotis* and *A. autumnalis* were potentially found in more fragmented areas. Given its sensitivity to habitat modifications, we expected to find *P. haematotis* within extensive areas of conserved forest with low levels of fragmentation. However, it was evident that both species avoided areas where agricultural activities take place, thus supporting the view that these particular species are sensitive to the effects of human activities (Ríos–Muñoz & Navarro–Sigüenza, 2009). The present potential presence areas of these species are a consequence of years of intense modifications to their natural habitats. Some species have adapted better than others to these changes and still persist in modified areas or are simply concentrated in remaining forest fragments. As these results are based on potential species distributions only, empirical corroborations in the field are essential.

Amazona oratrix is in a critical situation as its potential distribution is in the southwest of the Yucatan Peninsula, an area that has suffered a high level of modification due to extensive cattle ranching (Villalobos–Zapata et al., 2010). According to the obtained results, this species is found in very fragmented areas and its probability of occurrence is greater at sites with a high degree of anthropization, contradicting the existing literature on the ecology of this parrot species (Enkerlin–Hoeflich, 2000). Macías–Caballero & Iñigo–Elías (2003) established that one of the most abundant populations of *A. oratrix* in Mexico was found in this region of the Yucatan Peninsula and this has been confirmed by previous visits to the area (Plasencia–Vázquez & Escalona–Segura, 2014a).

It is interesting that *A. oratrix*, characterized by a low reproductive rate and a preference for high and medium tropical forest, is present in large numbers within an area of significant human activity, making it highly vulnerable to illegal trafficking and deforestation (Enkerlin–Hoeflich, 2000). The historical component was not taken into account during this study and may be a determining factor in explaining the observed patterns. Very little is known about this species in the Yucatan Peninsula and there are no data on its abundance prior to the deforestation of most of the forest that dominated the distribution area of this species. Many parrots have high longevity (Munshi–South & Wilkinson, 2006), thus many of the individuals observed in this region of the Yucatan Peninsula are adults that have managed to survive in a suboptimum environment. Furthermore, adult parrots normally remain in the forest canopy, making them difficult to capture, while parrot chicks suffer the highest capture rates by poachers as they are more easily found and removed from their nests. Therefore, despite low reproduction rates and high predation, large numbers of *A. oratrix* individuals continue to be observed (Plasencia–Vázquez & Escalona–Segura, 2014a). Unfortunately, in the near future, the *A. oratrix* population will be composed of mainly old individuals, with very low productivity and on the verge of a potentially catastrophic population crash.

In general, the present study revealed that the existing potential geographical distribution of psittacine species in the Yucatan Peninsula is determined by the interaction between variables that represent forest fragmentation, vegetation cover, and land uses. At present, the areas occupied by the parrot species are determined by the combined effect of natural and anthropogenic factors. As forest perturbation in the Yucatan Peninsula increases, anthropogenic factors become increasingly determinative in psittacine distribution.

At the scale of this study, many human activities carried out within the Yucatan Peninsula do not occupy extensive areas and therefore do not appear to affect psittacine populations. Smaller scale studies are required to assess how and to what extent activities such as agriculture or the expansion of urban areas and human settlements impact parrot populations. The results presented in this study on parrot species present in the Yucatan Peninsula are just an approximation of reality, analysed under determined conditions. There is therefore a need for future research conducted at a smaller, more local scale and including other environmental variables.

Acknowledgments

We are grateful to CONACyT (National Commission for Science and Technology) for the financial support provided through grants 239499 and 21467 and to INEGI (National Institute for Geography and Statistics, Mexico) for providing the Land Use and Vegetation map. We would like to thank Y. Ferrer–Sánchez, K. Renton and D. Denis–Ávila for their valuable comments and Idea Wild for the donated equipment. We appreciate the suggestions and comments provided by the reviewers.

References

Antogiovanni, M. & Metzger, J. P., 2005. Influence of matrix habitat on the occurrence of insectivorous bird species in Amazonian forest fragments. *Biological Conservation*, 122: 441–451.

Barrera, A., 1962. La Península de Yucatán como provincia biótica. *Revista de la Sociedad Mexicana de Historia Natural*, 23: 71–150.

Berovides, V. & Cañizares, M., 2004. Diagnóstico del decline de los psitácidos cubanos y su posible solución. *Biología*, 18(2): 109–112.

Bregman, T., Sekercioglu, P. C. H. & Tobias, J. A., 2014. Global patterns and predictors of bird species responses to forest fragmentation: Implications for ecosystem function and conservation. *Biological Conservation*, 169: 372–383.

Brotons, L., Mönkkönen, M. & Martin, J. L., 2002. Are fragments islands? Landscape context and density–area relationships in boreal forest birds. *The American Naturalist*, 162: 343–357.

Bucher, E., 1992. Neotropical parrots as agricultural pests. In: *New World parrots in crisis. Solutions from conservation biology*: 201–219 (S. R. Beissinger & N. F. R. Snyder, Eds.). Smithsonian Inst. Press, New York.

Carrascal, L. M., 2004. Distribución y abundancia de las aves en la Península Ibérica. Una aproximación biogeográfica y macroecológica. In: *La Ornitología hoy. Homenaje al Profesor Francisco Bernis*: 155–189 (J. L. Tellería, Ed.). Editorial Complutense, Universidad Complutense de Madrid, Madrid.

Carrascal, L. M., Galván, I. & Gordo, O., 2009. Partial least squares regression as an alternative to current regression methods used in ecology. *Oikos*, 118: 681–690.

Céspedes–Flores, S. E. & Moreno–Sánchez, E., 2010. Estimación del valor de la pérdida de recurso forestal y su relación con la reforestación en las entidades federativas de México. *Investigación ambiental*, 2(2): 5–13.

Donovan, T. M. & Lamberson, R. H., 2001. Area–sensitive distributions counteract negative effects of habitat fragmentation on breeding birds. *Ecology*, 82(4): 1170–1179.

Dormann, C. F., Elith, J., Bacher, S. Buchmann, C., Carl, G,. Carré, G., García–Marquéz, J. R., Gruber, B., Lafourcade, B., Leitão, P. J., Münkemüller, T., McClean, C., Osborne, P. E., Reineking, B., Schröder, B., Skidmore, A. K., Zurell, D. & Lautenbach, S., 2012. Collinearity: a review of methods to deal with it and a simulation study evaluating their performance. *Ecography*, 35: 1–20.

Enkerlin–Hoeflich, E. C., 2000. Loro cabeza amarilla. In: *Las Aves de México en Peligro de Extinción*: 230–234 (G. Ceballos & L. Marquez Valdelamar, Eds.). CONABIO/UNAM, México D.F.

ESRI, 1999. ArcView 3.2. ESRI (Environmental Scientific Research Institute). Redlands, California, USA.

Evans, B. E. I., Ashley, J. & Marsden, S. J., 2005. Abundance, habitat use, and movements of blue–winged macaws (*Primolius maracana*) and other parrots in and around an Atlantic forest reserve. *The Wilson Bulletin*, 117(2): 154–164.

Fahrig, L., 2003. Effects of habitat fragmentation on biodiversity. *Annual Review of Ecology, Evolution and Systematics*, 34: 487–515.

Fitzpatrick, M. C., Gotelli, N. J. & Ellison, A. M., 2013. MaxEnt versus MaxLike: empirical comparisons with ant species distributions. *Ecosphere*, 4(5): 55. Doi: http://dx.doi.org/10.1890/ES13–00066.1

Forshaw, J. M., 2006. *Parrots of the World: An Identification Guide*. Princeton University Press, Princeton, New Jersey, USA.

Guariguata, M., Chazdon, R. Denslow, J. & Dupuy, J., 1997. Structure and floristics of secondary and old–growth forest stands in lowland Costa Rica. *Plant Ecology*, 132: 107–120.

Gurrutxaga, M., 2006. Efectos de la fragmentación de hábitats y pérdida de conectividad ecológica dentro de la dinámica territorial. *Polígonos*, 16: 35–54.

Hanski, I., Zurita, G. A. Bellocq, M. I. & Rybicki, J., 2013. Species–fragmented area relationship. *Proceedings of the National Academy of Sciences*, 110(31): 12715–12720.

Hartley, M. J. & Hunter, M. L. Jr., 1998. A meta–analysis of forest cover, edge effects and artificial nest predation. *Conservation Biology*, 12: 465–469.

Hernandez, P. A., Graham, C. H. Master, L. L. & Albert, D. L., 2006. The effect of sample size and species characteristics on performance of different species distribution modeling methods. *Ecography*, 29: 773–785.

Howell, S. N. G. & Webb, S., 1995. *A guide to the birds of Mexico and northern Central America*. Oxford Univ. Press, London, UK.

INEGI, 2010. *Conjunto de datos vectoriales de la carta de uso del suelo y vegetación: escala 1:250 000. Serie IV (continuo nacional). Aguascalientes, México*. Instituto Nacional de Estadística y Geografía.

Lugo–Hubp, J., Acevedes–Quesada, J. F. & Espinosa–Pereña, R., 1992 Rasgos geomorfológicos mayores de la Península de Yucatán. *Revista Mexicana de Ciencias Geológicas*, 10: 143–150.

Macías–Caballero, C. M. & Iñigo–Elías, E. E., 2003. *Evaluación del estado de conservación actual de las poblaciones del loro cabeza amarilla (Amazona oratrix) en México*. México D.F., Comisión Nacional para el Conocimiento y Uso de la Biodiversidad.

MacKinnon, H. B., 2005. *Birds of Yucatan Peninsula*. Amigos de Sian Ka'an, AC. Cancún, Quintana Roo, México.

Marín–Togo, M. C., Monterrubio–Rico, T. C., Renton, K., Rubio–Rocha, Y., Macías–Caballero, C., Ortega–Rodríguez, J. M. & Cancino–Murillo, R., 2012. Reduced current distribution of Psittacidae on the Mexican Pacific coast: potential impacts of habitat loss and capture for trade. *Biodiversity and Conservation*, 21: 451–473.

Monterrubio–Rico, T. C., Renton, K., Ortega–Rodríguez, J. M., Pérez–Arteaga, A. & Cancino–Murillo, R., 2010. The Endangered yellow–headed parrot *Amazona oratrix* along the Pacific coast of Mexico. *Oryx*, 44: 602–609.

Monterrubio–Rico, T. C., De Labra Hernández, M. A., Ortega–Rodríguez, J. M., Cancino–Murillo, R. & Villaseñor–Gómez, J. F., 2011. Distribución actual y potencial de la guacamaya verde en Michoacán, México. *Revista Mexicana de Biodiversidad*, 82: 1311–1319.

Morales–Pérez, L., 2005. Evaluación de la abundancia poblacional y recursos alimenticios para tres géneros de psitácidos en hábitats conservados y perturbados de la costa de Jalisco, México. Tesis de maestría, Posgrado en Ciencias Biológicas, Universidad Nacional Autónoma de México, México, D.F.

Moran, P. A. P., 1950. Notes on continuous stochastic phenomena. *Biometrika*, 37: 17–23.

Munshi–South, J. & Wilkinson, G. S., 2006. Diet influences life span in parrots (Psittaciformes). *The Auk*, 123(1): 108–118.

Norma Oficial Mexicana, 2010. *NOM–059–SEMARNAT–2010. Protección ambiental–Especies nativas de México de flora y fauna silvestres–Categorías de riesgo y especificaciones para su inclusión, exclusión o cambio–Lista de especies en riesgo.* México, D.F., Diario Oficial de la Federación.

Ortiz–Maciel, S. G., Hori–Ochoa, C. & Enkerlin–Hoeflich, E., 2010. Maroon–Fronted Parrot (*Rhynchopsitta terrisi*) breeding home range and habitat selection in the Northern Sierra Madre Oriental, Mexico. *The Wilson Journal of Ornithology*, 122(3): 513–517.

Pennington, T. D. & Sarukhán, J., 1998. *Árboles tropicales de México. Manual para la identificación de las principales especies.* 2a. ed. Universidad Nacional Autónoma de México–Fondo de Cultura Económica, México.

Peterson, R. T. & Chalif, E. L., 1998. *Aves de México: guía de campo.* Diana, México, D.F.

Peterson, A. T., Soberón, J., Pearson, R. G., Anderson, R. P., Martínez–Meyer, E., Nakamura, M. & Araujo, M. B., 2011. *Ecological niches and geographic distributions.* Princeton University Press, Princeton.

Phillips, S. J., Anderson, R. P. & Schapire, R. E., 2006. Maximum entropy modeling of species geographic distributions. *Ecological Modelling*, 190: 231–259.

Plasencia–Vázquez, A. H., Escalona–Segura, G. & Esparza–Olguín, L. G., 2014. Modelación de la distribución geográfica potencial de dos especies de psitácidos neotropicales utilizando variables climáticas y topográficas. *Acta Zoológica Mexicana (nueva serie)*, 30(3): 471–490.

Plasencia–Vázquez, A. H. & Escalona–Segura, G., 2014a. Relative abundance of parrots throughout the Yucatan Peninsula: implications for their conservation. *The Wilson Journal of Ornithology*, 126(4): 759–766.

– 2014b. Caracterización del área de distribución geográfica potencial de las especies de aves psitácidas de la Península de Yucatán, México. *Biología Tropical*, 62(4): 1509–1522.

Raman, T. R. S., 2004. *Effects of landscape matrix and plantations on birds in tropical rainforest fragments of the Western Ghats, India.* CERC Technical Report No. 9. Nature Conservation Foundation, Mysore, India.

Reino, L., Beja, P., Osborne, P. E., Morgado, R., Fabião, A. & Rotenberry, J. T., 2009. Distance to edges, edge contrast and landscape fragmentation: interactions affecting farmland birds around forest plantations. *Biological Conservation*, 142(4), 824–838.

Rey Benayas, J. M., Bullock, J. M. & Newton, A., 2008. Creating woodland islets to reconcile ecological restoration, conservation, and agricultural land use. *Frontiers in Ecology and the Environment*, 6. Doi: 10.1890/070057.

Rice, J., 1999. Reintroduction of parrots in the Greater Antilles and South America. *Restoration and Reclamation Review*, 4(4): 1–6. http://conservancy. umn.edu/bitstream/59302/1/4.4.Rice.pdf

Ríos–Muñoz, C. A. & Navarro–Sigüenza, A. G., 2009. Efectos del cambio de uso de suelo en la disponibilidad hipotética de hábitat para los psitácidos de México. *Ornitología Neotropical*, 20: 491–509.

Romero–Balderas, K. G, Naranjo, E. J., Morales, H. H. & Nigh, R. B., 2006. Daños ocasionados por vertebrados silvestres al cultivo de maíz en la selva lacandona, Chiapas, México. *Interciencia*, 31(4): 276–283. http://www.scielo.org.ve/scielo.php?script=sci_ arttext&pid=S0378–18442006000400007&lng=es

Sawada, M., 1999. Technological tools. Rookcase: an Excel 97/2000 visual basic (VB) add–in for exploring global and local spatial autocorrelation. *Bulletin of the Ecological Society of America*, 80: 231–234.

Tamungang, S. A., Ayodele, I. A. & Akum, Z. E., 2001. Basic home range for the conservation of the African Grey Parrot in Korup National Park, Cameroon. *Journal of Cameroon Academy of Science*, 1(3): 155–158.

Villalobos–Zapata, G. J. & Mendoza Vega, J., 2010. *La Biodiversidad en Campeche: Estudio de Estado. Comisión Nacional para el Conocimiento y Uso de la Biodiversidad.* (CONABIO), Gobierno del Estado de Campeche, Universidad Autónoma de Campeche, El Colegio de la Frontera Sur, México.

Waltert, M., Mardiastuti, A. & Mühlenberg, M., 2005. Effects of deforestation and forest modification on understorey birds in Central Sulawesi, Indonesia. *Bird Conservation International*, 15: 257–273.

Weterings, M. J. A., Weterings–Schonck, S. M., Vester, H. F. M. & Calmé, S., 2008. Senescence of *Manilkara zapota* trees and implications for large frugivorous birds in the Sothern Yucatan Peninsula, Mexico. *Forest Ecology and Management*, 256: 1604–1611.

White, T. H. Jr., Collazo, J. A., Vilella, F. J. & Guerrero, S. A., 2005. Effects of hurricane George on habitat use by captive–reared Hispaniolan parrots (*Amazona ventralis*) released in the Dominican Republic. *Neotropical Ornithology*, 16, 405–417.

Willson, M. F., Morrison, J., Sieving, K. E., De Santo, T. L., Santisteban, L. & Díaz, I., 2001. Patterns of predation risk and survival of bird nests in a chilean agricultural landscape. *Conservation Biology*, 15: 447–456.

XLSTAT, 2014. Software, Version 2014.1. Copyright Addinsoft 1995–2014. http://www.xlstat.com/es/

Appendix 1. Descriptive statistics of the variables comprising the third level interactions that had greatest influence on presence/absence and distribution probability of psittacine species in the Yucatan Peninsula (mean ± standard error, minimum–maximum, sample size in brackets: TFA. Total forest area; TE. Total edge; MPE. Mean patch edge; MPAR. Mean perimeter–area ratio; Liv. Livestock raising; TSdMhF. Tropical semi–deciduous medium–height forest; TSeMhF. Tropical semi–evergreen medium–height forest; DF. Distance between fragments.

	Species			
Variables	A. albifrons	A. autumnalis	A. farinosa	E. nana
TFA (km²)	19.89 ± 0.90	19.11 ± 0.86	20.37 ± 0.81	18.26 ± 0.91
	0–29.40	0–29.33	0–29.33	0–29.27
	(100)	(100)	(100)	(100)
TE (km)	25.97 ± 0.96	27.95 ± 1.01	28.40 ± 1.02	27.11 ± 1.09
	0–59.70	0–53.60	0–56.53	0–53.74
	(100)	(100)	(100)	(100)
MPE (km)	19.48 ± 0.90	16.96 ± 0.91	17.70 ± 0.89	17.08 ± 0.94
	3.29–41.23	2.42–43.24	2.42–43.24	2.42–43.24
	(98)	(99)	(99)	(98)
MPAR (km)	6.58 ± 1.21	9.14 ± 1.96	10.98 ± 2.39	9.28 ± 1.84
	0.69–62.57	0.69–125.25	0.69–130.47	0.69–125.25
	(97)	(94)	(94)	(94)
Liv (km²)	6.16 ± 1.01	8.59 ±1.20	7.94 ±1.20	9.26 ± 1.21
	0.01–27.67	0.01–28.97	0.01–28.97	0.01–28.97
	(52)	(50)	(50)	(52)
TSdMhF (km²)	21.44 ± 1.48	18.50 ± 1.92	18.23 ± 1.80	17.66 ± 1.87
	1.91–29.04	0.003–29.21	0.18–29.21	0.003–29.21
	(32)	(28)	(27)	(30)
TSeMhF (km²)	17.81 ± 1.33	17.33 ± 1.07	18.16 ± 1.07	16.31 ± 1.22
	0.01–29.21	0.70–29.33	0.67–29.33	0.56–29.22
	(55)	(69)	(66)	(62)
DF (km)	8.34 ± 1.98	22.22 ± 6.22	18.29 ± 5.80	19.26 ± 4.69
	0–109.09	0–466.80	0–466.80	0–295.18
	(98)	(99)	99	(98)

Apéndice 1. Estadísticos descriptivos de las variables integradas en interacciones del tercer orden que más influyeron en la presencia y la ausencia y en la probabilidad de distribución de las especies de loros de la península de Yucatán (media ± error estándar, mínimo–máximo, tamaño de muestra entre paréntesis). TFA. Superficie forestal total; TE. Longitud total de los perímetros de los fragmentos; MPE. Media del perímetro de los fragmentos; MPAR. Media de la proporción entre la superficie y el perímetro; Liv. Ganadería; TSdMhF. Bosque tropical subcaducifolio de altura mediana; TSeMhF. Bosque tropical subperennifolio de altura mediana; DF. Distancia entre los fragmentos.

	Species		
A. oratrix	*A. xantholora*	*P. haematotis*	*P. senilis*
17.08 ± 0.98	19.09 ± 0.89	19.26 ± 0.87	20.04 ± 0.81
0–29.27	0–29.27	0–29.33	0–29.33
(100)	(100)	(100)	(100)
25.49 ± 1.17	26.80 ± 1.04	27.68 ± 1.00	28.28 ± 0.98
0–53.60	0–53.60	0–53.60	0–53.60
(100)	(100)	(100)	(100)
15.66 ± 0.94	17.86 ± 0.94	16.98 ± 0.90	17.71 ± 0.93
1.42–43.24	2.42–43.24	2.42–43.24	2.78–48.24
(98)	(98)	(99)	(99)
11.73 ± 2.36	8.76 ± 1.82	9.17 ± 1.96	8.50 ± 1.96
0.69–133.12	0.69–125.25	0.69–125.25	0.69–125.25
(94)	(95)	(94)	(93)
8.65 ± 1.11	8.63 ± 1.21	8.55 ± 1.21	7.45 ± 1.16
0.01–28.97	0.01–28.97	0.01–28.97	0.01–28.97
(56)	(51)	(50)	(48)
17.65 ± 1.91	17.98 ± 1.84	18.50 ± 1.92	17.94 ± 2.16
0.003–29.21	0.003–9.21	0.003–29.21	0.003–29.21
(27)	(31)	(28)	(24)
15.92 ± 1.32	17.69 ± 1.17	17.05 ± 1.11	17.98 ± 1.02
0–29.22	0.70–29.22	0.67–29.33	0.67–29.33
(58)	(61)	(68)	(74)
20.46 ± 6.13	13.87 ± 3.75	22.54 ± 6.22	19.88 ± 5.92
0–487.26	0–295.18	0–466.80	0–466.80
(98)	(98)	(99)	(99)

Bird mortality related to collisions with ski–lift cables: do we estimate just the tip of the iceberg?

N. Bech, S. Beltran, J. Boissier, J. F. Allienne, J. Resseguier & C. Novoa

Bech, N., Beltran, S., Boissier, J., Allienne, J. F., Resseguier, J. & Novoa, C., 2012. Bird mortality related to collisions with ski–lift cables: do we estimate just the tip of the iceberg? *Animal Biodiversity and Conservation*, 35.1: 95–98.

Abstract

Bird mortality related to collisions with ski–lift cables: do we estimate just the tip of the iceberg?— Collisions with ski–lift cables are an important cause of death for grouse species living close to alpine ski resorts. As several biases may reduce the detection probability of bird carcasses, the mortality rates related to these collisions are generally underestimated. The possibility that injured birds may continue flying for some distance after striking cables represents a major source of error, known as crippling bias. Estimating the crippling losses resulting from birds dying far from the ski–lift corridors is difficult and it is usually assessed by systematic searches of carcasses on both sides of the ski–lifts. Using molecular tracking, we were able to demonstrate that a rock ptarmigan hen flew up to 600 m after striking a ski–lift cable, a distance preventing its detection by traditional carcasses surveys. Given the difficulty in conducting systematic searches over large areas surrounding the ski–lifts, only an experiment using radio–tagged birds would allow us to estimate the real mortality rate associated with cable collision.

Key words: Bird collision, Crippling bias, Forensic approach, Human infrastructure, Rock ptarmigan, Ski–lift wires.

Resumen

Mortalidad de aves causada por colisión con los cables de los remontes de las pistas de esquí: ¿sólo vemos la punta del iceberg?— Las colisiones con los cables de los remontes son una importante causa de mortalidad para las diversas especies de galliformes que habitan en las inmediaciones de las estaciones de esquí alpino. Las tasas de mortalidad asociadas a esta causa resultan frecuentemente subestimadas como consecuencia de los diversos factores que pueden reducir la probabilidad de detección de los cadáveres. Así, por ejemplo, el posible desplazamiento de las aves heridas después de la colisión representa una importante causa de error llamada sesgo por mutilación (crippling bias). La estima de las pérdidas resultantes correspondientes a aves que mueren lejos de la vertical de los remontes es un aspecto de difícil cuantificación que normalmente se evalúa mediante búsquedas sistematizadas de restos a ambos lados de estas estructuras lineales. Mediante el uso de técnicas de rastreo molecular hemos sido capaces de detectar que una hembra de lagópodo alpino voló unos 600 m después de colisionar con un cable remontador, distancia a la cual el ave no hubiese sido detectada mediante los métodos de muestreo tradicionales. Debido a la dificultad de llevar a cabo búsquedas sistemáticas sobre superficies extensas alrededor de los remontes, tan solo el radio seguimiento de las aves permitiría una estima consistente de la tasa real de mortalidad asociada a la colisión con cables.

Palabras clave: Colisiones de aves, Sesgo por mutilación, Enfoque forense, Infraestructuras humanas, Lagópodo alpino, Cable remontador.

Nicolas Bech, Sophie Beltran, Univ. of Perpignan, Via Domitia, CNRS, UMR 5244, Evolutionary and Ecology of Interactions (2EI), Perpignan, F–66860, France. Present address: Equipe Ecologie, Evolution Symbiose, Lab. EBI Ecologie & Biologie des Interactions, Univ. de Poitiers–UFR Sciences Fondamentales et Appliquées, UMR CNRS 7267, Bât. B8, 40 avenue du Recteur Pineau, F–86022 Poitiers Cedex, France. – Jérôme Boissier, Jean François Allienne, Univ. of Perpignan, Via Domitia, CNRS, UMR 5244, Evolutionary and Ecology of Interactions (2EI), Perpignan, F–66860, France.– Jean Resseguier & Claude Novoa, ONCFS, Dept of Studies and Research, F–66500, Prades, France.

Corresponding author: Nicolas Bech: nicolas.bech@univ-poitiers.fr

Introduction

The skiing activities in recent decades have had a profound impact on the alpine landscapes of many European mountains. In the French Alps and the Pyrenees, a recent survey conducted by the Mountain Game Observatory counted a total of 252 ski resorts with 3,117 ski or chair lifts for a total length of some 2,575 km (Observatoire des Galliformes de Montagne, 2006). Several studies have reported adverse effects derived from the development of winter tourism on alpine wildlife and particularly on grouse species (Miquet, 1990; Menoni & Magnani, 1998; Martin, 2001; Watson & Moss, 2004; Arlettaz et al., 2007; Thiel et al., 2008; Patthey et al., 2008). Among these negative effects, the mortality of birds colliding with cables has been considered as one of the main causes of the decline of grouse abundance close to ski resorts. Indeed, grouse species seem particularly vulnerable to strike–wire mortality (Bevanger, 1995) and a recent survey of 252 French ski resorts confirmed this point (Observatoire des Galliformes de Montagne, 2006; Buffet & Dumont–Dayot, in press). Generally, mortality resulting from human infrastructures such as ski–lift cables, power lines or fences is estimated by counting animals found dead underneath or close to the human infrastructures (Bevanger & Brøseth, 2001; Barrios & Rodriguez, 2004). However, some injured birds may continue moving for some distance after striking and death can occur several hundred meters further away. This bias, known as the crippling bias, refers to animals which can be found dead far from the infrastructures and therefore not counted in estimates of mortality rate related to the structures. Widening the searching areas on both sides of ski–lift corridors would clearly help to produce more realistic estimates of collision mortality rates (Bevanger, 1999), but the width of the searching zone is not easy to define. In this paper, we report on a case of rock ptarmigan (*Lagopus muta muta*) mortality related to a ski–lift collision at a Pyrenean ski resort. Using molecular tracking, we demonstrated that this bird travelled several hundred meters after striking a ski–lift. Finally, we discuss the difficulty in estimating the different biases associated with cable collision.

Material and methods

The study was carried out in the Err–Puigmal ski resort located in the eastern French Pyrenees (N 42° 23' – E 2° 05'). This small ski resort ranges from 1,830 m to 2,700 m above sea level and includes two chair–lifts and seven ski–lifts. Five cases of rock ptarmigan mortality related to ski–lift collisions have been observed following the building in 2005 of a new ski–lift, called The Montserrat. On 10 VI 2008, we found a severely injured female rock ptarmigan at about 600 m from the Montserrat ski–lift (fig. 1). The bird died a few minutes after capture and a post–mortem examination revealed that its right wing and right leg were broken. During a survey carried out three days later under the ski–lift, we found some rock ptarmigan feathers just under the cables (fig. 1). Genomic DNA of both

samples was extracted from the tip (~3 mm) of feather samples and from muscle tissue in the case of the corpse, using silica columns (e.Z.N.A Kit of OMEGA BIO–TEK) and following the manufacturer's protocol. Both samples were tested with 14 microsatellite markers. The relevant DNA fragments were amplified using Polymerase Chain Reaction (PCR). To maximise efficiency and minimize costs, these PCRs were performed in four multiplexes using the QIAGEN multiplex kit following the manufacturer's protocol. Thus, PCR were performed in a 10 μL volume and contained 1μL of extraction product, 1x of 'Qiagen Multiplex PCR Master Mix' and 0.2 μM of primers. Cycling protocol contained an initial denaturation of 15 min at 95°C followed by 30 cycles of 30s at 94°C, 90s at the annealing temperature of 57°C (for all markers) and 1 min at 72°C, followed by a final extension of 30 min at 60°C. PCR products were electrophoresed on an automatic sequencer (CEQ™ 8000, Beckman Coulter) and genotypes were determined using the fragment analyzer package from Beckman Coulter. We genotyped the two samples six times in order to reach a genotype consensus and to avoid genotype errors due to false or null alleles for the samples of low quality, *i.e.* feathers (Valière et al., 2007).

Deviation from Hardy–Weinberg expectancies and linkage disequilibrium were tested using the global tests in FSTATv.2.9.3.2 (Goudet, 2001). The level of significance was adjusted for multiple testing using Bonferroni correction. Furthermore, polymorphism was estimated over all loci using the allelic richness (AR), expected heterozygosity (H_e), and F_{IS} computed with FSTAT v.2.9.3.2. This was performed with 34 supplementary DNA samples. These 34 supplementary DNA samples were isolated from moult feather or muscle removed from birds found dead during census or radio tracking. These samples, found in a radius 60 km around our ski–lift, had already been used in a genetic study of a metapopulation of the species focusing on the Pyrenees (Bech et al., 2009). This study indicated that the samples belonged to a single population (named 'Carança–Puigmal') (Bech et al., 2009). In order to check the power of our markers, we computed the probability of identity (PI) index corrected for small samples of individuals (PI unbiased) using GIMLET v.1.3.2 software (Valière, 2002). This index indicates the probability that two unrelated individuals in a population share a multilocus genotype. Very low PI values (PI values of $2.0 \cdot 10^{-5}$ are considered as low (Paetkau & Strobeck, 1994)) suggest high power to distinguish between two individuals. We then matched the genotypes obtained from the samples to check whether they belonged to the same individual.

Results

After Bonferroni correction we did not detect any deviation from Hardy–Weinberg expectations or any linkage disequilibrium. Our microsatellite panel revealed a high polymorphism degree characterised by a mean allelic richness (AR) of 5.457, a mean expected

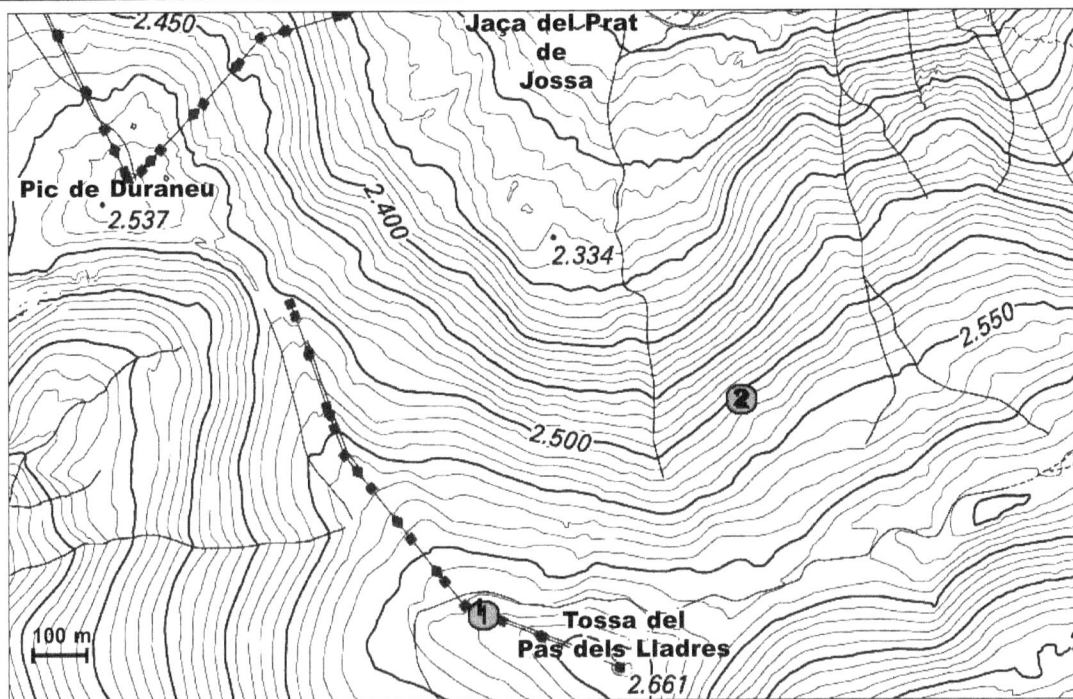

Fig. 1. Topographic map of the upper part of Err–Puigmal ski resort. The lines with diamonds figure the ski–lifts: 1. Location where the feathers were found (wire–collision); 2. Location of the rock ptarmigan hen carcase (map from Institut Cartografic de Catalunya).

Fig. 1. Mapa topográfico de la parte superior de la estación de esquí alpino de Err–Puigmal. Las trazas con rombos representan los remontes: 1. Lugar donde se encontraron las plumas (colisión con cable); 2. Localización del cadáver de la hembra de lagópodo alpino (mapa del Institut Cartogràfic de Catalunya).

heterozygosity (H_e) of 0.637, and a F_{IS} value of 0.148. The probability of identity was $1.216 \cdot 10^{-12}$, indicating that our microsatellite markers are polymorphic enough to distinguish between two individuals from our sampled population. For each sample, the six genotypic profiles obtained were identical, indicating that there were no false alleles and no null alleles. The genotype comparison between samples (*i.e.* collected under the ski–lift and on the rock ptarmigan corpse) gave strict identical genotype profiles for all microsatellite markers. Feathers found under the ski–lift thus came from the female rock ptarmigan corpse. By inference, this result shows that the injuries had been caused by the collision. The steep slope under the ski lift probably facilitated the distance travelled by the hen after the collision (fig. 1).

Discussion

Mortality related to collisions with wires, cables or fences is of major concern for the conservation of tetraonid populations (Bevanger, 1995; Moss, 2001). Out of 835 bird deaths associated with such collisions observed from 1997 to 2009 on 225 ski resorts of the French Alps and Pyrenees, 771 involved galliformes species (Buffet & Dumont–Dayot, in press). These numbers must be considered as minimal because cases of death associated with collisions are generally underestimated because only birds found dead directly underneath or close to the ski–lift are taken into account when estimating mortality rates. Correction factors must therefore be applied to produce more realistic estimates of collision mortality rates (Bevanger, 1999). Such factors correct for different biases such as season (Bevanger & Brøseth, 2004), removal by scavengers (Bevanger & Brøseth, 2000; Morrison, 2002; Barrios & Rodriguez, 2004; Kikuchi, 2008), search efficiency (Morrison, 2002; Kikuchi, 2008), habitat structure (Bevanger & Brøseth, 2004) or crippling losses (Bevanger & Brøseth, 2004). The latter, known as crippling bias, refers to victims of collision found dead outside the search zone. The rock ptarmigan mortality case reported in this study clearly fits this definition. Whereas our study shows the usefulness of a forensic approach for a better identification of the mortality cause, we cannot evaluate the magnitude of the underestimation related to crippling bias. The long distance reported in this study between the corpse and the ski–lift suggests that underestimation associated with crippling losses could be much greater than usually thought. As suggested by Bevanger (1995), an experiment using radio–tagged birds would allow a better estimation of crippling losses beyond the direct mortality rate associated with cable collision.

Acknowledgments

The authors thank Renaud Rabastens and Ramon Martinez for their valuable help. We are especially grateful to N. J. Aebsicher and Rosa Agudo for their constructive comments on the manuscript. This research was supported by the Bureau des Ressources Génétiques, the Office National de la Chasse et de la Faune Sauvage, the French Ministère de l'Enseignement Supérieur et de la Recherche Scientifique and the Centre National de la Recherche Scientifique.

References

Arlettaz, R., Patthey, P., Baltic, M., Leu, T., Schaub, M., Palme, R. & Jenni–Eiermann, S., 2007. Spreading free–riding snow sports represent a novel serious threat for wildlife. *P. Roy. Soc. B.–Biol. Sci.*, 274: 1219–1224.

Barrios, L. & Rodriguez, A., 2004. Behavioural and environmental correlates of soaring–bird mortality at on–shore wind turbines. *J. Appl. Ecol.*, 41: 72–81.

Bech, N., Boissier, J., Drovetski, S. & Novoa, C., 2009. Population genetic structure of rock ptarmigan in the 'sky islands' of French Pyrenees: implications for conservation. *Anim. Conserv.*, 12: 138–146.

Bevanger, K., 1995. Estimates and population consequences of tetraonid mortality caused by collisions with high tension power lines in Norway. *J. Appl. Ecol.*, 32: 745–753.

– 1999. Estimating bird mortality caused by collision with power lines and electrocution, a review of methodology. In: *Birds and power lines, collision, electrocution and breeding*: 29–56 (M. Ferrer & G. F. E. Janss, Eds.). *Quercus*, Madrid.

Bevanger, K. & Brøseth, H., 2000. Reindeer *Rangifer tarandus* fences as a mortality factor for ptarmigan *Lagopus* spp. *Wildlife Biol.*, 6: 121–126.

– 2001. Bird collisions with power lines – an experiment with ptarmigan *Lagopus* spp. *Biol. Conserv.*, 99: 341–36.

– 2004. Impact of power lines on bird mortality in a subalpine area. *Anim. Biodivers. Conserv.*, 27: 67–77.

Buffet, N. & Dumont–Dayot, E., in press. *Bird collision with overhead ski–cables: a source of mortality which can be reduced.* Betham Science Publishers.

Goudet, J., 2001. FSTAT, a program to estimate and test gene diversities and fixation indices (version 2.9.3). Available at http://www2.unil.ch/popgen/softwares/fstat.html.

Kikuchi, R., 2008. Adverse impacts of wind power generation on collision behaviour of birds and anti–predator behaviour of squirrels. *J. Nat. Conserv.*, 16: 44–55.

Martin, K., 2001. Wildlife communities in alpine and sub–alpine habitats. In: *Wildlife–habitat relationships in Oregon and Washington*: 285–310 (D. H. Johnson & T. A. O'Neil, Eds.). Oregon State University Press, Corvallis, OR.

Menoni, E. & Magnani, Y., 1998. Human disturbance of grouse in France. *Grouse News*, 15: 4–8.

Miquet, A., 1990. Mortality in Black grouse *Tetrao tetrix* due to elevated cables. *Biol. Conserv.*, 54: 349–355

Morrison, M., 2002. Searcher bias and scavenging rates in bird/wind energy studies. *Subcontractor Report*.

Moss, R., 2001. Second extinction of capercaillie (*Tetrao urogallus*) in Scotland ? *Biol. Conserv.*, 101: 255–57.

Observatoire des Galliformes de Montagne, 2006. [Percussion des oiseaux dans les câbles aériens des domaines skiables]. Zoom n°4, Sevrier, France (in French).

Patthey, P., Wirthner, S., Signorell, N. & Arlettaz, R., 2008. Impact of outdoor winter sports on the abundance of a key indicator species of alpine ecosystems. *J. Appl. Ecol.*, 45: 1704–1711.

Paetkau, D. & Strobeck, C., 1994. Microsatellite analysis of genetic variation in black bear populations. *Mol. Ecol.*, 3: 489–495.

Thiel, D., Jenni–Eiermann, S., Braunisch, V., Palme, R. & Jenni, L., 2008. Ski tourism affects habitat use and evokes a physiological stress response in capercaillie *Tetrao urogallus*: a new methodological approach. *J. Appl. Ecol.*, 45: 845–853.

Valière, N., 2002. GIMLET: a computer program for analysing genetic individual identification data. *Mol. Ecol. Notes*, 2: 377–379.

Valière, N., Bonenfant, C., Toïgo, C., Luikart, G., Gaillard, J. M. & Klein, F., 2007. Importance of a pilot study for non–invasive genetic sampling: genotyping errors and population size estimation in red deer. *Conserv. Genet.*, 8: 69–78.

Watson, A. & Moss, R., 2004. Impacts of ski–development on ptarmigan (*Lagopus mutus*) at Cairn Gorm, Scotland. *Biol. Conserv.*, 116: 267–275.

17

Improving the reviewing process in Ecology and Evolutionary Biology

G. D. Grossman

Grossman, G. D., 2014. Improving the reviewing process in Ecology and Evolutionary Biology. *Animal Biodiversity and Conservation*, 37.1: 101–105.

Abstract

Improving the reviewing process in Ecology and Evolutionary Biology.— I discuss current issues in reviewing and editorial practices in ecology and evolutionary biology and suggest possible solutions for current problems. The reviewing crisis is unlikely to change unless steps are taken by journals to provide greater inclusiveness and incentives to reviewers. In addition, both journals and institutions should reduce their emphasis on publication numbers (least publishable units) and impact factors and focus instead on article synthesis and quality which will require longer publications. Academic and research institutions should consider reviewing manuscripts and editorial positions an important part of a researcher's professional activities and reward them accordingly. Rewarding reviewers either monetarily or via other incentives such as free journal subscriptions may encourage participation in the reviewing process for both profit and non–profit journals. Reviewer performance will likely be improved by measures that increase inclusiveness, such as sending reviews and decision letters to reviewers. Journals may be able to evaluate the efficacy of their reviewing process by comparing citations of rejected but subsequently published papers with those published within the journal at similar times. Finally, constructive reviews: 1) identify important shortcomings and suggest solutions when possible, 2) distinguish trivial from non–trivial problems, and 3) include editor's evaluations of the reviews including identification of trivial versus substantive comments (*i.e.*, those that must be addressed).

Key words: Publication process, Reviewing, Editorial, Editors.

Resumen

Mejora del proceso de revisión de artículos en ecología y biología evolutiva.— Se debaten los problemas actuales de la revisión y las prácticas editoriales en los campos de la ecología y la biología evolutiva, y se sugieren posibles soluciones para los mismos. La crisis por la que está pasando la revisión no cambiará a menos que las revistas tomen medidas para aumentar la inclusividad de los revisores y los incentivos a los mismos. Asimismo, tanto las revistas como las instituciones deberían prestar menos atención a las cifras relativas a la publicación (las unidades mínimas publicables) y los factores de impacto, y centrar el interés en la síntesis y la calidad de los artículos, lo que exigirá que las publicaciones sean más largas. Las instituciones académicas y de investigación deberían considerar la revisión de los manuscritos y las posturas de las editoriales como una parte importante de las actividades profesionales de un investigador, y compensarlas en consecuencia. Recompensar a los revisores, ya sea económicamente o con otros incentivos, como suscripciones gratuitas a revistas, puede alentar la participación en el proceso de revisión, para las revistas con y sin ánimo de lucro. Probablemente pueda mejorarse el rendimiento de los revisores con medidas que aumenten la inclusividad, como el envío a los revisores de las revisiones y las notificaciones de las decisiones adoptadas. Las revistas tal vez puedan evaluar la eficacia de sus procesos de revisión comparando las citas de los artículos rechazados que se hayan publicado posteriormente con las de los que se publicaron en la revista en el mismo momento. Por último, las revisiones constructivas deben: 1) determinar las deficiencias importantes y sugerir soluciones siempre que sea posible, 2) distinguir los problemas triviales de los que no lo sean y 3) contener las evaluaciones que el editor haga de las revisiones, incluida la determinación de las observaciones triviales y las sustantivas (las que deben abordarse).

Palabras clave: Proceso de publicación, Revisión, Editorial, Editores.

Gary D. Grossman, Warnell School of Forestry & Natural Resources, Univ. of Georgia, Athens GA 30602, USA.

Discussion of shortcomings in the peer review process and editorial practices within scientific journals likely started with publication of the first journal and employment of the first editor. Recently, multiple aspects of this topic have been described in publications dealing with ecology and evolutionary biology (EEB) (Hochberg et al., 2009; Mesnard, 2010; Statzner & Resh, 2010; Albuquerque, 2011; Rohr & Martin, 2012a) and the subject has received considerable attention in the biomedical research community (Smith, 2006; Tite & Schroter, 2007). Multiple critical issues of the peer review process have been raised including: 1) the difficulties of finding good reviewers*, 2) the lack of reward for reviewing, 3) the increased number of manuscripts submitted to journals exacerbating issue 1, and 4) negative institutional policies reduce incentives for participating in the editorial process. Although a number of potential solutions to the reviewer crisis have been suggested, there is little consensus regarding what should be done (DeVries et al., 2009; Montesinos, 2012; Rohr & Martin, 2012b; Duffy, 2013) and there appear to be few changes in editorial practices by journals (Grod et al., 2010). In this paper I will discuss additional issues contributing to the reviewer crisis and propose several additional solutions. Much of what I report is based on my own experiences as an author of 110+ papers and also as a reviewer/editorial board member/associate editor for five different journals run by both scientific societies and commercial publishers.

Publication proliferation

There is no doubt that the number of manuscripts submitted for publication in scientific journals has increased substantially in the last few decades, primarily due to an increase in the number of scientists. In addition, the pressures of promotion and high competition for jobs in the last four decades contribute to the pressure to 'slice' publications into what historically has been known as the Least Publishable Unit (LPU) or 'salami tactic'. The combination of increasing publication frequency and decreasing publication length was recognized decades ago (Broad, 1981; Lyman, 2013) and is one of the main factors contributing to the reviewer crises in EEB. There is no doubt that many will judge a scientist's performance based on publication quantity rather than quality, and this is likely true for most scientific fields. The phenomenon itself is most easily observed in discussions of faculty search or tenure/promotion committees. It is clear that overall productivity (*i.e.*, number of publications) should play a role in evaluations, but first assessments (and cuts) typically are made using simple criteria such as 'number of publications in refereed journals'. This criterion is easy, quick and may even be correlated with quality, but it also encourages vita padding. It is easily gamed by dividing larger potential research publications into LPUs, which contribute significantly to the editorial burden of the EEB community. Nonetheless, I doubt

that publication frequency will ever disappear as an assessment criterion, but perhaps journal editors and referees should be more stringent in accepting papers that clearly are small slices of a complete pie.

The LPU syndrome has been exacerbated by the proliferation of journals in EEB (Statzner & Resh, 2010); including the explosion of 'Letters' (*i.e.*, short format) and open–access journals (Bohannon, 2013), all of which require enough papers to regularly fill issues. Some researchers appear to think that the publication process is slower than it was 25 years ago (Statzner & Resh, 2010), but recent studies provide surprising answers to that question. For example, there has been no demonstrable increase in average review time for journals in either behavioral sciences or natural history between 1980 and 2012 (Pautasso & Schaefer, 2010; Lyman, 2013). In addition, although there is a positive correlation between impact factor of a journal and the number of manuscripts submitted, there also is a negative trend between impact factor and time to acceptance (Pautasso & Schaefer, 2010). Hence, higher number of submissions does not necessarily result in more extensive editorial delays (Pautasso & Schaefer, 2010). It is possible; however, that the latter result is a consequence of many papers being rejected by journals without review (Pautasso & Schaefer, 2010) as has been the policy of a number of prominent EEB journals. This practice, although providing a quick turn–around for a manuscript, is quite susceptible to bias and cliquishness in publication, as noted in1974 (VanValen & Pitelka, 1974) and still in evidence today (Arnqvist, 2013). Nonetheless, in contrast to the results of Pautasso and Schaefer (2010) a recent survey of EEB editors showed a negative relationship between the number of papers handled and the proportion rejected without review (McPeek et al., 2009).

The referee pool

Given the increasing number of both journals and submissions, coupled with a pool of experienced referees that while increasing, still is insufficient to handle the current load (Hauser & Fehr, 2007; Statzner & Resh, 2010; Arnqvist, 2013; Duffy, 2013), it is obvious that the EEB community has yet to effectively deal with the 'reviewer crises'. Several investigators have suggested ideas for dealing with the decreased willingness of referees to perform reviews, the high number of review requests received by 'good referees', and issues of review quality (Hauser & Fehr, 2007; Fox & Petchey, 2010; Rohr & Martin, 2012a; Duffy, 2013). These suggestions involve punishing slow reviewers and rewarding timely referees who provide thorough reviews, but as all authors admit, these solutions may do little to prevent some scientists employing 'cheater' strategies. Nonetheless, they all are right that changes are necessary to improve the current status of reviewing.

Perhaps referees are no more nor less altruistic than they have been in the past, but what has changed

---

* I will use the term reviewer and referee interchangeably.

in the last 30 years are the external constraints on a researcher's time. Most university researchers, at least in the United States, are now faced with a plethora of administrative responsibilities from both their own universities and governmental sources (*e.g.,* faculty committees, training sessions for compliance with laws such as the Family Educational Rights and Privacy Act [FERPA], monthly documentation of graduate student performance, Institutional Animal Use and Care Committee (IACUC) requirements and training, federal data accessibility requirements, etc.). Concomitantly, both university and federal research budgets have been slashed in the United States and other countries; consequently researchers must devote much more time to seeking research funding than they have in the past. This is one of the major reasons referees are slow or reluctant to review papers; simply put, there is little time or energy left after performing one's daily research responsibilities (Statzner & Resh, 2010). At the same time, the qualifications needed to obtain a research or faculty position are increasing (Statzner & Resh, 2010). Hence, even if someone is a 'good Samaritan' (McPeek et al., 2009), there are strong selective pressures acting against altruism, even if they are merely perceived rather than real.

There is no solution to this problem until reviewing manuscripts, and editorial work in general, are viewed as normative responsibilities, with appropriate recognition and rewards from administrators. I suspect that in most institutions, editorial board membership or extensive reviewing rarely results in raises, increased release time or help from support staff. My supposition is that administrators resort to claims like 'well everyone does that so we can just assume that it is a constant across faculty' but the current crises suggest that reviewing and editorial work are not constant across faculty. In addition, an erroneous assumption by administrators that reviewing is equal across faculty promotes 'cheaters' who do no reviewing and devote all their time to writing grants or papers instead, especially when promotion decisions are made on a comparative basis. Faculty must become more proactive in demanding that incentives be provided for highly active and competent reviewers and associate editors, and managing editors should support them in this quest. Hopefully, this will result in administrators providing substantive rewards for participation in the editorial process as well as penalties for faculty who do not participate.

How can journals and editors improve the situation

At present, there appear to be few journals that provide incentives for reviewers. A few journals provide free access to online versions of the journal although frequently this only extends over a month or two. Certainly one perquisite for reviewers that could increase referee responsiveness would be to give a free online subscription to the journal after a given number of reviews in a year. Even non–profit scientific societies could employ this incentive because it is not costly. Incentives could be provided on a graduated scale

where it might take four reviews in a year to obtain free access for a year, and a single review might earn only three months access. Of course this may penalize members of scientific societies who already receive a journal subscription, but they still might not have online access or they could be rewarded with free access in the next year or access to a journal they do not receive (many scientific societies publish multiple journals). No incentive scheme is perfect but it seems that some experimentation is called for at the present time, given the repeatedly voiced concerns by both editors and authors.

It is possible that paying referees for reviews could improve both referee participation and performance, but its discussion mostly has occurred on online forums. I have found no published evaluation of this practice in EEB, although EEB outside examiners are paid by universities for dissertation reviews in both Australia and New Zealand and likely other countries. In addition, multiple European countries (Ireland, Poland and Spain) pay for proposal reviews, as does at least one commercial publisher for editorial board work. Nonetheless, a study of biomedical reviewers found that reviewers had mixed opinions regarding the positive impacts of financial rewards on the reviewing process (Tite & Schroter, 2007). The biggest objections to payment for services involve the end of volunteerism, and the assumption that financial rewards will bias the reviewing process, or pull referees away from journals that cannot provide incentives. There is a lack of evidence but I suspect this is unlikely. From a philosophical perspective, I deplore the loss of the volunteer ethic in science; however, the current crisis seems immune to philosophical regrets and perhaps represents the triumph of the market economy even in science. One of my goals is to suggest possible approaches leading to data on potential strategies to resolve the reviewer crisis. It would be useful for an EEB journal or society to conduct an experiment in which some reviewers are paid and others not and then compare the quality, timing, and responsiveness of the two reviewer treatment groups. There is no doubt that such an experiment would require a sophisticated design and still likely present logistical hurdles, however, it should aid in determining whether financial rewards would improve the reviewing process in EEB. Finally, it is true that payment for reviewers and editorial work may present logistical and financial difficulties for non–profit journals, however these obstacles are mostly irrelevant for the many journals published by highly profitable commercial publishers or open–access journals with high publication fees.

One of the reasons for the poor performance of reviewers is that too many journals fail to cultivate a culture of inclusion in the editorial process. I suspect reviewer performance would be substantively improved if journals practiced a few simple steps that demonstrated the importance of individual reviews in the overall editorial process. For example, although some journals provide a reviewer with all reviews of a manuscript and editor's decision letter, too many do not. Reading the comments of other reviewers and the editor always is an educational experience and is an excellent mechanism for less experienced reviewers to learn from more experienced reviewers. In addition, it would be beneficial for everyone involved if editors explained their reasoning when they overrule a referee. Finally, I wonder how much effort is expended by journals in evaluating whether their editorial practices are efficient and unbiased, or whether the prevailing attitude is one of *laissez faire* (Grod et al., 2010). Certainly one way that journals could evaluate the accuracy of their reviewing practices would be to compare citation frequencies of a random sample of articles rejected by the journal but subsequently published in other journals with a sample of articles accepted in that same year. Although citation frequencies are not a perfect metric of quality, they are easily obtained and certainly indicative of quality if the citations are positive. Such an analysis should be conducted with historic data, for example volumes published 10, 7 and 4 years previously. If no difference exists between citation frequencies of the two sets of papers, and assuming that the rejected papers were appropriate subject matter for both journals, then it would be cause for examining historic editorial practices, or to determine if specific associate editors were the cause of these rejections. Of course the citations would have to be checked randomly to assure that the citations were comparable (*i.e.*, to avoid the case where total citations are equal but one paper has all positive citations and the other has all negative citations). It also might help identify continuing trends in problematic editorial practices. In addition, recent work has shown that factors such as journal impact factors may affect reviews independent of manuscript quality, and that reviewer ratings of the same manuscript may not be highly correlated (Eyre–Walker & Stoletzki, 2013). It is likely that there is little formal or quantitative evaluation of associate editors for many journals, except where an editor's behavior becomes intolerable, such as failing to act on multiple manuscripts. These issues all call for journals to evaluate the accuracy and precision of their reviewing policies.

Improving reviews and reviewing

There is no doubt that high quality referees and editors are both typically overworked. Nonetheless, if editors believe that reviewers should not use this excuse, then neither should they. My own experience suggests there has been a decline in the quality of review interpretation and decisions made by editors as well as a general decline in review quality. I have already mentioned fostering a sense of inclusion for referees in the editorial process and (Statzner & Resh, 2010) have covered many of the current negative trends in the editorial process. Having published my first paper in 1977, I have seen just about every constructive and inane comment possible, typically with no comments from the editor on inappropriate or obviously erroneous comments. I believe that it is an editor's responsibility to ensure that an editorial decision letter does not come back to an author without commentary on the reviews. At the very least, editors should identify reviewer's comments that must be addressed versus those that are optional.

Nonetheless, the evaluation of reviewer's comments by editors certainly is not general policy for scientific journals. Given the complaints by editors regarding the poor quality of many reviews, this is not a trivial issue, yet most editors provide an author with little guidance other than 'all comments must be addressed, especially revisions that you do not incorporate'. But how much detail must be provided by an author when a comment clearly is erroneous: a not infrequent situation? This can be particularly problematical for young scientists, especially given the many picayunish negative comments written by reviews of today. Frankly, if editors are actually reading reviews closely, as they should, then it does not take much more time to identify which comments need to be addressed and which do not. After all, how can an editor reach an informed decision without evaluating reviews, even when both ratings are reject? Every author deserves at least this much from an editor. An additional problem of today is that the category of 'accepted with revision' seems to have disappeared from many journals and instead the author is told that their manuscript has landed in the large gray category called 'not acceptable in this form'. I have spoken with many researchers, especially young researchers, who have interpreted this as a rejection, when in fact it really is just code for 'significant revision'. Nonetheless, some editors have justified this change by saying that it was difficult to obtain substantive revisions from authors once the term 'accepted' had been used.

What constitutes a good review?

A thorough discussion of the reviewing process is provided by DeVries et al. (2009), an article that is particularly useful for young scientists. An interesting psychological question for both editors and reviewers is whether a paper should be viewed as acceptable until a sufficient number of problems render it unacceptable, or whether papers should be viewed as unacceptable until a sufficient number of positive points are identified so that it becomes acceptable. I favor the first view point, mainly because I believe it leads to more constructive reviewing and hopefully a more positive experience for the authors, even when a paper is rejected. Many journals do not have review templates that ask a reviewer to specifically identify both the strengths and weaknesses of the manuscript but this would lead to more objective reviewing and improved editorial decisions.

For both referees and editors, clearly the criterion for any comment is whether or not it is truly constructive. Probably the most significant improvement would be to require referees to reference their criticisms. I have seen comments ranging from 'this simply is wrong' to the 'literature review was inadequate' without any subsequent explanation of why a given technique was wrong nor any subsequent listing of missing papers. Such comments are completely unhelpful to the author and certainly do not fall under the rubric of 'constructive criticisms'. It is not the reviewer's responsibility to rewrite an author's manuscript; nonetheless, unconstructive comments and reviews help no one and eventually result in a bad reputation for a journal. I know more than one scientist who simply has stopped submitting manuscripts to journals that have persistently poor reviewing policies even when they have high impact factors. Nonetheless, clearly this is a luxury of the full professor, not the untenured assistant professor. A final comment on writing style is warranted, given that many current referees seem to have little tolerance for a style different from their own. I have received reviews stating that a manuscript is poorly written without any description of how this judgment was reached, let alone an 'example' paragraph that was rewritten to demonstrate good writing. In addition it is not uncommon to receive reviews in which one reviewer ranks the paper as well written while another says it is poorly written. Once again, this is the type of comment that should prompt an editor's intervention but this is rare in my experience. Consequently, if you cannot identify specific problems in grammar, clarity or verbosity accompanied by examples of how this can be corrected, then it is likely that you and the author have different writing styles, and it should be left at that. An even more problematical stylistic issue is that of non–native English writers, and the level of grammatical 'stretch' that should be allowed in such manuscripts (Clavero, 2010)

As with any large volunteer enterprise, problems exist with the current peer review system and whether or not they will be fixed depends on the EEB community itself. Nonetheless, I hope that the suggestions made in this paper are helpful, even if they only lead to small improvements in the overall EEB editorial system. Most importantly, journals should begin conducting experiments regarding changes in editorial practices that may improve the various aspects of the 'reviewing crisis', and ultimately communicate the results of these experiments to the EEB community.

Acknowledgements

I apologize in advance for any omissions contained in this article and to any journals, administrators, etc. who already employ the suggestions in this article. I am sure they are out there and should be congratulated. I would like to thank D. DeVries, E. Garcia–Berthou, A. Hildrew, D. Jackson, M. McCallum, V. Resh, J.–C. Senar, J. Schaefer and J. Rohr for thoughtful comments on the manuscript, and my family for their ever present support. In addition, the reviewers: Mario Diaz and Sara Schroter provided insightful commentary on the ms. Conceptual stimulation for this paper was aided by Jittery Joe's and Two Story. Finally, the Warnell School of Forestry and Natural Resources provided material support for this paper.

References

Albuquerque, U. P., 2011. The tragedy of the common reviewers – the peer review process. *Rev. Bras. Farmacogn. Braz. J. Pharmacogn.*, 21: 1–3.

Arnqvist, G., 2013. Editorial rejects? Novelty, schnovelty! *Trends Ecol. Evol.*, 28: 448–449.

Bohannon, J., 2013. Who's Afraid of Peer Review? *Science*, 342: 60–65.

Broad, W. J., 1981. The publishing game: getting more for less. *Science*, 211: 1137–1139.

Clavero, M., 2010. 'Awkward wording, rephrase': linguistic injustice in ecological journals. *TREE*, 25: 552–553.

DeVries, D. R., Marschall, E. A. & Stein, R. A., 2009. Exploring the peer review process: what is it does it work and can it be improved? *Fisheries*, 34: 270–279.

Duffy, D. C., 2013. Reviewing reviewers. *The Scientist* http://www.the–scientist.com/?articles.view/articleNo/36575/title/Opinion—Reviewing–Reviewers/

Eyre–Walker, A. & Stoletzki, N., 2013. The Assessment of Science: The Relative Merits of Post–Publication Review, the Impact Factor, and the Number of Citations. *PLoS Biol*, 11(10): e1001675. doi:10.1371/journal.pbio.1001675.

Fox, J. & Petchey, O. L., 2010. Pubcreds: fixing the peer review process by 'privatizing' the reviewer commons. *Bull. Ecol. Soc. Am.*, 91: 325–333.

Grod, O. N., Lortie, C. J. & Budden, A. E., 2010. Behind the shroud: a survey of editors in ecology and evolution. *Front. Ecol. Environ.*, 8: 187–192.

Hauser, M. & Fehr, E., 2007. An incentive solution to the peer review problem. *PLoS Biol.*, 5: 703.

Hochberg, M. E., Chase, J. M., Gotelli, N. J., Hastings, A. & Naeem, S., 2009. The tragedy of the reviewer commons. *Ecol. Lett.*, 12: 2–4

Lyman, R. L., 2013. Three–Decade History of the Duration of Peer Review. *J. Scholarly Pub.*, 44: 211–220.

McPeek, M. A., DeAngelis, D. L, Shaw, R. G., Moore, A. J., Rausher, M. D., Strong, D. R., Ellison, A. M., Barrett, L., Rieseberg, L., Breed, M. D., Sullivan, J., Osenberg, C. W., Holyoak, M. & Elgar, M. A., 2009. The golden rule of reviewing. *Am. Nat.*, 173(5): E155–E158.

Mesnard, L., 2010. On Hochberg et al.'s 'the tragedy of the reviewer commons'. *Scientometrics*, 84: 903–917.

Montesinos, D., 2012. Type I error hinders recycling: a response to Rohr and Martin. *Trends Ecol. Evol.*, 27: 311–312.

Pautasso, M. & Schaefer, H., 2010. Peer review delay and selectivity in ecology journals. *Scientometrics*, 84: 307–315.

Rohr, J. R. & Martin, L. B., 2012a. Reduce, reuse, recycle scientific reviews. *Trends Ecol. Evol.*, 27: 192–193.

– 2012b. Type I error is unlikely to hinder review recycling: a reply to Montesinos. *Trends Ecol. Evol.*, 27: 312–313.

Smith, R., 2006. The trouble with medical journals. *J. R. Soc. Med.*, 99: 115–119.

Statzner, B. & Resh, V. H., 2010. Negative changes in the scientific publication process in ecology: potential causes and consequences. *Freshwat. Biol.*, 55: 2639–2653.

Tite, L. & Schroter, S., 2007. Why do peer reviewers decline to review? A survey. *J. Epidem. Comm. Health*, 61: 9–12.

VanValen, L. & Pitelka, F., 1974. Intellectual Censorship in Ecology. *Ecology*, 55: 925–926.

18

Effect of wild ungulate density on invertebrates in a Mediterranean ecosystem

A. J. Carpio, J. Castro–López,
J. Guerrero–Casado, L. Ruiz–Aizpurua,
J. Vicente & F. S. Tortosa

Carpio, A. J., Castro–López, J., Guerrero–Casado, J., Ruiz–Aizpurua, L., Vicente, J. & Tortosa, F. S., 2014. Effect of wild ungulate density on invertebrates in a Mediterranean ecosystem. *Animal Biodiversity and Conservation*, 37.2: 115–125.

Abstract

Effect of wild ungulate density on invertebrates in a Mediterranean ecosystem.— In recent decades, the abundance and distribution of certain big game species, particularly red deer (*Cervus elaphus*) and wild boar (*Sus scrofa*), have increased in south central Spain as a result of hunting management strategies. The high density of these ungulate species may affect the abundance of epigeous invertebrates. We tested the relationships between big game abundance and biodiversity, taxon richness, the biomass of invertebrates and their frequency on nine hunting estates and in comparison to ungulate exclusion areas. Ungulate exclusion itself affected invertebrate richness, since lower values were found in the open plots, whereas the highest differences in invertebrate diversity between fenced and open plots was found in areas with high wild boar density. Where wild boar densities were high, the number of invertebrates decreased, while where they were low, red deer had a positive effect on invertebrate abundance. Fenced plots thus seemed to provide refuge for invertebrates, particularly where wild boar were abundant. This study supports the idea that the structure of fauna communities is damaged by high density populations of ungulates, probably due to decreased food availability owing to overgrazing, modified conditions of ecological microniches and direct predation. However, the effects depended on the group of invertebrates, since saprophytic species could benefit from high ungulate abundance. Our findings reflect the need to control ungulate population density under Mediterranean conditions in south–western Europe and to implement ungulate exclusion plots.

Key words: Biodiversity, Invertebrates, Overabundance, Red deer, Wild boar

Resumen

Efecto de la densidad de ungulados silvestres sobre los invertebrados en un ecosistema Mediterráneo.— En las últimas décadas, la abundancia y distribución de determinadas especies de caza mayor, especialmente el ciervo rojo (*Cervus elaphus*) y el jabalí (*Sus scrofa*), han aumentado en la zona centromeridional de España como resultado de las estrategias de gestión cinegética. La alta densidad de estas especies de ungulados puede afectar a la abundancia de los invertebrados epigeos. Estudiamos la relación entre la abundancia de las especies de caza mayor y la biodiversidad, la riqueza de taxones, la biomasa de invertebrados y su frecuencia en nueve fincas de caza, y se comparó con las zonas de exclusión de ungulados. De por sí, la exclusión de ungulados afectó a la riqueza de invertebrados, ya que se encontraron valores más bajos en las parcelas abiertas, mientras que las mayores diferencias en la diversidad de invertebrados entre parcelas abiertas y cercadas se encontraron en zonas con una alta densidad de jabalíes. Donde la densidad de jabalíes era alta, el número de invertebrados disminuyó, mientras que donde era baja, el ciervo rojo tuvo un efecto positivo en la abundancia de invertebrados. Así, las parcelas cercadas parecían ofrecer refugio a los invertebrados, sobre todo donde los jabalíes eran abundantes. Este estudio apoya la idea de que las poblaciones con una alta densidad de ungulados perjudican a la estructura de las comunidades faunísticas, probablemente debido a la disminución de la disponibilidad de alimentos como consecuencia del sobrepastoreo, la modificación de las condiciones de los micronichos ecológicos y la depredación directa. Sin embargo, los efectos dependieron del grupo de invertebrados, ya que las especies saprofitas podrían beneficiarse de la alta abundancia de ungulados. Nuestros resultados reflejan la necesidad de controlar la densidad de las poblaciones de ungulados en condiciones mediterráneas en el suroeste de Europa y de establecer parcelas de exclusión de ungulados.

Palabras clave: Biodiversidad, Invertebrados, Sobreabundancia, Ciervo rojo, Jabalí

Antonio J. Carpio, Jesús Castro–López, José Guerrero–Casado, Leire Ruiz–Aizpurua & Francisco S. Tortosa, Dept of Zoology, Univ. of Cordoba, Campus de Rabanales Ed. Darwin, 14071 Córdoba, Spain.– Joaquín Vicente, Inst. en Investigación de Recursos Cinegéticos, IREC (CSIC, UCLM, JCCM), Ronda de Toledo s/n., 13071, Ciudad Real, Spain.

Corresponding author: A. J. Carpio. E–mail address: b42carca@uco.es

Introduction

The soil invertebrate community participates actively in ecological processes that are essential for substrate soil fertility and plant succession (Hedlund & Öhrn, 2000; Osler & Sommerkorn, 2007). Sources of soil disturbance and their effect on invertebrates, including the use of pesticides, phytosanitary treatment and other measure, have been thoroughly studied in agricultural ecosystems (Vickery et al., 2009; Raebel et al., 2012). However, knowledge of the factors affecting invertebrate communities in forest ecosystems is scarce (McIntyre, 2000).

Ungulate density and range has increased throughout Europe and North America over the last century (Clutton–Brock & Albon, 1992; Côté et al., 2004; Gordon et al., 2004; Sarasa & Sarasa, 2013) as a result of the extirpation of large predators (Breitenmoser, 1998), changes in sylviculture and agriculture, and the intensification of game management (Apollonio et al., 2010). This increase in wild ungulate populations may have a strong impact on soil nutrient status and biota due to grazing, rooting, trampling and dunging, and changes in plant community due to herbivory can also affect invertebrate community structure (see Spalinger et al., 2012), but specific studies on these relationships are scarce. High densities of either livestock (Rosa–García et al., 2009) or wild ungulates (Côté et al., 2004; Mohr et al., 2005) are known to affect epigeous invertebrate communities, which are useful bioindicators (Gerlach et al., 2013) and important food resources for many species of birds, including the red–legged partridge, a key prey for many predators and the most important game bird in Spain (Wilson et al., 1999). Previous studies on the effect of ungulates on invertebrates have been conducted in areas in which ungulates are invasive (Cuevas et al., 2010, 2012) and in temperate climates, focusing on deciduous forests (Côté et al., 2004; Mohr et al., 2005; Mizuki et al., 2010). However, few studies have reported the effect of native ungulates on invertebrate soil diversity in semiarid areas, such as Mediterranean habitats (Gebeyehu & Samways, 2006). Red deer (*Cervus elaphus*) and wild boar (*Sus scrofa*) are the principal wild ungulate species in Southern European Mediterranean habitats, reaching very high abundances when intensive hunting management is performed (Vicente et al., 2007), ranging between 0.04 to 66.77 deer/km² (mean = 19.51; n = 22 populations) (Acevedo et al., 2008). In fact, the red deer is considered by some authors to be among the most invasive species in the world (Lowe et al., 2000) and its negative effect on some arthropod taxa such us Orthoptera or other phytophagous insect has been reported in subalpine grasslands (Goméz et al., 2004; Spalinger et al., 2012).

A high abundance of wild boar has also been reported to have a strong impact on edaphic fauna through disturbance (Herrero et al., 2006; Giménez–Anaya et al., 2008), rooting, and the direct consumption of meso– and macroinvertebrates (Cuevas et al., 2010). However, despite the large increase in the densities of wild boar and deer, little is known about the

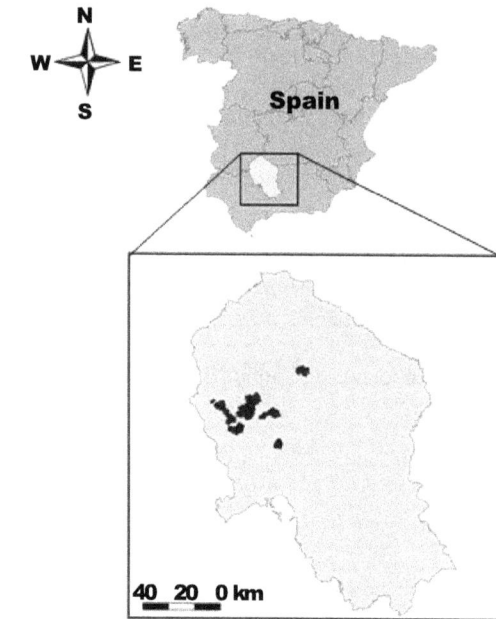

Fig. 1. Map of Spain showing the location of the sampling sites (Córdoba province in light grey).

Fig. 1. Mapa de España que muestra la ubicación de los sitios de muestreo (provincia de Córdoba en gris claro).

ecological impact of their overabundance on Mediterranean ecosystems (Barrios–García & Ballari, 2012; Carpio et al., 2014b) and particularly on the epigeous invertebrate assemblage, essential elements in the diet of many birds (Holland et al., 2006).

The aim of this study was to determine the impact of wild boar and red deer on diversity, richness and biomass of epigeous invertebrates in a semiarid Mediterranean environment from south central Spain, within the native distribution range of these two ungulate species.

Material and methods

Study area

Data were collected on 9 different hunting estates, which had an average area of 2,470 hectares (range 1,480–3,600 ha), located in southern Spain. The altitude ranged from 400 to 800 m.a.s.l. The dominant vegetation included tree species such as holm oak (*Quercus ilex*) and cork oak (*Quercus suber*), pine plantations (*Pinus pinea* and *Pinus pinaster*), shrub species such as *Cystus* spp., *Erica* spp., *Pistacia* spp., *Phyllirea* spp. and *Rosmarinus officinalis,* and scattered pastures and small areas of crops (Vicente et al., 2007). These savannah–like landscape units are called 'dehesas'. The study sites are mainly devoted to recreational hunting for wild boar and red deer.

Estimating red deer and wild boar abundance

Deer population size was estimated at hunting estate level, the estates being considered as discrete management units. Two spotlight counting events between September and October 2011 were used to estimate the deer population size at each estate. Transects (mean length = 20.3 km ± 2.34 SE) were driven at 10–15 km/h (Carpio et al., 2014a). The distance from the observer to the centre of a deer group was measured, and compass bearings were taken to determine the angle between deer, or deer groups, and the transect line. The distance between the observer and the deer was measured with a Leica LRF 1200 Scan telemeter (Solms, Germany) (range 15–1,100 m; precision ±1m/±0.1%). The abundance of the deer populations was estimated by distance sampling (Buckland et al., 2004, Distance 5.0 software). Half–normal, uniform and hazard rate models for the detection function were fitted against the data using cosine, hermite polynomial and simple polynomial adjustment terms, which were fitted sequentially. The selection of the best model and adjustment term were based on Akaike's Information Criterion (AIC). The best relative fit of the model and adjustment term for distance–sampling was the hazard–rate cosine based on the lowest AIC score. However, this census method suffers significant variations depending on the type of game mode that is practiced (hunts or stalking).

Two 4 km transects per site were sampled for signs of wild boar activity following the guidelines of Acevedo et al. (2007). Each transect consisted of 40 segments of 100 m in length and 1 m in width. Every 100 m segment was divided into 10 sectors of 10 m in length. Sign frequency was defined as the average number of 10–m sectors containing droppings per 100–m transect (Carpio et al., 2014b), and a single average value of wild boar abundance was calculated per estate.

Experimental plots

We used five ungulate proof fences in each one of the nine hunting states. These fenced plots (hereafter FP) were constructed three to five years prior to data collection and they were constructed from steel. Each FP was 0.5 ha, with a mesh size of 150 mm × 100 mm in order to prevent the ungulates access, although they were accessible to other animals (Carpio et al., 2014b). Two pitfall traps were randomly placed in each FP, resulting in a total of 90 traps where ungulates were excluded. Another two pitfall traps were placed 100m outside of each FP as controls (Open Plots, OP), resulting in 180 pitfall traps in total.

We conducted two surveys of invertebrates. The pitfall traps consisted of plastic receptacles, with a capacity of 0.75 litres and an opening diameter of 12 cm, buried at ground level (Paschetta et al., 2013). These were half filled with a solution of salts (to preserve the specimens caught) and soap (to break the water surface tension). The trapped invertebrates were collected 14 days after the traps had been set (Allombert et al., 2005). The contents of the receptacles were passed through a sieve. The invertebrates were preserved in 100 ml plastic containers with 70% alcohol and later identified by stereomicroscope in the laboratory. Specimens were identified to order level (Barrientos & Abelló, 2004), as in some previous studies on the diet of farmland birds (Holland et al., 2006).

We studied the diversity and structure of invertebrate orders larger than 0.02 mm (mesofauna and macrofauna) present in our study area, excluding microfauna (less than 0.02 mm) (Swift et al., 1979). We therefore studied the most important groups in the diet of red–legged partridge chicks (Holland et al., 2006; Aebischer & Ewald, 2012). We excluded pitfall traps containing necrophagous insects (11% of placed traps) and also those in which more than 50% of individuals belonged to the order Hymenoptera (13% of placed traps) owing to the proximity of ant nests as these could exert a repellent effect on other arthropods (Blum, 1978).

For each sampling point, we calculated the invertebrate dry weight (B), taxon richness (S) and the Shannon index (Shannon, 1948).

To obtain the dry weight, the contents of the pitfall traps were dehydrated in an oven at 80ºC for 24 h. A precision scale (0.001 g) was used. We calculated the values for each variable from the average of the two pitfall traps in each pair of sampling periods (OP and FP).

Vegetation structure

The vegetation structure was described by creating a buffer area of a 25 m radius around each pitfall trap and the percentage of grass, scrub and woodland cover was estimated by eye, following similar protocols for general habitat–species studies (Morrison et al., 1992). All the estimates of vegetation structure were performed by the same observer (A.J.C).

The amount of plant biomass was assessed from cuttings in an area of 25 cm² of herbaceous vegetation. Two sampling points were randomly selected in both the fenced and the open plots. The sampled vegetation was dried in a drying oven with hot air circulation at 60ºC until a constant weight was obtained. An electric balance (precision: 0.01 g) was used.

Statistical analysis

The relationships between ungulate abundance (separately for red deer and wild boar, respectively) on invertebrate richness, dry mass, the Shannon index and absolute frequency (number of invertebrates per sample) were tested using generalized linear mixed models (GLMMs). With regard to the absolute frequency models, the analyses were carried out separately for each of the four taxonomic groups into which the samples had been pooled. The taxonomic categories were 'Hymenoptera' (n = 1,120), 'Insecta' other than Hymenoptera (16 orders, n = 1,743), class 'Arachnida' (including orders Araneida, Acari, Opiliones, Scorpionida, Pseudoescorpionida and

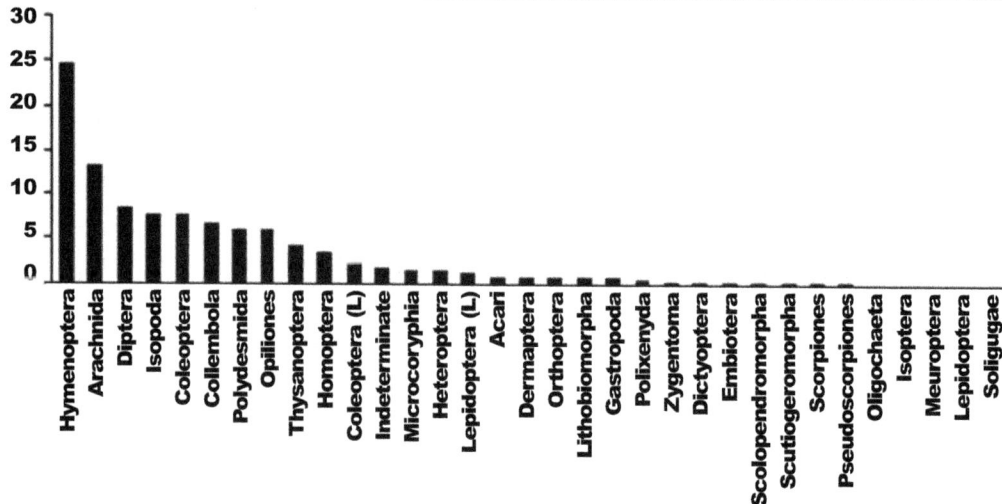

Fig. 2. Percentage of trapped invertebrates belonging to each order as regards the total.

Fig. 2. Porcentaje de invertebrados atrapados pertenecientes a cada orden con respecto al total.

Solifugae; six orders, n = 906), and 'others' (including the subphylum Myriapoda, order Isopoda, and classes Oligochaeta and Gastropoda; nine taxa, n = 787).

Treatment (two levels: open vs. fenced plots) was included in the model as the factor, whereas red deer and wild boar abundances, in addition to the vegetal biomass (g) and percentage of grass, shrub and tree covers, were included as co–variables. We also included the interaction between the treatment and the abundances of ungulates and the interaction between deer and wild boar density. The estate was included (nine levels) as a random factor. Since every plot was sampled twice, the sampling dates were included in the model as repeated measures.

A normal distribution function and an identity link were used for dry mass, and the Shannon index, and a Poisson function and log–link function were used for richness and absolute frequency models. Rather than using criteria based on parsimony to select the 'best model' (which favour precision vs. bias) we used the full models: (i) because our models had high degrees of freedom (nine explanatory variables) and there was no need to guard against over–fitting, (ii) to protect from the bias of regression coefficients, and (iii) to preserve the accuracy of confidence intervals while using other non–collinear factors for control purposes (multiplicity adjustment, while our understanding of the underlying biological processes led us to believe that the important variables to control for had been included). The assumptions of normality, homogeneity and independence in the residuals were assessed in models with normal distribution function (Zuur et al., 2009). Statistical analyses were performed using InfoStats and SAS 9.0 statistical software. The significant p–value was set at p = 0.05.

Table 1. Mean and standard deviations for the variables quantifying invertebrate abundance: OP. Open plots; FP. Fenced plots.

Tabla 1. Media y desviaciones estándar de las variables que cuantifican la abundancia de invertebrados: OP. Parcelas abiertas; FP. Parcelas cerradas.

	March collection		April collection		March + April	
	OP	FP	OP	FP	OP	FP
Shannon index	1.56 ± 0.36	1.75 ± 0.28	1.57 ± 0.37	1.65 ± 0.32	1.56 ± 0.36	1.66 ± 0.31
Taxon richness	7.06 ± 2.31	8.13 ± 1.89	7.3 ± 2.24	8.22 ± 2.41	7.11 ± .45	8.32 ± 2.15
Weight arthropods	0.15 ± 0.18	0.13 ± 0.11	0.08 ± 0.08	0.16 ± 0.18	0.12 ± 0.15	0.14 ± 0.15
Absolute frequency	36.44 ± 5.1	37.8 ± 3.07	43 ± 4.91	56.09 ± 5.67	39.22 ± 3.5	47.05 ± 3.4

Table 2. Full model on the effects of ungulates on invertebrate richness, dry mass and Shannon diversity index ($^*p < 0.05$).

Tabla 2. Modelo completo sobre los efectos de los ungulados en la riqueza, el peso seco y el índice de diversidad de Shannon de los invertebrados (*p < 0,05).

	Taxon richness		Dry mass		Shannon index	
	F	β	F	β	F	β
Treatment (T)	7.8*	1.21	1.62	0.42	0.07	1.01
Deer density (Dd)	0.01	2.18	0.44	−2.76	0.03	0.63
Wild boar abundance (Wba)	0.01	−3.36	0.54	−0.95	0.03	−0.82
Shrub	0.1	0.0017	0.05	0.0006	1.10	0.0009
Dd*Wba	0.01	0.59	0.28	3.67	0.01	0.36
T*Dd	0.61	−3.25	0.63	1.76	1.57	−0.81
Plant biomass	0.17	−0.15	0.01	0.0052	4.31	0.03
Wooded	1.1	0.017	0.05	−0.0017	3.1	0.0045
Grass	0.43	−0.0004	3.62	1.9	1.65	−0.0001
T*Wba	1.78	4.92	1.33	−2.23	4.3*	1.17

Results

The best relative fit of the model and adjustment term for distance–sampling was the hazard–rate cosine based on the lowest AIC score. The average red deer density, expressed as the number of deer per 100 ha, ranged from 25 to 68 (average 39 ± 14 SD). The coefficients of variation of distance–sampling estimates ranged from 2.95% to 38.86%. The abundance indices for wild boar ranged from 0.04 to 0.47 (average 0.26 ± SD 0.15).

We identified 5781 invertebrates, 3,201 of which were captured in FP and 2,580 in OP (table 1). They were spread over 33 taxa (17 insect orders, six Arachnida orders, six Myriapoda orders, one Crustacean order, one Gastropoda class, one Oligochaeta class and a group corresponding to indeterminate individuals; fig. 2).

The invertebrate dry mass was marginally significant and positively associated with the percentage of grass cover (table 2, $F_{1,123} = 3.62$, $p = 0.059$), whereas invertebrate richness differed statistically between treatments, with the values for the OP being lower than those for the FP ($F_{1,123} = 7.8$, $p < 0.05$). The Shannon Index was statistically related to the interaction between treatment and wild boar abundance, meaning that the differences in arthropod diversity were only evidenced when high wild boar densities occurred ($F_{1,123} = 4.31$, $p < 0.05$; table 2). This was mainly due to an increase in the diversity index in the FP with high densities of wild boar (fig. 3), with diversity remaining similar in the OP.

Table 3 shows the models concerning the relationships between invertebrate numbers on the surface (absolute abundance) and ungulate densities, both overall and separately for each taxonomic group: Insecta (no Hymenoptera), Hymenoptera, Arachnida and 'others'. The percentage of shrubs was statistically and negatively related to both Hymenoptera counts and the total amount of arthropods. Interestingly, the interaction between deer and wild boar abundances was statistically related to the total invertebrate counts (fig. 4A) and the number of invertebrates included in the 'others' group (fig. 4B). Independently of red deer abundance, when high wild boar densities occurred the number of invertebrates decreased, although at low wild boar abundance a positive association between red deer density and the number of invertebrates was recorded. Those invertebrates included in the 'others' group were more frequent in areas with high abundance of both red deer and wild boar. A positive relationship between red deer density and the absolute frequency of trapped invertebrates was also found (fig. 5A, 5B).

Discussion

Our main results were that (i) higher values of invertebrates richness were found in ungulate exclusion areas, and (ii) the high densities of wild boar had a particularly negative effect on invertebrates diversity. These findings support the negative relationships between high wild boar abundance and invertebrates in Mediterranean ecosystems, which may be considered to be arthropod hotspots (Hernandez–Manrique et al., 2012).

The higher abundance of invertebrates in the FP may be caused by a local attraction effect, since invertebrates might seek refuge in fenced patches

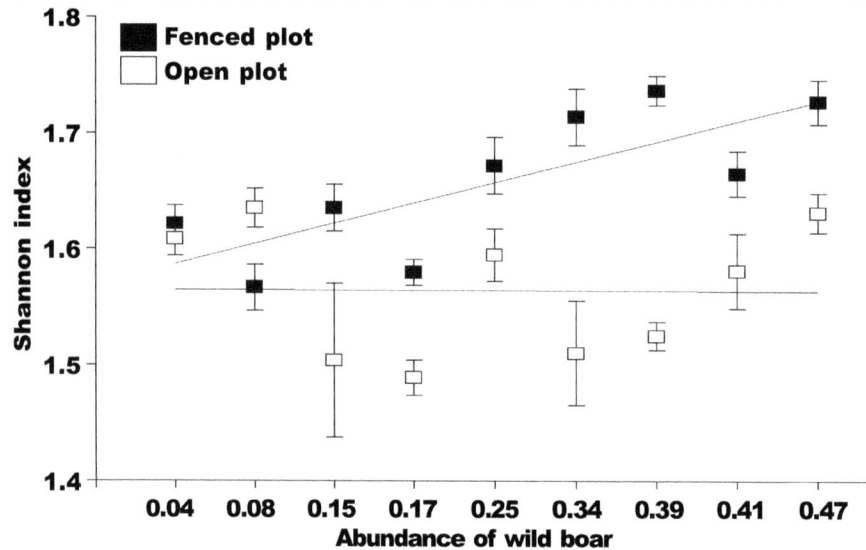

Fig. 3. Shannon index as a function of wild boar abundance index per estate (mean ± SE).

Fig. 3. Índice de Shannon en función del índice de abundancia de jabalíes por finca (media ± EE).

in which they actively look for the conditions inside the plots where no wild board predation (Grayson & Hassall, 1985) or overgrazing occurs. Overgrazing is known to cause a decrease in the food that is available to the edaphic fauna (Dennis et al., 2001, 2008;

Rosa–García et al., 2009, 2010) and suitable places for egg production, laying and incubation. Moreover, inside the fenced plots, the invertebrates would avoid disturbance from wild boar and red deer, which strongly affect soil compaction/structure through trampling and

Table 3. Full models on the effects of ungulates on the number of invertebrates (Insecta, Hymenoptera, Arachnida, others and total, respectively): ** $p < 0.01$; * $p < 0.05$.

Tabla 3. Modelos completos sobre los efectos de los ungulados en el número de invertebrados (Insecta, Hymenoptera, Arácnida, otros y total, respectivamente): ** p < 0,01; * p < 0,05.

	Insecta		Hymenoptera		Arachnida		Others		Total	
	F	β	F	β	F	β	F	β	F	β
Intercept	1.1	0.44	1.57*	1.48	0.14	1.78	1.17	−10.2	2.86**	3.18
Treatment (T)	0.43	0.14	0.22	0.16	0.22	0.27	0.025	−1.26	0.39	0.21
Deer density (Dd)	2.2	2.10	0.36	−0.75	0.004	0.47	5.99*	80.53	5.19*	3.75
Wild boar abundance (Wba)	0.26	0.17	0	0.095	0.009	−0.45	3.33	50.54	2.13	1.65
Shrub	0.5	0.025	7.31**	−0.13	0.02	0.012	2.12	−1.96	9.88**	−0.17
Dd*Wba	0.96	−2.82	0.023	0.49	0.005	0.39	5.36*	−177.4	4.29*	7.59
T*Dd	2.31	−0.63	0.003	−0.061	0.14	−0.67	0.14	−9.1	0.42	−0.79
Plant biomass	0.58	0.049	0.07	−0.026	0.06	−0.04	1.17	2.51	0.37	0.19
Wooded	2.5	−0.084	0.43	−0.049	0.02	0.017	0.004	0.12	0.31	−0.086
Grass	0.54	0.001	0.78	−0.002	0.16	−0.001	0.06	−0.015	0.43	−0.002
T*Wba	2.32	0.88	0.07	−0.22	0.09	0.022	0.45	13.98	0.89	0.94

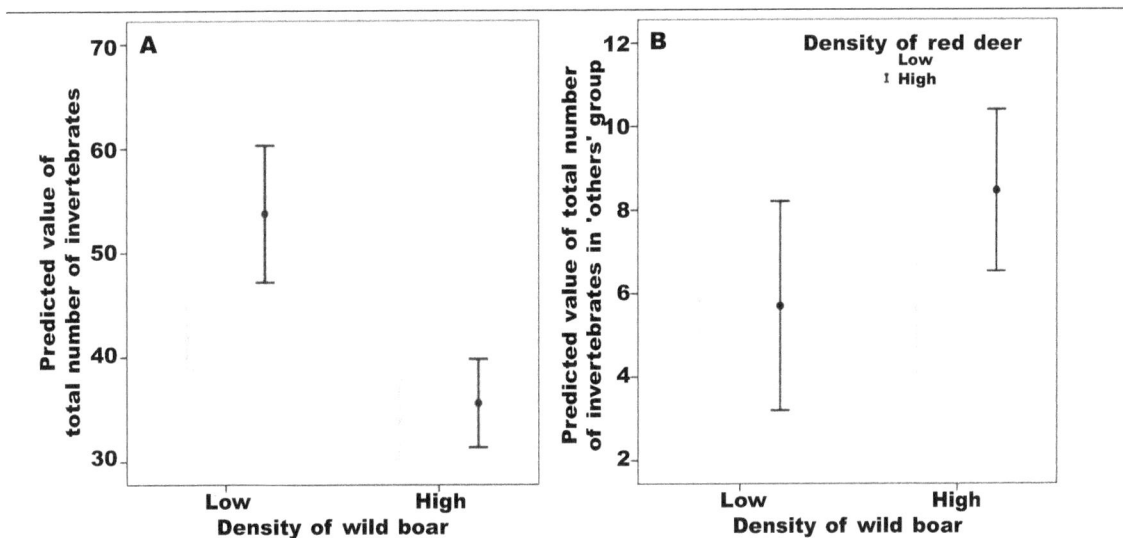

Fig. 4. Total number of invertebrates (A) and number of invertebrates included in 'others' group (B) as a function of interaction of wild boar and red deer abundance index groups (categorized according to the median ± 95% CI of the abundance indexes).

Fig. 4. Número total de invertebrados (A) y número de invertebrados incluidos en el grupo otros (B) en función de la interacción de los índices de abundancia de jabalíes y de ciervos rojos (agrupados según la mediana ± IC del 95% de los índices de abundancia).

rooting activities (Massei & Genov, 2004; Bueno, 2011). This could alter the establishment of a range of invertebrate species with different ecological requirements (Thiele–Bruhn et al., 2012), thus reducing the diversity of invertebrates. Our study supports previous findings in other environments showing that the overabundance of wild boar damages the structure of fauna communities (Côté et al., 2004; Allombert et al., 2005; Mohr et al., 2005; Albon & Brewer, 2007; Cuevas et al., 2012; Wirthner et al., 2012). However, in our study, the principal predictor of the invertebrate dry mass was the percentage of pasture cover, probably because pasture cover benefits certain abundant species more than others, and the ungulate effect is not appreciated in terms of invertebrate biomass.

Moreover, the differences on invertebrates diversity (Shannon Index) between fenced and open areas were higher in hunting states with higher wild boar density. In other words, the values of Shannon Diversity Index were much higher in ungulate proof areas than in open areas characterized by high wild boar densities. This may be due to the less favourable habitat in the surroundings as a consequence of overgrazing and rooting activity, possibly attracting more invertebrates to undisturbed patches (Gardiner & Hassall, 2009). Indeed, the wild boar diet includes not only vegetation but also many meso– and macro–invertebrates (Cuevas et al., 2010). Therefore, high wild boar densities may cause an intense disturbance of edaphic fauna, and invertebrates from the area tend to aggregate more in FP than in areas with lower wild boar abundance.

Interestingly, the interaction between deer and wild boar abundances was statistically related to the total counts of invertebrates and the number of invertebrates included in the 'others' group. Independently of red deer abundance, when wild boar densities were high, the number of invertebrates decreased, indicating that the wild boar, at high densities, have an overall negative impact on invertebrates. However, when wild boar abundance was low, a positive association between red deer density and the number of invertebrates was evident. We observed a positive relationship between red deer density and the absolute frequency of trapped invertebrates and the 'others' category, which must be explained in terms of the interaction between red deer and wild boar abundances (also significant, see discussion below). In contrast, as figure 5A shows, the high absolute frequency of invertebrates was recorded at intermediate values of red deer density, which is in agreement with previous studies that suggest a positive effect of moderate grazing pressure (Gómez et al., 2004).

Our results further suggest that Isopoda and Myriapoda groups, the most abundant taxa found in the 'other' group, could benefit from high red deer abundances (fig. 5B). These groups have phytophagous but also important saprophytic diets and may therefore benefit from the removal of bushes and the presence of the layer of grass, which provides an increased amount of organic plant matter, and therefore an increased source of food (Bugalho & Milne, 2003; Côté et al., 2004). Furthermore, ungulate faeces attract invertebrates that consume the dung and gain moisture from it or consume microbes within it (Stewart, 2001).

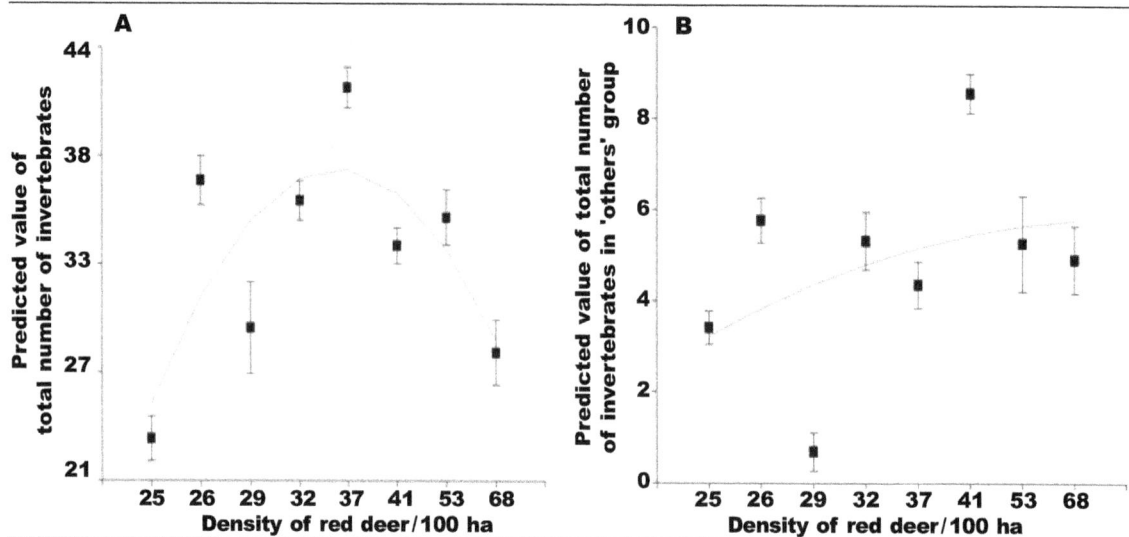

Fig. 5. Total number of invertebrates (A) and number of invertebrates included in 'others' group (B) as a function of red deer density per estate (mean ± SE).

Fig. 5. Número total de invertebrados (A) y número de invertebrados incluidos en el grupo otros (B) en función de la densidad de ciervos rojos por finca (media ± EE).

With regard to the Arachnida and Insecta category, we found no differences in abundance either inside or outside the fenced plots, although grass favoured the presence of the Araneida order (Rosa–García et al., 2009). The composition of the habitat and the development of pastures as a result of moderate deer grazing may benefit the presence of animals included in the Arachnida category (Dennis et al., 2001; Paschetta et al., 2013). On the other hand, our results show that the percentage of shrub cover has negative effects on the abundance of Hymenoptera. A study carried out by Azcarate & Peco (2012) in a Mediterranean ecosystem led them to conclude that the generation of a more hete-rogeneous environment at the smaller scales increased the species diversity of ants. However, the reasons for the negative influence of shrubs on Hymenoptera remain unclear and more research on the type of ecological relationships that exist between them are therefore necessary as few studies have focused on discovering these relationships in a Mediterranean environment.

General conclusions

This research has evidenced the relationships between ungulate abundance (in high density areas) and edaphic invertebrate abundance and richness under Mediterra-nean constraints. Overall, this study supports the notion that high density populations of wild boar may damage the structure of soil fauna communities as a result of a decrease in food availability owing to overgrazing, soil disturbance by rooting, and direct predation. The conservation applications of this study refer to wild boar population density control under Mediterranean

conditions where big game hunting has become an important industry. In particular, high densities of wild boar have a strong impact on invertebrates when com-pared to red deer, and a positive association was even noted in regard to the number of trapped invertebrates. Furthermore, since fenced plots evidenced a local scale effect, playing a role as refuges, the implementation of ungulate proof exclusion fences is desirable in order to maintain invertebrate communities, which would in turn enhance the food availability for many birds, including the red legged–partridge. However, more studies are needed to develop field protocols (*e.g.* the size and location of such fenced patches) and to assess population control effects on the invertebrate community.

Acknowledgements

We should like to thank E. López and the students Oscar, David and Macarena for their help in collecting field data, and two anonymous reviewers who contri-buted to the improvement of the manuscript. We are also grateful to the estate keepers for their hospitality and assistance in the field work, and to the Andalucía Autonomous Government's Environmental Council for financing this work via a project to improve the habitat of the black vulture in Córdoba province, thanks to which we have been able to carry out this work.

References

Acevedo, P., Vicente, J., Höfle, U., Cassinello, J., Ruiz–Fons, F. & Gortázar, C., 2007. Estimation of European wild boar relative abundance and aggregation: a novel method in epidemiological risk assessment. *Epidemiology and Infection,* 135: 519–527.

Acevedo, P., Ruiz–Fons, F., Vicente, J., Reyes–García, A. R., Alzaga, V. & Gortázar, C., 2008. Estimating red deer abundance in a wide range of management situations in Mediterranean habitats. *Journal of Zoology,* 276: 37–47.

Aebischer, N. J. & Ewald, J. A., 2012. The grey partridge in the UK: popiulation status, research, policy and prospects. *Animal Biodiversity and Conservation,* 35.2: 203–212.

Albon, S. D. & Brewer, M. J., 2007. Quantifying the grazing impacts associated with different herbivores on rangelands. *Journal of Applied Ecology,* 44: 1176–1187.

Allombert, S., Stockton, S. & Martin, J. L., 2005. A Natural Experiment on the Impact of Overabundant Deer on Forest Invertebrates. *Conservation Biology,* 19: 1917–1929.

Apollonio, M., Anderson, R. & Putman, R., 2010. *European Ungulates and their Management in the 21st Century.* Cambridge Univ. Press, Cambridge.

Azcarate, F. M. & Peco, B., 2012. Abandonment of grazing in a Mediterranean grassland area: consequences for ant assemblages. *Insect Conservation and Diversity,* 5: 279–288.

Barrientos, J. A. & Abelló, P., 2004. *Curso práctico de entomología.* Univ. Autónoma de Barcelona, Barcelona.

Barrios–García, M. N. & Ballari, S. A., 2012. Impact of wild boar (*Sus scrofa*) in its introduced and native range: a review. *Biological Invasions,* 14: 2283–2300.

Blum, M. S., 1978. Biochemical defenses of insects. In: *Biochemistry of insects*: 465–513 (M. Rockstein, Ed.). Academic Press, New York.

Breitenmoser, U., 1998. Large predators in the Alps: the fall and rise of man's competitors. *Biological Conservation,* 83: 279–289.

Buckland, S. T., Anderson, D. R., Burnham, K. P., Laake, J. L., Borchers, D. L. & Thomas, L., 2004. Introduction to distance sampling: estimating abundance of biological population. Oxford Univ. Press, Oxford.

Bueno, C. G., 2011. Impacto de las perturbaciones del jabalí en los pastos supraforestales de los Pirineos Centrales. Ph. D. Thesis, Univ. de Zaragoza, Instituto Pirenaico de Ecología (CSIC).

Bugalho, M. N. & Milne, J. A., 2003. The composition of the diet of red deer (*Cervus elaphus*) in a Mediterranean environment: a case of summer nutritional constraint? *Forest Ecology and Management,* 181: 23–29.

Carpio, A. J., Guerrero–Casado, J., Ruiz–Aizpurua, L., Vicente, J., Tortosa, F. S., 2014a. The high abundance of wild ungulates in a Mediterranean region: is this compatible with the European rabbit? *Wildlife Biology,* 20: 161–166.

Carpio, A. J., Guerrero–Casado, J., Tortosa, F. S. & Vicente, J., 2014b. Predation of simulated red–legged partridge nests in big game estates from

South Central Spain. *European Journal of Wildlife Research,* 60: 391–394.

Clutton–Brock, T. H. & Albon, S. D., 1992. Trial and error in the Highlands. *Nature,* 358: 11–12.

Côté, S. D., Rooney, T. P., Trembley, J. P., Dussault, C. & Waller, D. M., 2004. Ecological impacts of deer overabundance. *Annual Review of Ecology, Evolution, and Systematics*, 35: 113–147.

Cuevas, M. F., Mastrantonio, L., Ojeda, R. A. & Jaksic, F. M., 2012. Effects of wild boar disturbance on vegetation and soil properties in the Monte Desert, Argentina. *Mammalian Biology,* 77: 299–306.

Cuevas, M. F., Novillo, A., Campos, C., Dacar, M. A. & Ojeda, R. A., 2010. Food habits and impact of rooting behaviour of the invasive wild boar, *Sus scrofa,* in a protected area of the Monte Desert, Argentina. *Journal of Arid Environment,* 74: 1582–1585.

Dennis, P., Skartveit, J., McCracken, D. I., Pakeman, R. J., Beaton, K., Kunaver, A. & Evans, D. M., 2008. The effects of livestock grazing on foliar arthropods associated with bird diet in upland grasslands of Scotland. *Journal of Applied Ecology*, 45: 279–287.

Dennis, P., Young, M. R. & Bentley, C., 2001. The effects of varied grazing management on epigeal spiders harvestmen and pseudoscorpions of *Nardus stricta* grassland in upland Scotland. *Agriculture, Ecosystems and Environment*, 86: 39–57.

Gardiner, T. & Hassall, M., 2009. Does microclimate affect grasshopper populations after cutting of hay in improved grassland? *Journal of Insect Conservation,* 13: 97–102.

Gebeyehu, S. & Samways, M. J., 2006. Conservation refugium value of a large mesa for grasshoppers in South Africa. *Biodiversity and Conservation,* 15: 717–734.

Gerlach, J., Samways, M. & Pryke, J., 2013. Terrestrial invertebrates as bioindicators: an overview of available taxonomic groups. *Journal of Insect Conservation,* 17: 831–850.

Giménez–Anaya, A., Herrero, J., Rosell, C., Couto, S. & García–Serrano, A., 2008. Food habits of wild boars (*Sus scrofa*) in a Mediterranean Coastal Wetland. *Wetlands,* 28: 197–203.

Gómez, J. M., Gonzáles–Megías, A. & Sanchez–Pinero, F., 2004. Effects of ungulates on epigeal arthropods in Sierra Nevada National Park (Southeast Spain). *Biodiversity and Conservation,* 13: 733–752.

Gordon, I. J., Hester, A. J. & Festa–Bianchet, M., 2004. The management of wild large herbivores to meet economic, conservation and environmental objectives. *Journal of Applied Ecology,* 41: 1021–1031.

Grayson, F. W. L. & Hassall, M., 1985. Effects of rabbit grazing on population variables of *Chorthippus brunneus* (Orthoptera). *Oikos,* 44: 27–34.

Hedlund, K. & Öhrn, M. S., 2000. Tritrophic interactions in a soil community enhance decomposition rates. *Oikos,* 88: 585–591.

Hernandez–Manrique, O. L., Numa, C., Verdu, J. R., Galante, E. & Lobo, J. M., 2012. Current protected sites do not allow the representation of endangered invertebrates: the Spanish case. *Insect Conservation and Diversity,* 5: 414–421.

Herrero, J., García–Serrano, A., Couto, S., Ortuño, V. M. & García–González, R., 2006. Diet of wild boar *Sus scrofa* L. and crop damage in an intensive agroecosystem. *European Journal of Wildlife Research*, 52: 245–250.

Holland, J. M., Hutchison, M. A. S., Smith, B. & Aebischer, N. J., 2006. A review of invertebrates and seed–bearing plants as food for farmland birds in Europe. *Annals of Applied Biology*, 148: 49–71.

Lowe, S., Browne, M., Boudjelas, S. & De Poorter, M., 2000. *100 of the World's Worst Invasive Alien Species A selection from the Global Invasive Species Database*. The Invasive Species Specialist Group (ISSG). Auckland, New Zealand.

Massei, G. & Genov, P., 2004. The environmental impact of wild boar. *Galemys*, 16: 135–145.

McIntyre, N. E., 2000. Ecology of urban arthropods: a review and a call to action. *Annals of the Entomological Society of America*, 93: 825–835.

Mizuki, I., Yamasaki, M., Kakutani, T. & Isagi, Y., 2010. Negligible impact of deer–induced habitat degradation on the genetic diversity of extant *Bombus diversus* populations in comparison with museum specimens. *Journal of Insect Conservation*, 14: 191–198.

Mohr, D., Cohnstaedt, L. & Topp, W., 2005. Wild boar and red deer affect soil nutrients and soil biota in steep oak stands of the Eifel. *Soil Biology and Biochemistry*, 37: 693–700.

Morrison, M. L., Marcot, B. G. & Mannan, R. W., 1992. *Wildlife–habitat relationship. Concepts and applications*. Univ. of Wisconsin Press, Madison.

Osler, G. H. R. & Sommerkorn, M., 2007. Toward a complete soil C and N cycle: incorporating the soil fauna. *Ecology*, 88: 1611–1621.

Paschetta, M., La Morgia, V., Masante, D., Negro, M., Rolando, A. & Isaia, M., 2013. Grazing history influences biodiversity: a case study on ground–dwelling arachnids (*Arachnida: Araneae, Opiliones*) in the Natural Park of Alpi Marittime (NW Italy). *Journal of Insect Conservation*, 17: 339–356.

Raebel, E. M., Merckx, T., Feber, R. E., Riordan, P., Macdonald, D. W. & Thompson, D. J., 2012. Identifying high–quality pond habitats for Odonata in lowland England: implications for agri–environment schemes. *Insect Conservation and Diversity*, 5: 422–432.

Rosa–García, R., Jáuregui, B. M., García, U., Osoro, K. & Celaya, R., 2009. Effects of livestock breed and grazing pressure on ground–dwelling arthropods in Cantabrian heathlands. *Ecological Entomology*, 34: 466–475.

Rosa–García, R., Ocharan, F. J., García, U., Osoro, K. & Celaya, R., 2010. Arthropod fauna on grassland–heathland associations under different grazing managements with domestic ruminants. *Comptes Rendus Biologies*, 333: 226–234.

Shannon, C. E., 1948. A mathematical theory of communication. *Bell System Technical Journal*, 27: 379–423.

Sarasa, M. & Sarasa, J. A., 2013. Intensive monitoring suggests population oscillations and migration in wild boar *Sus scrofa* in the Pyrenees. *Animal Biodiversity and Conservation*, 36: 79–88.

Spalinger, L. C., Haynes, A. G., Schutz, M. & Risch, A. C., 2012. Impact of wild ungulate grazing on Orthoptera abundance and diversity in subalpine grasslands. *Insect Conservation and Diversity*, 5: 444–452.

Stewart, A. J. A., 2001. The impact of deer on lowland woodland invertebrates: a review of the evidence and priorities for future research. *Forestry*, 74: 259–270.

Swift, M. J., Heal, O. W. & Anderson, J. M., 1979. *Decomposition in terrestrial ecosystems*. Blackwell Scientific, Oxford, UK.

Thiele–Bruhn, S., Bloem, J., Vries, F. Td., Kalbitz, K. & Wagg, C., 2012. Linking soil biodiversity and agricultural soil management. *Current Opinion in Environmental Sustainability*, 4: 523–528.

Vicente, J., Höfle, U., Fernández–de–Mera, I. G. & Gortázar, C., 2007. The importance of parasites life histories and host density in predicting the impact of infections in red deer. *Oecologia*, 152: 655–664.

Vickery, J., Feber, R. & Fuller, R., 2009. Arable field margins managed for biodiversity conservation: A review of food resource provision for farmland birds. *Agriculture, Ecosystems and Environment*, 133: 1–13.

Wilson, J. D., Morris, A. J., Arroyo, B. E., Clark, S. C. & Bradbury, R. B., 1999. A review of the abundance and diversity of invertebrate and plant foods of granivorous birds in northern Europe in relation to agricultural change. *Agriculture, Ecosystems and Environment*, 75: 13–30.

Wirthner, S., Schütz, M., Page–Dumroese, D. S., Busse, M. D., Kirchner, J. W. & Risch, A. C., 2012. Do changes in soil properties after rooting by wild boars (*Sus scrofa*) affect understory vegetation in Swiss hardwood forests? *Canadian Journal of Forest Research*, 42: 585–592.

Zuur, A. F., Ieno, E. N., Walker, N. J., Saveliev, A. A. & Smith, G. M., 2009. *Mixed Effects Models and Extensions in Ecology with R*. New York, Springer.

19

An assessment of wetland nature reserves and the protection of China's vertebrate diversity

R. Sun, Y. Zheng, T. Lei & G. Cui

Sun, R., Zheng, Y. Lei, T. & Cui, G., 2014. An assessment of wetland nature reserves and the protection of China's vertebrate diversity. *Animal Biodiversity and Conservation*, 37.2: 217–225.

Abstract

An assessment of wetland nature reserves and the protection of China's vertebrate diversity.— We assessed 148 wetland nature reserves in China and the distribution of the four taxa of endemic and threatened terrestrial vertebrates, reptiles, amphibians, birds and mammals. Assessment of the wetland nature reserves was combined with the governmental list of the endemic and threatened vertebrates to identify the richness of the species. Species richness was scored as high, medium or low using a factor analysis method, and 31 wetland ecosystems were marked as high protection areas. The relationship between the threatened species and the endemic species in the reserves was also analyzed. We found that both richness patterns were similar. Based on the richness study, a nature reserve classification system with corresponding management is expected to be established in the future to protect species diversity in China.

Key words: Wetland nature reserve, Vertebrate diversity, Threatened species, Endemic species, Factor analysis

Resumen

Evaluación de los humedales declarados reserva natural y protección de la diversidad de vertebrados de China.— Se evaluaron las reservas naturales existentes en China y se estudió la distribución de los cuatro taxones de vertebrados terrestres endémicos y amenazados, que comprenden los reptiles, los anfibios, las aves y los mamíferos. Con vistas a determinar la riqueza de las especies, se combinó la evaluación anterior de los 148 humedales declarados reserva natural con la lista de los vertebrados endémicos y amenazados elaborada por el gobierno. Dicha riqueza de especies se clasificó en tres categorías: alta, media y baja mediante un análisis factorial y se seleccionaron 31 ecosistemas de humedales como zonas de alta protección. Asimismo, se analizó la relación existente entre las especies amenazadas y las endémicas en las reservas, y se observó que ambos modelos de riqueza eran parecidos. Se espera que, sobre la base del estudio de la riqueza, se pueda crear un sistema de clasificación de las reservas naturales, con su gestión correspondiente, con el objetivo de proteger la diversidad de las especies en China.

Palabras clave: Humedal declarado reserva natural, Diversidad de vertebrados, Especies amenazadas, Especies endémicas, Análisis factorial

Rui Sun, College of Food and Biology Engineering, Qilu Univ. of Technology, Jinan, 250353, China.– Yaomin Zheng, Inst. of Remote Sensing Applications, Chinese Academy of Sciences, Beijing 100101, China.– Ting Lei & Guofa Cui, College of Nature Conservation, Beijing Forestry Univ., Beijing 100083, China.

Corresponding author: Rui Sun. E–mail: sunrui_qut@163.com

Introduction

Wetland ecosystems are considered key habitat environments for mammals, birds, reptiles, amphibians, and aquatic plants (Gibbs, 2000; Gibbons, 2003; Roe et al., 2006). In order to protect biodiversity in wetland ecosystems, in its biotic and abiotic dimensions, the establishment of protected areas, such as nature reserves, has been widely adopted worldwide (Soule, 1991; Margules & Pressey, 2000). Undoubtedly, it is necessary to determine an effective way to protect all types and species of wildlife.

As stated by Margules & Pressey (2000), the main role of reserves is to separate elements of biodiversity from processes that threaten their existence in the wild. However, for social, economic or political reasons, some reserves could be at risk of degradation or revocation of status. Biodiversity loss may occur in wetland systems, for example, due to land use changes, habitat destruction, pollution, exploitation of resources, or invasive species. The human impact is most apparent in wetland systems covering a relatively small area. Establishing protected area networks (PANs) can improve the management of protected areas and reduce human impact. Protected area networks can be defined as collections of at least two individual protected areas or reserves that are managed spatially, economically or socially in a cooperative and synergistic manner. Well–designed, protected area networks can improve species persistence in a certain region, representing the richness of overall species (Bonn et al., 2002). In a protected area, choosing appropriate surrogate species as effective indicators is critical to identify conversation priorities and establish conservation planning and management. Many studies have focused on biodiversity using surrogates of species richness (Wiens et al., 2008; Loyola et al., 2007; Caro & O'Doherty, 1999; Pinto et al., 2008) and the approach has established the relationship between species richness and taxa in some reserves (Plumptre et al., 2007; Lund & Rahbek, 2002; Negi & Gadgil, 2002; Qian, 2007; Wolters et al., 2006). Based on this relationship, the cross–taxon congruence of species diversity in the reserves has been applied to conservation strategies. Furthermore, biodiversity conservation can be planned based on such analysis. Conservation prioritization of all species, especially threatened and endemic vertebrate species, is considered essential for planning (Brooks et al., 2004; Grenyer et al., 2006; Rodrigues et al., 2004).

The main challenges to biodiversity conservation are related to the implementation of conservation plans (Pressey, 1998). Improper management may damage reserves and cause further loss of diversity. To avoid such an outcome, various protection policies have been developed for management of nature reserves (Margules & Pressey, 2000; Margules et al., 2002; Goodman, 2003; Wiersma & Nudds, 2009). National or international reserve systems should be central to biodiversity conservation. In China, endemic and threatened terrestrial vertebrates in the wetland reserves have been considered and protected, and a classification policy was issued by the government in 1994. However, systematic management does not seem to have been effectively implemented, particularly at a national level (Liu et al., 2003; Cui, 2004). There is a pressing need to solve many issues related to wetland conservation and biodiversity protection in China. The first of these is related to a severe decrease in the area of wetlands. In 2005, the area of wetlands was about 660,000 km^2 (Liu & Diamond, 2005) but in 2009 it had decreased to 359,478 km^2 (Niu et al., 2009), causing a sharp decline in habitat for species. Second, numerous threatened and endemic wetland vertebrates might not be registered in the reserves (Li et al., 2002; Wang & Xie, 2004; Yan et al., 2005), so conversation priorities could be incorrect. Third, the existing reserves or PANs are influenced by regional restrictions or distribution of rare species. Some of these reserves were established for maximum return of investment. To optimize reserve resources, it is therefore necessary to develop a new, effective and well–designed network for China's nature reserves, based on others' successful experiences (Reid, 1998; Scott et al., 2001; Cabeza & Moilanen, 2001). Unfortunately, threatened and endemic species in protected wetland areas in China have not been analyzed to date on a large scale to meet such an objective.

In this work, we analyzed the ranking of the protected wetland areas in China based on threatened and endemic vertebrate species richness. The species in those protected areas were identified and their protection value was evaluated. This work could provide useful data and new insights to establish and improve management systems for nature reserves in China.

Methods

Data collection

The data on species richness in wetland nature reserves were provided by the Beijing Forestry University's nature reserve database that includes scientific research reports, management plans, and project reports. The database was originally established as a reference for national and provincial reserve committees. The vertebrates in the database were classified into four taxa: mammals, birds, reptiles, and amphibians.

One hundred and forty–eight wetland nature reserves in China (77 national reserves and 71 local reserves), and 22 international wetlands were involved in this work. The classification was based on the State Bureau of Technical Supervision, 1994. The samples are the most important protected wetland areas administered by the State Forestry Administration of the People's Republic of China. The species data for each of these reserves were collected, as reported in references (Zheng et al., 2012).

Because of differences in information or category, some species that are not in the threatened species list in the IUCN Red List are listed in the China Species Red List. Thus, the threatened species in this work refer to those whose taxonomic level is considered threatened, including critically endangered (CR), endangered (EN), and vulnerable (VU) species, in the IUCN red list, the China Red Data Book of Endangered Animals, or the

China Species Red List. The national key protected species of China are also involved. The endemic species are indexed in the China Species Red List, referring to the species only found on mainland China. The basic information for endemic birds and mammals was collected from the literature (Mackinnon & Phillipps, 2000; Smith & Xie, 2008). The numbers of threatened and endemic species were in accordance with the data in the literature. Therefore, the database was used for all species in each reserve and the sampling covered all provinces in China except Gansu and Chongqing.

Ranking analysis of wetland protected areas

The ranking of the wetland protected areas was analyzed using the method developed by Plumptre et al, 2007. In short, the protected wetland areas were ranked in terms of the number of the endemic and threatened species for four vertebrate taxa. The weight on each taxon was assumed to be the same. The rank scores were standardized for each protected area by dividing by the maximum rank score to account for the fact that some sites do not have data for all taxa, and to account for the varying number of the endemic or threatened species between taxa. A mean rank was then calculated for both the endemic and threatened species for each site across the four taxa by summing ranks from each taxon and dividing this by the number of taxa for which there were survey data. These final standardized scores were then used to rank all the wetland protected areas.

Factor analysis

The following procedure was used to analyze the richness of species, namely factor scores. The richness of four vertebrate taxa was separately analyzed into the following eight classes: threatened amphibians, threatened reptiles, threatened birds, threatened mammals, endemic amphibians, endemic reptiles, endemic birds, and endemic mammals, referred to as indicators of X1, X2, X3, X4, X5, X6, X7 and X8, respectively.

Species richness in each class in all nature reserves was corrected using the following correction formula:

$$\ln(1 + X) / \ln(area)$$

where x is the number of species and area refers to the area of reserve sites.

The Kaiser–Meryer–Olkin test (KMO) and Bartlett sphericity test were performed, based on the correction results, to check whether the correlation matrix can be presumed to be identified. The corrected richness values were then subjected to the factor analysis in the following procedure. The raw data were first standardized and the principal components were analyzed. The variance of the cumulative contribution rate to eigenvalues was calculated by the varimax orthogonal rotation transform method. A factor score was then calculated by the regression method. Finally, the weighted average of composite scores was calculated based on the contribution rate of the characteristic roots. The following equation was used to calculate the composition scores:

$$CS = \lambda_1 f_1 + \lambda_2 f_2 + + \lambda_n f_n$$

where CS is the composite scores; f_1, f_2 ... f_n are the factor values of the first, second, and n^{th} factors, respectively; λ_1, λ_2 ... λ_n are the contribution rates to the eigenvalues for the first, second, and n^{th} factors.

Identification of the conservation priority

SPSS 17.0 software was used to calculate the single variable frequency distribution. Three levels with the composition scores of 1–49, 50–98, and 99–148 were designated as low, medium, or high, respectively, for all reserves. The levels were in accordance with the conservation priorities of the wetland nature reserves.

Correlation analysis

The relationship between the numbers of threatened and endemic species and the ranking of wetland ecosystem was studied. Pearson correlation coefficients were used to interpret the strength of the relationship. The significance of the correlation coefficient for all tests was P < 0.01, as shown in table 1. In short, the coefficient of the high correlation was above 0.50. The range of moderate correlation was between 0.30 and 0.10. The coefficient of the low correlation was below 0.10. (Lamoreux et al., 2005).

Results

Correlation analysis and identification of common factors

The value in the KMO test was 0.711, slightly higher than 0.7. The concomitant probability obtained by the Bartlett sphericity test was 0, considerably lower than the significance level of 0.05, fitting the adequacy to enter a factor analysis (Ferguson & Cox, 2007)

Table 1 summarizes the Pearson correlation coefficients for eight diversity indicators. A positive correlation was observed for every pair of indicators, whether cross–taxon or intra–taxon. The correlation coefficient of 11 in 28 cross–taxon comparisons was larger than 0.5, suggesting that they are in moderate correlation. There was a high correlation between the threatened species and endemic species for a certain taxon.

For the threatened species, there was a high correlation between amphibians and reptiles and a moderate correlation between reptiles and birds or mammals. On the other hand, for the endemic species, a strong correlation was found between birds and mammals. However, the richness of threatened birds was lower than that of threatened amphibians and mammals if the indicator values were corrected in the reserve area.

Table 2 shows the correlation matrix between the eigenvalues and eigenvectors. Two eigenvalues, which were larger than 1, were considered as the significant common factors. One of the common factors, referred as to f_1, with an eigenvalue of 4.421, was characterized as the terrestrial species factor for amphibians, reptiles, and mammals. The other common factor, referred to as f_2, with an eigenvalue of 1.198, was characterized as the avian species factor for the birds.

Table 1. Pearson correlation coefficients for eight diversity indicators: * Significant correlation at 0.01 level (2–tailed).

*Tabla 1. Coeficientes de correlación de Pearson para ocho indicadores de la diversidad: * La correlación es significativa al nivel 0,01 (bilateral).*

Taxon	X2	X3	X4	X5	X6	X7	X8
X1	0.647	0.384	0.416	0.714	0.626	0.479	0.465
X2		0.481	0.533	0.698	0.738	0.405	0.344
X3			0.267	0.258	0.418	0.773	0.154
X4				0.412	0.427	0.346	0.722
X5					0.583	0.303	0.412
X6						0.549	0.524
X7							0.449

Reserve ranking and species richness analysis

Results showed 70.77% of the bird species and 59.31% of the mammal species in China were recorded in the 148 protected areas. The percentages for the reptile species and amphibian species were 47.6% and 35.5%, respectively (fig. 1). Except for amphibian species, the proportion of the threatened species taxa was higher than that of the endemic species. Thus, based on the rule of priority, the threatened species was placed on the higher priority than the endemic species in the China wetland protected areas. For the amphibians, only one species, which is *Andrias davidanus*, was in the protection list in a special reserve in China. The endemic amphibians should be considered as the indicator for the protection functions in the protected areas.

Figure 2 shows the distribution matrix of priority ranking for all wetland nature reserves. The horizontal axis is for threatened species and the vertical axis is for endemic species. The reserves of 31, as marked as dark box, were scored as high level for both threatened species and endemic species. Thus, these reserves should have the highest overall protection value. On the contrary, the reserves of 20 were scored as low level for both threatened species and endemic species. They were considered as the lowest overall protection value. Other reserves were between the highest and the lowest protection values.

Most of the 31 reserves with the highest protection value were located in central and southern China in the middle and lower reaches of the Yangtze River (fig. 3). The major wetland ecosystems in these reserves are lakes, rivers, and marshes. Thus, these 31 reserves may be representative for the diversity of threatened and endemic species.

Figure 4A shows the species richness, expressed in percentages, for the threatened species in the 31 reserves, and figure 4B shows the percentages for the endemic species.

The types of the wetland reserves included offshore (coast) areas, lakes, rivers, and marshes (Tang & Huang, 2003; Scott & Jones, 1995). For the 31 reserves with the highest protection value, the most common type was lake reserves (16 lakes), and the second most common type was river reserves (12 rivers reserves). Lakes can generally provide good habitats for migratory water birds. For instance, the Poyanghu reserve and

Table 2 A summary of factor analysis results: Ts. Terrestrial species; As. Avian species.

Tabla 2. Resumen de los resultados del análisis factorial: Ts Especies terrestres; As. Especies de aves.

Item	Ts (F1)	As (F2)
X1	0.722	0.380
X2	0.715	0.428
X3	0.103	0.940
X4	0.772	0.082
X5	0.779	0.195
X6	0.695	0.461
X7	0.277	0.858
X8	0.778	0.059
Eigen value	4.421	1.198
Variance contribution rate	42.648	27.593
Cumulative variance contribution rate		
	42.648	70.241

Table 3. Classification of the 31 reserves with the highest conservation value: * Location is shown in parenthesis; LP. Level of protection (N. National; L. Local); ** Wetlands of international importance (Wii).

Tabla 3. Clasificación de las 31 reservas con el valor de conservación más elevado. La ubicación se muestra entre paréntesis; LP. Grado de protección (N. Nacional; L. Local); ** Humedales de importancia internacional (Wii).*

Reserve*	Wetland type	Area (km²)	LP	Wii**
Poyanghu (Jiangxi)	Lake	224	N	✓
Yancheng (Jiangsu)	Offshore and seacoast	4,530	N	✓
Dongtinghu (Hunan)	Lake	1,900	N	✓
Xinfengjiang (Guangdong)	Lake	1,014	L	
Huidonglianhuashanbaipenzhu (Guangdong)	Lake	140.34	L	
Sanjiangyuan (Qinghai)	Marsh, lake	152,300	N	✓
Zhouzhiheihe (Shananxi)	River	131.25	L	
Anqing (Anhui)	Lake	987	L	
Longganhu (Hubei)	Lake	223.22	N	
Tonglingdanshuitun (Anhui)	Lake	519.50	N	
Nuoshuihe (Sichun)	River	94.80	L	
Longchuanfengshuba (Guangdong)	Lake	156.70	L	
Yangzie (Anhui)	Lake	185.65	N	
Shengjinhu (Anhui)	Lake	333.40	N	
Hanjiang (Shananxi)	River	336.05	L	
Bitahai (Yunnan)	Marsh, lake	330.70	L	✓
Longtan (Guangxi)	River	428.48	L	
Aibihu (Xinjiang)	Lake	267.09	N	
Kashahu (Sichuan)	Lake	317.00	L	
Baheliuyu (Sichuan)	River	492.60	L	
Huangheshidi (Henan)	River	227.80	N	
Weiningcaohai (Guizhou)	Lake	629.63	N	
Haizishan (Sichuan)	Lake	459.16	L	
Yangchengmanghe (Shanxi)	River	55.73	N	
Jiuzhaigou (Sichuan)	Lake	642.97	N	
Heyuanxingang (Guangdong)	Lake	75.13	L	
Yuncheng (Shanxi)	River	868.61	L	
Danjiangshidi (Henan)	River	837.38	N	
Pingnan Mandarin Duck and Macaque (Fujian)	River	14.57	L	
Yinghu (Shananxi)	River	198	L	
Neixiangtuanhe (Henan)	River	45.47	L	

Dongtinghu reserve are the most important protected areas in China for birds. Besides, they have served as habitats for many threatened mammals. For instance, the Yangzie reserve and Tonglingdanshuitun reserve play a primary role in the protection of Chinese alligator, such as *alligator sinensis*, and freshwater dolphin, such as *Lipotes vexillifer*.

The reserves which are not located around the Yangtze River are the Huidonglianhuashanbaipenzhu reserve, Xinfengjiang reserve, Longchuanfengshuba

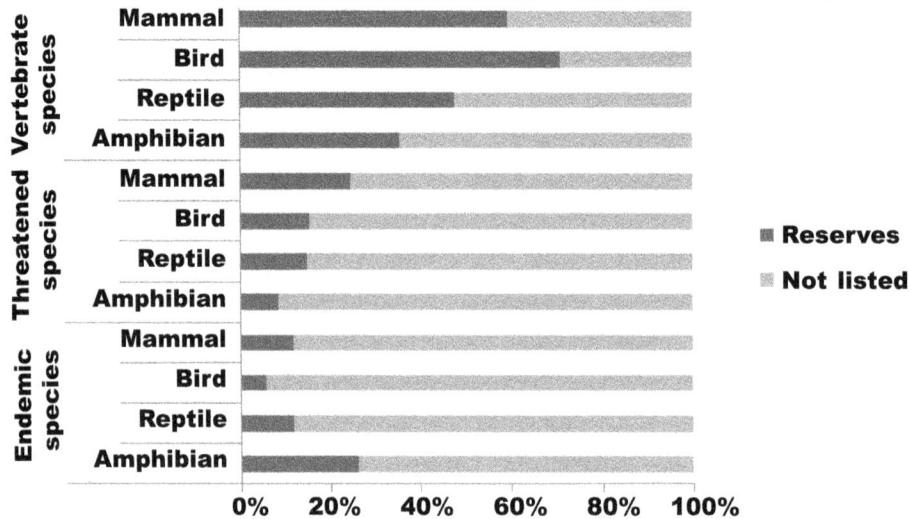

Fig. 1. The percentage of vertebrate species in reserves in China.

Fig. 1. Porcentaje de especies de vertebrados en reservas en China.

reserve, and Heyuanxingang reserve. These reserves are distributed mainly in Guangdong Province in southeast China. This type of reserve includes lakes, artificial reservoirs, and forests, providing ideal habitats for amphibians and reptiles.

The typical river reserves in the 31 reserves include Danjiangshidi reserve, Hanjiang reserve, and Huangheshidi reserve. All of them are highly important migrating sites for migratory birds, especially for waterbirds such as ciconia nigra, cygnus cygnus, grus grus, geese, and ducks. The regions of these reserves generally cover the forest ecosystems by rivers. The abundance of threatened amphibians, reptiles, and mammals living in these reserves is greater than those in other wetland reserves.

The Yancheng reserve is the only offshore type reserve, providing the largest winter habitat for *grus japonensis*. It also provides a comfortable habitat for migratory waterbirds. The Sanjangyuan reserve and the Bitahai reserve are typical plateau marsh type reserves in a plateau environment with high richness of endemic species.

The primary lake and river reserves have high richness and they are home to both threatened and endemic species. Nevertheless, some of the reserves have not received sufficient attention at the national level. For example, the Pingnan Mandarin Duck and Macaque reserve and the Nuoshuihe reserve are not well–protected because of their small size.

Discussion

The fundamental aim of nature reserves is to protect wildlife. Many reserves, however, are in areas of scenic interest, such as near like lakes and rivers, and it is foreseeable that local government develops scenic spots for the public. In doing so, reserves would lose their basic conservation function and biodiversity in these areas will decrease. Wildlife habitats are declining in many areas in China due to the expansive development and urban sprawl. High priority reserves therefore urgently require an adaptive management approach to protect the habitats of the species.

Endemic species \ Threatened species	High	Medium	Low
High	31	10	8
Medium	9	18	22
Low	9	21	20

Fig. 2. Matrix of priority levels in 148 wetland nature reserves in China.

Fig. 2. Matriz de grados de prioridad de los 148 humedales declarados reserva natural en China.

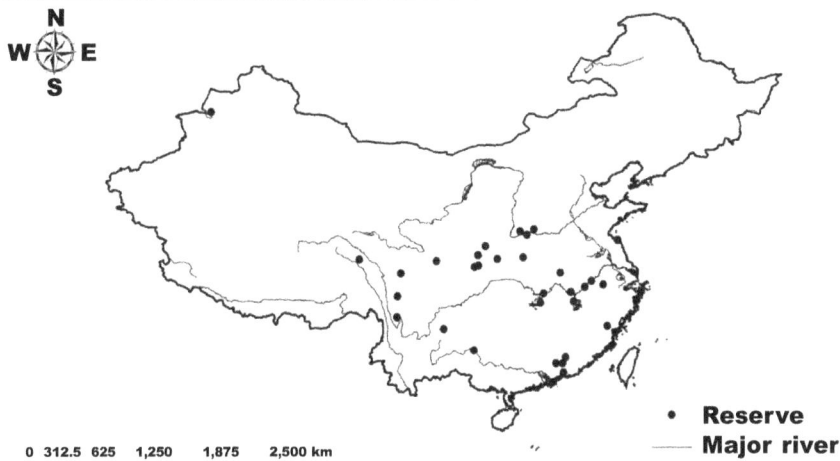

Fig. 3. The distribution of the 31 wetland nature reserves with high species protection value in Mainland China.

Fig. 3. Distribución de los 31 humedales declarados reserva natural que albergan especies con un alto valor de protección en la China continental.

The national protected areas are generally well–protected under Chinese law. However, many reserves are in a relatively low protection condition because of the lack of clear protection objectives and purpo-ses. For example, the Pingnan Mandarin Duck and Macaque reserve is only 14.57 km^2, but its level of species richness is relatively high. According to the island theory of biogeography, habitat size has a

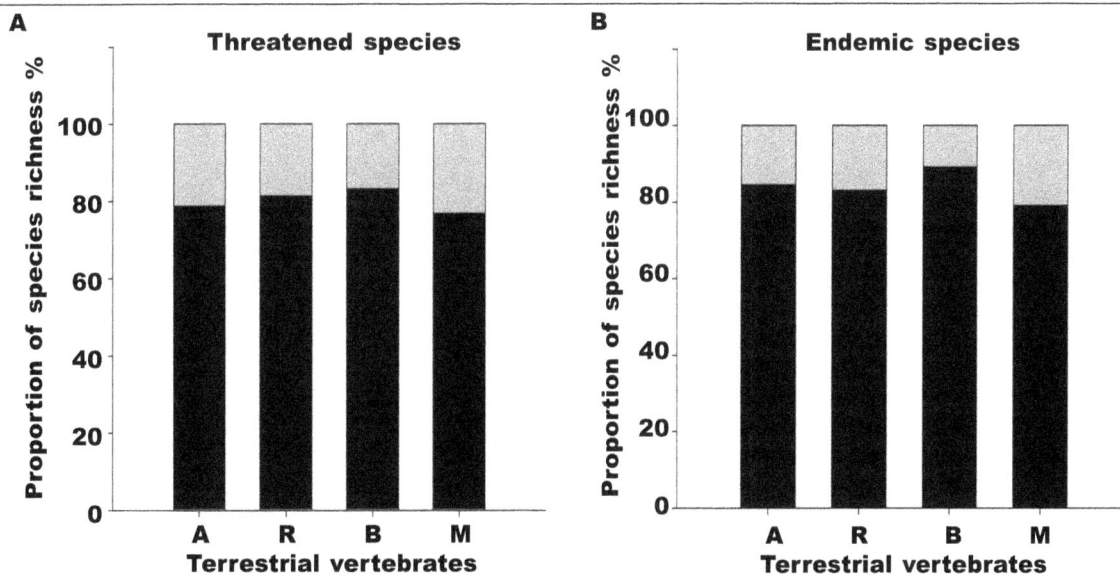

Fig. 4. Proportion of threatened species richness (A) and endemic species richness (B) of four vertebrate taxa in wetland nature reserves: A. Amphibians; R. Reptiles; B. Birds; M. Mammals; ■ Percentage of the 31 reserves with the highest conservation value; ▨ Percentage of the other 117 reserves.

Fig. 4. Proporción de la riqueza de especies amenazadas (A) y endémicas (B) de los cuatro taxones de vertebrados en los humedales declarados reserva natural: A. Anfibios; R. Reptiles; B. Aves: M. Mamíferos; ■ Porcentaje de las 31 reservas con el valor de conservación más elevado; ▨ Porcentaje de las otras 117 reservas.

significant, direct influence on species. A larger area should contain a larger diversity of species, and vice versa. Each single reserve is like an isolated island and is limited by local species size and management. If reserves are too small, the diversity of species will disappear. The current management mode in China is to rely on local capacity for protection and resources. If reserves have a highly scattered distribution, therefore, the island effect can be formed in the protection of diversity. To be a viable ecosystem for biodiversity, a protected reserve must be a certain size. Small reserves with a high richness, such as the Pingnan Mandarin Duck and Macaque Reserve, should thus be brought within the protected area network to avoid the disappearance of the species and conserve the regional biodiversity.

Regarding the 31 reserves which are mostly located near lakes and rivers, although their composite scores are high, human activity, such as that related to water supply, aquaculture, reclamation planting, and shipping, significantly influence the reserves. Excessive activity has a strong impact on vertebrate species and could lead to the serious depopulation of a species, such as the functional extinction of *Lipotes vexillifer* en Yangtze River in 2007.

Comprehensive management seems to be a need for all reserves. The key point is to make the ranking of reserves clear. All reserves can be ranked by conversation priority and an establishment policy can be made to protect wildlife and obtain funding. A protected area network should be developed rather than individual reserve protection based on the 31 high–score reserves. The results in this work indicate that the degree of concern of threatened terrestrial vertebrates in the wetland protected areas is obviously higher than that of endemic terrestrial vertebrates. The endemic species do not seem to have priority protection in a protected area as do threatened species, especially in a small–sized protected area. In the protected area network, the concern for the endemic terrestrial vertebrates should be increased. Biodiversity conservation planning should have solid roots in protection values of species and establishing a protected area network.

The classification and ranking results provide useful insight for the development of protection policies for vertebrate species for the government in China. The 31 national reserves should be included in a national reserve network. Furthermore, 15 provincial or municipal reserves which are also highly scored should be upgraded to a national level.

Conclusions

This work discusses the protection status of wildlife in wetland nature reserves in China. The composite score of the reserves was analyzed for both threatened and endemic species. The richness of species was correlated with the ranking of reserves and may help to address protection priorities. The characteristics of the highest scores in 31 reserves were analyzed. The results indicate that the management of reserve should be reinforced and a national reserve network should be established to protect the biodiversity in China. A nature reserve classification system with corresponding management will help to protect species diversity in China.

Acknowledgments

Financial support from the Forestry Department dedicated to essential industries (Project No. 201104029) is gratefully acknowledged. The authors thank the staff and management offices of the nature reserves for their help and support. Thanks are also given to Han Zhang for drawing the maps.

References

Bonn, A. Rodrigues, A. S. L. & Gaston, K. J., 2002. Threatened and endemic species: are they good indicators of patterns of biodiversity on a national scale. *Ecology Letters*, 5: 733–741.

Brooks, T. Fonseca, G. A. B. & Rodrigues, A. S. L., 2004. Species, data, and conservation planning. *Conservation Biology*, 18: 1682–1688.

Cabeza, M. & Moilanen, A., 2001. Design of reserve networks and the persistence of biodiversity. *Trends in ecology and evolution*, 16: 242–248.

Caro, T. M. & O'Doherty, G., 1999. On the use of surrogate species in conservation biology. *Conservation Biology*, 13: 805–814.

Cui, G., 2004. Special research fields and hot spots in science of nature reserves. *Journal of Beijing Forestry University*, 26: 102–105.

Ferguson, E. & Cox, T., 2007. Exploratory factor analysis: a users' guide. *International Journal of Selection and Assessment*, 1: 84–94.

Gibbons, J., 2003. Terrestrial habitat: a vital component for herpetofauna of isolated wetlands. *Wetlands*, 23: 630–635.

Gibbs, J., 2000. Wetland loss and biodiversity conservation. *Conservation Biology*, 14: 314–317.

Goodman, P. S., 2003. Assessing management effectiveness and setting priorities in protected areas in KwaZulu–Natal. *Bioscience*, 53: 843–850.

Grenyer, R., Orme, C. D. Jackson, S. F., Thomas, G. H., Davies, R. G., Davies, T. J., Jones, K. E., Olson, V. A., Ridgely, R. S., Rasmussen, P. C., Ding, T.-S., Bennett, P. M., Blackburn, T. M., Gaston, K. J., Gittleman, J. L. & Owens, I. P. F., 2006. Global distribution and conservation of rare and threatened vertebrates. *Nature*, 444: 93–96.

Lamoreux, J. F., Morrison, J. C., Ricketts, T. H., Olson, D. M., Dinerstein, E., McKnight, M. W. & Shugart, H. H., 2005. Global tests of biodiversity concordance and the importance of endemism. *Nature*, 440: 212–214.

Li, X. H. Li, D. M., Li, Y. M., Ma, Z. J. & Zhai, T. Q., 2002. Habitat evaluation for crested ibis: A GIS–based approach. *Ecological Research*, 17: 565–573.

Liu, J. & Diamond, J., 2005. China's environment in a globalizing world. *Nature*, 435 (7046): 1179–1186.

Liu, J. Ouyang, Z. Pimm, S. L., Raven, P. H., Wang,

X., Miao, H. & Han, N., 2003. Protecting China's biodiversity. *Science*, 300: 1240–1241.

Loyola, R. D. Kubota, U. & Lewinsohn, T. M., 2007. Endemic vertebrates are the most effective surrogates for identifying conservation priorities among Brazilian ecoregions. *Diversity and Distributions*, 13: 389–396.

Lund, M. & Rahbek, C., 2002. Cross–taxon congruence in complementarity and conservation of temperate biodiversity. *Animal Conservation*, 5: 163–171.

Mackinnon, J. & Phillipps, K., 2000. *A Field Guide of the Birds of China*. Oxford University Press, New York.

Margules, C. & Pressey, R., 2000. Systematic conservation planning. *Nature*, 405: 243–253.

Margules, C. Pressey, R. & Williams, P., 2002. Representing biodiversity: data and procedures for identifying priority areas for conservation. *Journal of Biosciences*, 27: 309–326.

Negi, H. & Gadgil, M., 2002. Cross–taxon surrogacy of biodiversity in the Indian Garhwal Himalaya. *Biological Conservation*, 105: 143–155.

Niu, Z. G., Gong, P., Cheng, X., Guo, J. H., Wang, L., Huang, H. B., Shen, S. Q., Wu, Y. Z., Wang, X. F., Wang, X. W., Ying, Q., Liang, L., Zhang, L. N., Wang, L., Yao, Q., Yang, Z. Z., Guo, Z. Q. & Dai, Y. J., , 2009. Remote sensing of wetland mapping and initial analysis of the relevant geographical features. *Science in China Series D: Earth Sciences* 39(2):188–203.

Pinto, M. P., José Alexandre Felizola Diniz-Filho, J. A. F., Bini, L. M., Blamires, D. & Rangel, T. F. L. V. B., 2008. Biodiversity surrogate groups and conservation priority areas: birds of the Brazilian Cerrado. *Diversity and Distributions*, 14: 78–86.

Plumptre, A. J., Davenport, T. R. B., Behanganac, M., Kityo, R., Eilu, G., Ssegawa, P., Ewango, C., Meirte, D., Kahindo, C., Herremans, M., Peterhans, J. K., Pilgrim, J. D., Wilson, M., Languy, M. & Moyer, D., 2007. The biodiversity of the Albertine Rift. *Biological Conservation*, 134: 178–194.

Pressey, R. L., 1998. Algorithms, politics and timber: an example of the role of science in a public, political negotiation process over new conservation areas in production forests. In: Ecology for everyone: communicating ecology to scientists, the public and politicians: 73–87 (R. T. Wills & R. J. Hobbs, Eds.). Surrey Beatty & Sons, Sydney.

Qian, H., 2007. Relationships between plant and animal species richness at a regional scale in China. *Conservation Biology*, 21: 937–944.

Reid, W., 1998. Biodiversity hotspots. *Trends in Ecology and Evolution*, 13: 275–280.

Rodrigues, A. S. L., Andelman, S. J., Bakarr, M. I., Boitani, L., Brooks, T. M., Cowling, R. M., Fishpool, L. D. C., Da Fonseca, G. A. B., Gaston, K. J., Hoffmann, M., Long, J. S., Marquet, P. A., Pilgrim, J. D., Pressey, R. L., Schipper, J., Sechrest, W., Stuart, S. N., Underhill, L. G., Waller, R. W., Watts, M. E. J. & Yan, V., 2004. Effectiveness of the global protected area network in representing species diversity. *Nature*, 428: 640–643.

Roe, J. Gibson, J. & Kingsbury, B., 2006. Beyond the wetland border: Estimating the impact of roads for two species of water snakes. *Biological Conservation*, 130: 161–168.

Scott, J. M., Davis, F. W., McGhie, R. G., Wright, R. G., Groves, C. & Estes, J., 2001. Nature reserves: do they capture the full range of America's biological diversity. *Ecological Applications*, 11: 999–1007.

Scott, D. & Jones, T., 1995. Classification and inventory of wetlands: A global overview. *Plant Ecology*, 118: 3–16.

Smith, A. & Xie, Y., 2008. *A guide to the mammals of China*. Princeton Univ Press, New Jersey.

Soule, M., 1991. Conservation: tactics for a constant crisis. *Science*, 253: 744–749.

Tang, X. & Huang, G., 2003. Study on classification system for wetland types in China. *Forest Research*, 16: 531–539.

Wang, S. & Xie, Y., 2004. *China species red list*. High Education Press, Beijing.

Wiens, J. A., Hayward, G. D., Holthausen, R. S. & Wisdom, M. J., 2008. Using surrogate species and groups for conservation planning and management. *Bioscience*, 58: 241–252.

Wiersma, Y. F. & Nudds, T. D., 2009. Efficiency and effectiveness in representative reserve design in Canada: The contribution of existing protected areas. *Biological Conservation*, 142: 1639–1646.

Wolters, V. Bengtsson, J. & Zaitsev, A., 2006. Relationship among the species richness of different taxa. *Ecology*, 87: 1886–1895.

Yan, P. Wu, X. B.. Shi, Y., Gu, C. M., Wang, R. P. & Wang, C. L., 2005. Identification of Chinese alligators (*Alligator sinensis*) meat by diagnostic PCR of the mitochondrial cytochrome b gene. *Biological Conservation*, 121: 45–51.

Zheng, Y. Zhang, H. Y., Niu, Z. G. & Gong, P., 2012. Protection efficacy of national wetland reserves in China. *Chinese Science Bulletin*, 57: 1116–1134.

20

On the taxonomical identity of some taxa of the Iberian endemic genus *Iberus* Montfort, 1810 (Gastropoda, Helicidae)

A. Martínez–Ortí & F. Robles

Martínez–Ortí, A. & Robles, F., 2012. On the taxonomical identity of some taxa of the Iberian endemic genus *Iberus* Montfort, 1810 (Gastropoda, Helicidae). *Animal Biodiversity and Conservation*, 35.1: 99–106.

Abstract

On the taxonomical identity of some taxa of the Iberian endemic genus Iberus *Montfort, 1810 (Gastropoda, Helicidae).*— To determine the taxonomic identity of some of the species included in the genus *Iberus*, we reviewed the type series described by Rossmässler (1854) and deposited at the Senckenberg Forschungsinstitut und Naturmuseum of Frankfurt (Germany). We concluded that *Helix alcarazana* (currently *Iberus alcarazanus*) is a junior synonym of *Iberus alonensis* (Férussac, 1821). The type localities of *Helix guiraoana* and *Helix alcarazana* were discussed and modified and the lectotypes of *Helix guiraoana*, *Helix guiraoana* var. *angustata*, *Helix alcarazana* and *Helix loxana* were designated. In addition, we reviewed the type series of *Helix alonensis* Férussac, 1821 (currently *Iberus alonensis*) deposited at the Muséum national d'Histoire naturelle of Paris, France and we selected its lectotype. The locality of Alicante has been designated as the restricted type locality of this species.

Key words: Mollusca, Helicidae, *Iberus*, Types, Synonym, Spain.

Resumen

Identidad taxonómica de algunos taxones del género endémico ibérico Iberus *Montfort, 1810 (Gastropoda, Helicidae).*— Se revisan las series tipo de las especies de *Iberus* descritas por Rossmässler (1854) y depositadas en el Senckenberg Forschungsinstitut und Naturmuseum de Frankfurt, Alemania. Se concluye que *Helix alcarazana* (actualmente *Iberus alcarazanus*) es sinónimo posterior de *Iberus alonensis* (Férussac, 1821). Se discuten y modifican las localidades tipo de *Helix* y de *Helix alcarazana* y se designan lectotipos para *Helix guiraoana*, *Helix guiraoana* var. *angustata*, *Helix alcarazana* y *Helix loxana*. Además, se revisa la serie tipo de *Helix alonensis* (Férussac, 1821) (actualmente *Iberus alonensis*) depositada en el Muséum national d'Histoire naturelle de París, Francia. Se selecciona el lectotipo y se designa "Alicante" como localidad tipo restringida.

Palabras clave: Mollusca, Helicidae, *Iberus*, Tipos, Sinonimia, España.

A. Martínez–Ortí, Museu Valencià d'Història Natural. Passeig de la Petxina 15, 46008 València and Dept. de Zoologia, Fac. de Biologia, Univ. de València, Av. Dr. Moliner 50, 46100 Burjassot, València (Spain).– F. Robles, Dept. de Geologia, Fac. de Biologia. Univ. de València, Av. Dr. Moliner 50, Burjassot, València (Spain)..

Corresponding author: A. Martínez–Ortí. E–mail: amorti@uv.es

Introduction

Iberus Montfort, 1810 is an endemic genus from the Iberian peninsula whose taxonomy has been largely controversial. Recent studies on molecular phylogeny conclude that the genus currently includes at least 16 taxa of specific or subspecific character, three of which remain unnamed (Elejalde et al., 2008a, 2008b). In order to clarify the taxonomical identity of some of these taxa, we sampled intensively the type localities of *Helix alcarazana* Rossmässler, 1854, *H. guiraoana* Rossmässler, 1854 and *H. guiraoana* var. *angustata* Rossmässler, 1854 and compared the material collected with the type series of these three species which are deposited in the Senckenberg Forschungsinstitut und Naturmuseum of Frankfurt am Main (Germany, SMF) collection (Martínez–Ortí et al., 2004b). The first two taxa, *I. guiraoanus* and *I. angustatus*, present conchological features that differ from the majority of the morphs of *Iberus*, such as flattened shells, with a rounded final lap, and a well–shaped umbilicus. However, as *I. alcarazanus* has a globose shell without umbilicus, differentiation of the latter species from the two other species can be clearly established. The original material of these species was collected by the Murcian naturalist Ángel Guirao Martínez (1817–1890) (Rossmässler, 1854; López Fernández, 2002). Taking advantage of the type series of *Helix loxana* Rossmässler, 1854 is present in the same collection and we designate herein its type specimen. The identity of these species was recently confirmed as a result of molecular studies; new studies should therefore be carried out to confirm whether it should be considered a subspecies of *I. marmoratus* (Férussac, 1821) or a valid species (Elejalde et al., 2008a),

We also revised the original material of *H. alonensis* Férussac, 1821, deposited in the collection of the Muséum national d'Histoire naturelle in Paris, France (MNHN). This material is a mixture of shells from a very wide area, extending from Castellón province, in the north of Spain, to Almeria province, in the south. As their morphologies are very different it has been considered necessary to designate both a lectotype and a restricted type locality to clarify which of the variants described by Férussac (1821) accurately represents *Helix alonensis* Férussac, 1821 (currently *Iberus alonensis*).

Material and methods

The taxa from the SMF are *Helix guiraoana* (SMF 7914/2) (currently *Iberus guiraoanus*) (figs. 1A–1B) which presents as *locus typicus* 'Castellon de la Plana' (Castellón, Valencian Community), *Helix guiraoana* var. *angustata* (SMF 7911 and SMF 7912) (currently *Iberus angustatus*) (figs. 1C–1F) 'La Sierra de los Dientes de la Vieja' (Granada, Andalusia), *H. alcarazana* (SMF 7915 and SMF 7916) (currently *I. alcarazanus*) (figs. 1G–1H) 'Sierra de Alcaraz (Sierra de Segura) in Alcaraz' (Albacete, Castilla–La Mancha, Spain) and *Helix loxana* (SMF 7879) (figs. 2G–2I) 'in Loja (Granada)'. We collected numerous samples

of the first three taxa in the type localities of taxa which have been compared them with specimens contained in the samples from SMF (Martínez–Ortí, 1999; Martínez–Ortí & Robles, 2003; Martínez–Ortí et al, 2004a).

The type series of *I. alonensis*, deposited in MNHN, consists of 21 specimens distributed in nine lots, five of them labelled with the locality 'Alicante', one with the label 'Castelnovo prés Segorbe' (correct spelling Castellnovo, Castellón province), both localities from the Valencian Community, while the other three lots come from Almería, in Andalusia (Martínez–Ortí et al., 2005).

Results

Types and localities of the *Iberus* species described by Rossmässler (1854)

Taxonomical position of *I. angustatus*

The type series of *Helix guiraoana* var. *angustata* is composed of two syntypes. The first presents the columelar edge of the peristome reflected on the umbilicus, giving the sensation that it is narrower, a character emphasised by Rossmässler in his diagnosis. This effect would be greater because part of the reflected edge seems to have disappeared as a result of a break (SMF 7911, figs. 1C–1D). The other syntype has a similar umbilicus of the *I. guiraoanus* (SMF 7912, figs. 1E–1F). Consequently, conchological differentiation between the two species cannot be established. By studying the molecular phylogeny of the species of *Iberus*, Elejalde et al. (2008a) separated several of them into well–characterized species, one of which (clade 4 in Elejalde et al., 2008a; fig. 1) is named *I. angustatus*.

The area of distribution of this species is contiguous to that of *I. guiraoanus*, the two being separated by the course of the Guadiana Menor River. The sampling density is very low, however, and can not exclude the possibility of areas of overlap between the two species, and the possibility of hybridization.

According to Arrébola (pers. comm.: e–mail 17 V 2012) *I. angustatus sensu* Elejalde et al. (2008a) may not have an open umbilicus, but rather half–covered or almost covered by the reflection of the peristome. There are areas where the specimens of *I. angustatus* show great variability in regard to the reflection of the peristome over the umbilicus, approaching the characteristics of *I. guiraoanus*.

As noted below, the type locality of *H. guiraoana* var. *angustata* is highly problematic. Together with the small number of specimens of the type series, consisting of only two specimens, it is therefore difficult to establish the taxonomic relationship between *H. guiraoana* of Rossmässler and its variety *angustata*. Until more intensive sampling is undertaken in the area of distribution of both taxa and until further investigations on the type locality of *H. g.* var. *angustata* provide new information on this topic, it seems prudent to retain the name of *I. angustatus* as that considered by Elejalde et al. (2008a).

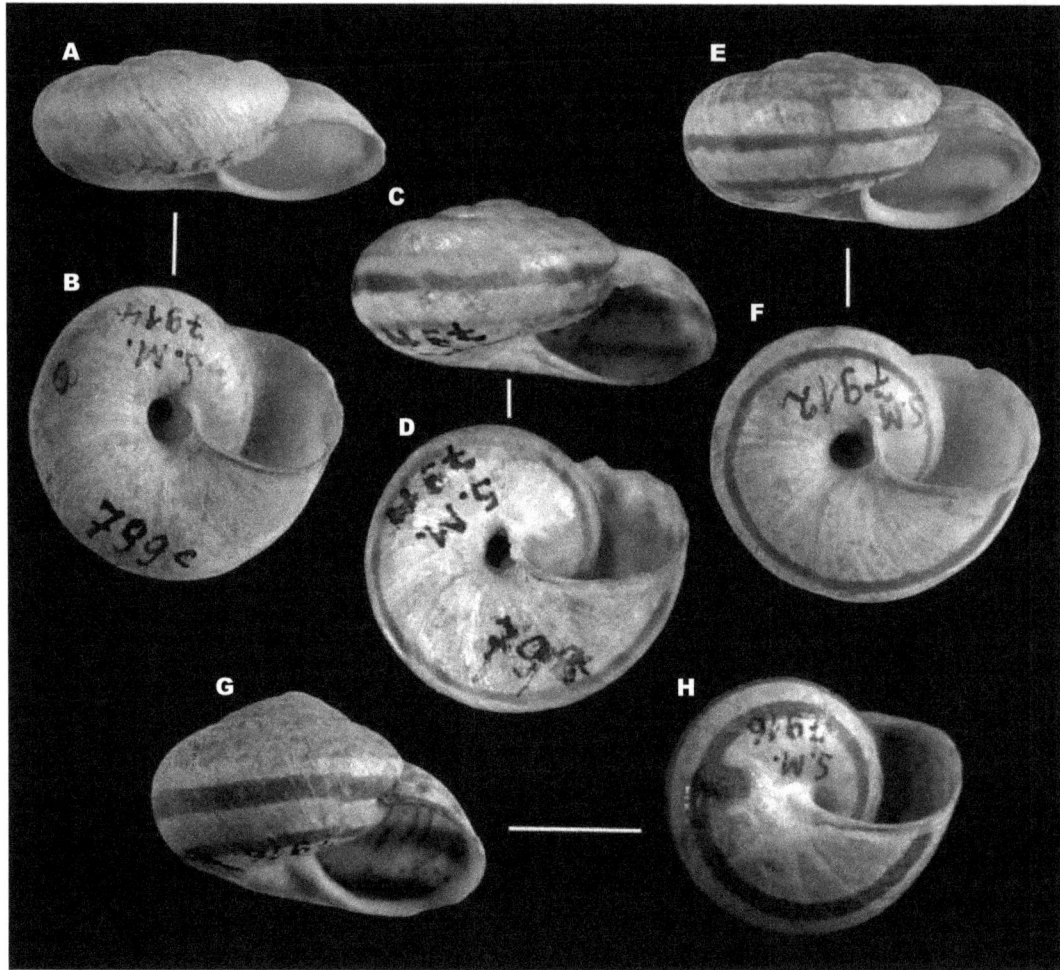

Fig. 1. A–B. Lectotype of *Iberus guiraoanus* (SMF 7914/2; 27.75 mm Ø); C–D. Lectotype of *I. angustatus* (SMF 7911; 21.8 mm Ø; phot. S. Hof, Naturmuseum Senckenberg); E–F. Paralectotype of *I. angustatus* (SMF 7912; 22.2 mm Ø); G–H. Lectotype of *I. alcarazanus* (SMF 7916/2; 22.7 mm Ø).

Fig. 1. A–B. Lectotipo de Iberus guiraoanus *(SMF 7914/2; 27,75 mm Ø); C–D. Lectotipo de* I. angustatus *(SMF 7911; 21,8 mm Ø; phot. S. Hof, Naturmuseum Senckenberg); E–F. Paralectotipo de* I. angustatus *(SMF 7912; 22,2 mm Ø); G–H. Lectotipo de* I. alcarazanus *(SMF 7916/2; 22,7 mm Ø).*

Type localities of *I. guiraoanus* and *I. alcarazanus*

The intensive sampling in both the type localities and the geographical areas close to those of *I. guiraoanus* and *I. alcarazanus* allowed us to ensure that *I. guiraoanus* does not live in 'Castellón' and that *I. alcarazanus* does not live in 'la Sierra de Alcaraz' (Albacete, Castilla–La Mancha, Spain), the two localities reported by Rossmässler (1854) in his original description (Martínez–Ortí, 1999; Martínez–Ortí & Robles, 2003; Martínez–Ortí et al., 2004b). On the contrary, *I. guiraoanus* is very abundant in 'Alcaraz' and *I. alcarazanus* lives in 'Castellón'. Therefore, in our view, it is clear that a mistake occurred. Taking into account that both samples were collected by

Guirao, prior to taxa descriptions by Rossmässler, it seems likely that this author received the samples incorrectly labelled. The other possibility is that Rossmässler mistakenly interchanged labels. In either case, according to the recommendation 76.A.2 of the International Code of Zoological Nomenclature (ICZN, 1999), we correct the declaration of the respective type localities: *I. guiraoanus* comes from the 'Sierra de Alcaraz' (Sierra de Segura) in Alcaraz (Albacete province) and *I. alcarazanus* comes from 'Castellón'.

Taxonomical position of *I. alcarazanus*

The conchological comparison of the specimens of *I. alcarazanus* from the Rossmässler collection with the shells of numerous individuals collected by the authors in the province of Castellón, which are attributed to *I. alonensis*, indicates that there are no significant differences between them. On the other hand, all *Iberus* samples whose DNA has been analyzed, from the northern half of the Iberian peninsula, which included the province of Castellón, have been identified as *I. alonensis* (Elejalde et al., 2008b). Therefore, the authors consider *I. alcarazanus* (Rossmässler, 1854) as a junior synonym of *I. alonensis* (Férussac, 1821).

Type locality of *I. angustatus*

The typical locality of *Helix guiraoana* var. *angustata* according to Rossmässler (1854) is 'under the blocks of limestone not far from the Sierra de los Dientes de la Vieja, in the province of Granada', where it was collected by Guirao. This place is very difficult to specify, because in the province of Granada and the rest of the Autonomous Community of Andalusia, there are several landforms that receive this designation. The authors visited a place known as 'Dientes de la Vieja', located near of Diezma (Granada), which due to its proximity to the area of distribution of *I. angustatus* sensu Elejalde et al. (2008a) seems the most appropriate, but we did not find specimens of this species. However, we found two different taxa, *I. loxanus* and *I. alonensis–like 02 sensu* Elejalde et al. (2008b). While it is possible that *I. angustatus* lived in this area from 'Dientes de la Vieja' and has become extinct since the mid–19th century, it cannot be ruled out that the material provided by Guirao to Rossmässler comes from other localities of the same name in the province of Granada. Consequently, and until new information is available on this matter, the situation of the type locality of *I. angustatus* remains doubtful.

Designation of types

Rossmässler (1854) did not designate types for their species and, according to the information we have, they have not been designated subsequently by other authors. As a result, and so as to stabilize the nomenclature, we proceed to designate lectotypes and paralectotypes from the syntypes.

Helix guiraoana Rossmässler, 1854. The type series is composed of two syntypes numbered SMF 79142/2. We designate as lectotype the specimen figured by Rossmässler (1854), figures 799, 799a and 799b and as paralectotype the specimen figured with the number 799c (figs. 1A–1B).

Helix guiraoana var. *angustata* Rossmässler, 1854. The series type is composed of two syntypes numbered SMF 7911 and SMF 7912. We designate as lectotype the specimen figured by Rossmässler (1854) having number 798 (SMF 7911, figs. 1C–1D) and as paralectotype the specimen SMF 7912 (figs. 1E–1F).

Helix alcarazana Rossmässler, 1854. The type series is composed of two syntypes: SMF 7915 and SMF 7916. According to information provided by Dr. Janssen (SMF), the specimen SMF 7915 is labelled as 'Typus' and considered as the original Ic. 795. It matches the size of the Ic. 795 but differs somewhat

in the colour pattern. Apparently there is no other specimen better matching Ic. 795 (Janssen, pers. comm. e–mail 16 I 12). Consequently, we designate as lectotype the specimen of the sample SMF 7915, figured by Rossmässler (1854) with the numbers 795, 795a and 795b, and as paralectotype the specimen of the sample 7916, formerly labelled in the collection of the SMF as 'Paratype' (figs. 1F–1G).

Helix loxana Rossmassler, 1854. Together with the species discussed above, Rossmässler (1854) described the species *Helix loxana*, with type locality 'in some mountains in the province of Granada, especially around Loja'. After molecular studies by Elejalde et al. (2008a) this species is considered a valid taxon (*Iberus loxanus*) with its own entity. Since the type specimen of the species has been not designated formally, we designate as lectotype the specimen of the figures 793, 793a and 793b (Rossmässler, 1854), deposited in the SMF having the number 7879 (figs. 2G–I).

Revision of the original material of *Iberus alonensis* from the Férussac collection (figs. 2A–2F; table 1)

Férussac in 1821 named and figured, without a formal description, the species *Helix (Helicogena) alonensis* (currently *Iberus alonensis*). The etymology of the name derives from 'Alone', an ancient Greek colony situated in the vicinity of Alicante, although its precise location is unknown (Herrero Alonso, 1986). According to the information provided by the website www.animalbase (see reference) the first mention of this species appears on page XIV of the Histoire Naturelle of Férussac & Deshayes (1819–1851), published on 6 IV 1821, as a name in the figure caption of the plate 39, which was published on 13 VII 1821. According to the article 10.1.1 of the ICZN the name *Helix alonensis* Férussac, 1821 does not appear until this second date (13 VII 1821). Between the publication of the taxon name and the figures, Férussac (1821–1822), in his Prodrome, cited these figures and varieties of *Helix alonensis* with reference to the afore mentioned plate 39 and plate 36A. The issue is more complex because he never published a plate 36A, as indicated by Kennard (1942). Indeed, this reference refers to the plate 39B that in figure 8 shows the shell and the living animal of *H. alonensis*. This mistake was corrected by Férussac in his annotated copy of the historical Edition of the Prodrome, deposited at the Muséum national d'Histoire naturelle in Paris and it was also marked by Coan et al. (2011) who require that the plate 39B ('originally to have been 36A') was published on 27 IX 1823.

The name *Helix alonensis* refers to figures 1 to 9 of the plate 39 of Férussac & Deshayes (1819–1851), but these figures represent five varieties (alpha to epsilon) from various localities, without criteria, in the original publication, to select one of them as most suitable to represent the species from the other figures. As a result, we revised the original material of Férussac in order to select a specimen of one of the variants to designate it as a lectotype, and from it, to designate a restricted type locality for the species.

The explanation of the Pl. 39 Férussac & Deshayes (1819–1851) identifies five varieties of *Helix*

Fig. 2. A–C. Lectotype of *Iberus alonensis* (MNHN, lot 1: 29.1 mm Ø); D. Paralectotype of *I. alonensis* (MNHN, lot 4: 26.8 mm Ø); E. Paralectotype of *I. alonensis* (MNHN, lot 4; 31.2 mm Ø); F. Paralectotype of *I. alonensis* (MNHN, lot 5; 33.4 mm Ø); G–I. Lectotype of *I. loxanus* (SMF 7879; 26.3 mm Ø; phot. E. Neubert, Naturmuseum Senckenberg).

Fig. 2. A–C. Lectotipo de Iberus alonensis *(MNHN, lote 1: 29,1 mm Ø); D. Paralectotipo de* I. alonensis *(MNHN, lote 4: 26,8 mm Ø); E. Paralectotipo of* I. alonensis *(MNHN, lote 4; 31,2 mm Ø); F. Paralectotipo of* I. alonensis *(MNHN, lote 5; 33,4 mm Ø); G–I. Lectotipo de* I. loxanus *(SMF 7879; 26,3 mm Ø; fotogr. E. Neubert, Naturmuseum Senckenberg).*

alonensis, with the locality of collection: (α) fig. 1, 2; from Alicante. (β) fig. 3; from Alicante (γ) fig. 4, 5; from Valencia. (δ) fig. 6; from Alicante. and (ε) fig. 7, 8, 9; from Almeria.

The material used by Férussac to name and figurate the species *H. alonensis* is deposited in the collection of malacology of the MNHN in Paris. In total it comprises 21 full shells in a good state of conservation, gathered in nine different lots. The three lots from Almeria (var. ε) (Pl. 39, figs. 7–9) contain five specimens in total. They seem to belong to a taxon currently under study and needing name (Martínez–Ortí et al., 2005), different from the taxon *alonensis*. As the five specimens from this locality (var. ε) belong to a new taxon, included by Elejalde et al. (2008b) in the clade named OTU *I. alonensis–like 01*, they are excluded from the syntypes of *H. alonensis*.

The remaining 16 specimens (syntypes), which form the series of *H. alonensis*, are distributed in the following lots (table 1, with dimensions):

Lot 1
Three specimens of the var. α. Locality: Alicante. Specimen 1–1 corresponds to the figs. 1 and 2 of the Pl. 39 Férussac & Deshayes (1819–1851) (figs. 2A–2C).

Lot 2
Three specimens of the var. α. Locality: Alicante.

Lot 3
One specimen of the var. α. Locality: Alicante.

Lot 4
Five specimens. Specimen 4–1 seems to correspond

Table 1. Conchological data of the type series of *Helix alonensis* deposited in the Muséum national d'Histoire naturelle of Paris (France): acr. Acronym; Hap. Aperture height; Hll. Height of the last lap; Lnº. Lap number; Øap. Aperture diameter; Of. Original figuration.

Tabla 1. Datos conquiológicos de la serie tipo de Helix alonensis *depositada en el Muséum national d'Histoire naturelle de París (Francia): acr. Sigla; Hap. Altura de la abertura; Hll. Altura de la última vuelta; Lnº. Número de vueltas; Øap. Diámetro de la abertura; Of. Figuración original.*

Lots

Acr	Var	Ømx	Hmx	Hll	Hap	Øap	Lnº	Locality	Of
Lot 1 (3 spec.)									
1–1	α	29.1	19.2	17.6	15.2	17.0	4½	Alicante (lectotype)	Figs. 1&2
1–2	α	28.4	18.4	17.6	13.8	16.5	4⅝	Alicante	–
1–3	α	28.8	19.4	18.2	15.9	17.8	4½	Alicante	–
Lot 2 (3 spec.)									
2–1	α	31.2	19.3	17.4	15.0	18.5	4⅛	Alicante	–
2–2	α	29.7	19.6	17.9	15.1	17.8	4¼	Alicante	–
2–3	α	33.4	21.0	19.9	16.8	19.5	4⅜	Alicante	–
Lot 3									
3–1	α	30.2	19.3	18.0	15.5	17.3	4¼	Alicante	–
Lot 4 (5 spec.)									
4–1	¿β?	26.8	16.6	15.6	13.4	15.7	4¼	Alicante	¿Fig. 3?
4–2	δ	31.2	21.0	18.1	15.4	17.2	4⅜	Alicante	Fig. 6
4–3	¿β?	32.0	20.5	18.3	15.8	19.1	4¼	Alicante	–
4–4	¿β?	32.0	20.4	19.3	16.0	18.2	4¼	Alicante	–
4–5	¿β?	33.4	22.5	21.0	16.4	19.0	4½	Alicante	–
Lot 5									
5–1	γ	33.4	23.9	21.1	17.2	20.0	4⅝	Castellnovo	Figs. 4&5
Lot 6 (3 spec.)									
3–1	?	27.7	17.5	16.5	15.0	15.1	4	Alicante	–
3–2	?	27.7	20.0	17.7	15.0	15.7	4½	Alicante	–
3–3	?	immature specimen not measured						Alicante	–

to the var. β , represented in fig. 3 of the Pl. 39 (fig. 2D). The specimen 4–2 presents the Greek letter δ recorded on the shell and it seems to be represented in fig. 6 of the Pl. 39 (fig. 2E).

Lot 5

One specimen of the var. γ represented in figures 4 and 5 of the Pl. 39 (fig. 2F). Although the figure caption of this plate indicates that its locality is Valencia, the label of the lot indicates 'dans les montagnes of Castelnovo (Castellnovo) près Segorbe', village located in the province of Castellón.

Lot 6

Three specimens without assignment of variety, one of them is immature. Locality: Alicante.

Designation of the lectotype of *I. alonensis*

All but one of the syntypes of *H. alonensis* proceed from Alicante. In fact, Férussac named this species in reference to its precedence because 'Alone' (gr. *als, alos*: salt) was a Greek colony situated in the vicinity of the current Alicante. The common name that Deshayes in Férussac & Deshayes (1819–1851) assigns to this species is, in fact, 'Hélice d'Alicante'. For these reasons, we designate a lectotype from this locality instead of Castellnovo.

Within the 15 revised syntypes of Alicante, we selected the first specimen cited by Férussac as lectotype (figs. 2A–2C), corresponding to the var. α and figured in the Pl. 39 figs. 1 and 2. This specimen has been identified in lot 1 and subsequently separated from the other syntypes, which should to be considered as paralectotypes. Therefore, the restricted type locality becomes 'Alicante' (Art. 76.2 ICZN; Recommendation 74E).

Conclusions

For an unknown reason, the labels of original samples of *I. guiraoanus* and *I. alcarazanus*, both collected by Guirao, were swapped. This likely occurred prior to the description made by Rossmässler (1854), and he probably received the samples with erroneous data. According to recommendation 76.A.2 of the ICZN, we corrected the designation of the respective type localities for each species: *I. guiraoanus* (Rossmässler, 1854), type locality: 'Sierra de Alcaraz, in Alcaraz' and *I. alcarazanus* (Rossmässler, 1854), type locality: 'Castellón'. Consequently, we considered *I. alcarazanus* as a junior synonym of *I. alonensis* (Férussac, 1821). As for the original sample of *I. alonensis*, deposited in the Férussac collection of the MNHN, we designated the specimen of the var. α figured in the Pl. 39, figs. 1 and 2 (Férussac & Deshayes, 1819–1851) as lectotype and, in consequence, we establish 'Alicante' as its restricted type locality.

Lastly, we designated as lectotype of *I. loxanus* the only syntype studied by Rossmässler (1854), which is deposited in the SMF.

Acknowledgements

This work is part of the project REN2002–00716 from the Ministry of Science and Innovation, and the Ministry of Science and Technology (Ref. CGL2008–01131) of Spain. We thank Dr. José Ramón Arrébola for his helpful comments, and also Virginie Héros, assistant curator of molluscs at the Muséum national d'Histoire naturelle of Paris, Dr. Ronald Janssen, curator of the Senckenberg Forschungsinstitut und Naturmuseum of Frankfurt am Main (SMF) for the transfer of the material type, and Dr. E. Neubert and S. Hof (both of the SMF) for providing photographs of specimens from the type series deposited in the SMF.

References

AnimalBase, *Helix alonensis* species taxon homepage: http://www.animalbase.uni–goettingen.de/zooweb/servlet/AnimalBase/home/ [Accessed 24 IV 2012].

Coan, E. V., Kabat, A. R. & Petit, R. E., 2011. 2,400 years of malacology, 8th ed., February 15, 2011, 936 pp. + 42 pp. [Annex of Collations]. *American Malacological Society*: http://www.malacological.org/publications/2400_malacology.php

Elejalde, M. A., Madeira, Mª J., Arrébola, J. R., Muñoz, B. & Gómez–Moliner, B. J., 2008a. Molecular phylogeny, taxonomy and evolution of the land snail genus *Iberus* (Pulmonata: Helicidae). *Journal of Zoological Systematics and Evolution Research*, 46(3): 193–202.

Elejalde, M. A., Madeira, Mª J., Muñoz, B., Arrébola, J. R. & Gómez–Moliner, J., 2008b. Mitochondrial DNA diversity and taxa delineation in the land snails of the *Iberus gualtieranus* (Pulmonata, Helicidae) complex. *Zoological Journal of the Linnean Society*, 154: 722–737.

Férussac, A. É. J. P. J. F. d'Audebard de, 1821–1822. *Tableaux systématique dés animaux mollusques suivis d'un Prodrome général pour tous les mollusques terrestres ou fluviatiles vivantes ou fossiles.* J.–B. Baillière, Paris.

Férussac, A. É. J. P. J. F. d'Audebard de & Deshayes, G. P., 1819–1851. *Histoire naturelle générale et particulière des mollusques terrestres et fluviatiles, tant des espèces que l'on trouve aujourd'hui vivantes, que des dépouilles fossiles de celles qui n'existent plus; classés d'après les caractères essentiels que présentent ces animaux et leurs coquilles*, 1: 8+184 pp.; 2(1): 402 pp.; 2(2) 260+22+16 pp.; Atlas 1: 70 pl.; 2: 166 + 5 pl. Paris, J.–B. Bailliere.

Herrero Alonso, A., 1986. Toponimia premusulmana de Alicante a través de la documentación medieval (II). Anales de la Universidad de Alicante. *Historia Medieval*, 4–5: 9–48.

International Commission of Zoological Nomenclature [ICZN], 1999. *International code of zoological nomenclature [the Code]*. Fourth edition. The International Trust for Zoological Nomenclature, c/o Natural History Museum, London. [online version at http://www.iczn.org/iczn/index.jsp].

Kennard, A. S., 1942. The Histoire and Prodrome of Férussac. Part II. –*Proceedings of the Malacological Society of London*, 25(3): 105–110.

López Fernández, C., 2002. *Ciencia y Enseñanza en algunas instituciones docentes murcianas. 1850–1936*, Ed. Univ. de Murcia.

Martínez–Ortí, A., 1999. *Moluscos terrestres testáceos de la Comunidad Valenciana*. Ph. D. Univ. de València.

Martínez–Ortí, A. & Robles, F., 2003. *Los Moluscos Continentales de la Comunidad Valenciana*. Conselleria de Territori i Habitatge, Colección Biodiversidad, 11. Valencia.

Martínez–Ortí, A., Robles, F. & Elejalde, A., 2004a. The Taxonomical Identity of Three Taxa of the Genus *Iberus* Monfort 1810: *Helix guiraoanus* Rossmässler 1854, *Helix guiraoana* var. *angustata* Rossmässler 1854 and *Helix alcarazana* Rossmässler 1854 (Gastropoda, Helicidae). *Molluscan Megadiversity: Sea, Land and Freshwater. 2nd World Congress of Malacology. F. Wells, Ed., Western Australian Museum, Perth (Australia)*: 95.

Martínez–Ortí, A., Aparicio, Mª T. & Robles, F., 2004b. La Malacofauna de la Sierra de Alcaraz (Albacete, España). *Iberus*, 22(2): 9–17.

Martínez–Ortí A., Robles F. & Gómez–Moliner, B., 2005. Estudio de la Serie Tipo del endemismo ibérico *Helix alonensis* Férussac, 1821 (Gastropoda, Helicidae). IV International Congress of the European Malacological Societies. *Notiziario (SIM)*, 23(5–8): 73.

Rossmässler, E. A., 1854. *Iconographie der Land–und Süsswasser–Mollusken Europa's, mit vorzüglicher Berücksichtigung kritischer und noch nicht abgebildeten Arten*, 3(1–2): 1–31. H. Costenoble Ed., Leipzig.

21

Habitat preference of the endangered Ethiopian walia ibex (*Capra walie*) in the Simien Mountains National Park, Ethiopia

D. Ejigu, A. Bekele, L. Powell & J.–M. Lernould

Ejigu, D., Bekele, A., Powell, L. & Lernould, J.–M., 2015. Habitat preference of the endangered Ethiopian walia ibex (*Capra walie*) in the Simien Mountains National Park, Ethiopia. *Animal Biodiversity and Conservation*, 38.1: 1–10.

Abstract

Habitat preference of the endangered Ethiopian walia ibex (Capra walie) in the Simien Mountains National Park, Ethiopia.— Walia ibex (*Capra walie*) is an endangered and endemic species restricted to the Simien Mountains National Park, Ethiopia. Recent expansion of human populations and livestock grazing in the park has prompted concerns that the range and habitats used by walia ibex have changed. We performed observations of walia ibex, conducted pellet counts of walia ibex and livestock, and measured vegetation and classified habitat characteristics at sample points during wet and dry seasons from October 2009 to November 2011. We assessed the effect of habitat characteristics on the presence of pellets of walia ibex, and then used a spatial model to create a predictive map to determine areas of high potential to support walia ibex. Rocky and shrubby habitats were more preferred than herbaceous habitats. Pellet distribution indicated that livestock and walia ibex were not usually found at the same sample point (*i.e.* 70% of quadrats with walia pellets were without livestock droppings; 73% of quadrats with livestock droppings did not have walia pellets). The best model to describe probability of presence of walia pellets included effects of herb cover (β = 0.047), shrub cover (β = 0.030), distance to cliff (β = –0.001), distance to road (β = 0.001), and altitude (β = 0.004). Walia ibexes have shifted to the eastern, steeper areas of the park, appearing to coincide with the occurrence of more intense, human–related activities in lowlands. Our study shows the complexities of managing areas that support human populations while also serving as a critical habitat for species of conservation concern.

Key words: Endemic, Ethiopia, Habitat preference, Simien Mountains, Walia Ibex

Resumen

Preferencia de hábitat del íbice de Etiopía (Capra wallie)*, en peligro de extinción, en el Parque Nacional de las Montañas Simien, en Etiopía.*— El íbice de Etiopía (*Capra wallie*) es una especie en peligro de extinción endémica del Parque Nacional de las Montañas Simien, en Etiopía. La reciente expansión de las poblaciones humanas y el pastoreo de ganado en el parque han suscitado preocupación por que los límites y los hábitats utilizados por el íbice de Etiopía hayan cambiado. Se realizaron observaciones del íbice de Etiopía y conteos de excrementos de íbice y de ganado, asimismo, se describió la vegetación y se clasificaron las características del hábitat en los puntos muestrales durante las estaciones seca y húmeda, desde octubre de 2009 hasta noviembre de 2011. Se evaluó la influencia de las características del hábitat en la presencia de excrementos de íbice y posteriormente se utilizó un modelo espacial para crear un mapa predictivo de las zonas con mayor probabilidad de albergar a esta especie. Los hábitats preferidos fueron los rocosos y arbustivos en comparación con los herbáceos. La distribución de los excrementos indicaba que el ganado y el íbice de Etiopía no solían encontrarse en el mismo punto muestral (el 70% de los cuadrados que contenían excrementos de íbice carecían de defecaciones de ganado y el 73% de los cuadrados con defecaciones de ganado no contenían excrementos de íbice). El mejor modelo para describir la probabilidad de presencia del íbice tomaba en consideración el efecto de la cubierta herbácea (β = 0,047), la cubierta arbustiva (β = 0,030), la distancia a un acantilado (β = –0,001), la distancia a una carretera (β = 0,001) y la altitud (β = 0,004). Los íbices de Etiopía se han trasladado hacia las zonas más orientales y abruptas del parque, lo que parece estar relacionado con la concentración de actividades humanas más intensas en las tierras bajas. Nuestro estudio pone de manifiesto la complejidad de gestionar zonas habitadas por poblaciones humanas y que a la vez constituyen un hábitat fundamental para las especies en conservación.

Palabras clave: Endémico, Etiopía, Preferencia de hábitat, Montañas Simien, Íbice de Etiopía

Dessalegn Ejigu, Dept. of Biology, College of Sciences, Bahir Dar Univ., P. O. Box 79, Bahir Dar, Ethiopia.– Afework Bekele, Dept. of Zoology, College of Natural Sciences, Addis Ababa Univ., Ethiopia.– Larkin Powell, School of Natural Resources, Univ. of Nebraska–Lincoln, Lincoln, Nebraska, USA.– Jean–Marc Lernould, Conservation des Especesetdes Populations Animales, Schlierbach, France.

Corresponding author: Dessalegn Ejigu. E–mail: dessalegn_ejigu@yahoo.com

Introduction

Walia ibex (*Capra walie* Ruppell, 1835) is a species of conservation concern and one of the Palearctic ibex species in Ethiopia (Nievergelt, 1981; Last, 1982). Distribution of Caprinae has been influenced mainly by rapid environmental changes caused by glaciation (Geist, 1971). Simien Mountains National Park (SMNP) is the southern limit of the natural range of ibexes in the world and the only place where walia ibex occurs (Nievergelt, 1981; Gebremedhin et al., 2009). Walia ibex lives at higher altitudes and is adapted to partial forest life in the SMNP. Thus, it lives in areas with different habitats compared to other ibex species occurring in the other regions of the world (Nievergelt, 1981; Fiorenza, 1983; Yalden & Largen, 1992).

Ibexes, in general, prefer areas with steep slope and cliffs and avoid grasslands and flat hillsides (Feng et al., 2007). The presence of livestock usually has a negative effect on their relative abundance and distribution. Livestock act as a disturbance, and ibex retreat to less suitable habitats (Namgail, 2006; Pelayo et al., 2007). The behavioural responses are key to understanding animal–habitat interactions; the way individuals obtain food, seek shelter, escape from predators, find mates, and care for the young can provide clues to the effect of disturbances (Hickman et al., 1993).

Walia ibex has a restricted habitat, and its main distribution range is in the steep, rocky and topographically heterogeneous habitats of the mountains of Ethiopia (Nievergelt, 1981; Yalden & Largen, 1992).The walia ibex is an outstanding rock climber on steep cliffs, and it prefers to live in mountainous areas, sub–afroalpine grasslands, and areas with low vegetation cover (Last, 1982; Yalden & Largen, 1992; Hurni & Ludi, 2000). The distribution of walia ibex in the SMNP has shifted towards the east since the 1970s, and intensified use of the park for livestock grazing has contributed significantly to such changes in walia ibex distribution (Hurni & Ludi, 2000). Low protection efficiency of wildlife habitat is the main conservation problem in the park (Ludi, 2005).Thus, walia ibex prefer areas with little or no disturbances and occupy the most remote and inaccessible habitats (Hurni & Ludi, 2000). Simien Mountains National Park is heavily affected by livestock grazing, fuel wood collection and timber cutting, and crop cultivation (Hurni & Ludi, 2000; Ludi, 2005).

Habitat preference models for a species can be used effectively in their conservation and management (Krausman & Morrison, 2003; Doswald et al., 2007). Such models provide information to determine the species' ecological niche through the relationship between observed species locations and habitat variables that restrict or drive their distribution (Hirzel & Le Lay, 2008). Factors such as competition, predation, human disturbances and the type of habitat patches can affect the species' habitat preference (Ottaviani et al., 2004; Rhodes et al., 2005). Habitat loss is a critical threat to most endangered species and the problem becomes significant in the SMNP where walia ibex occurs. Identification of suitable habitats is an essential step to ensure sustainable conservation of species such as walia ibex (Huettmann & Calgary, 2003; Jean Desbiez et al., 2009).

Unless resources are abundant, two populations cannot occupy the same niche at the same place and time (Hardin, 1960). Some degree of competition can occur in natural populations (Namgail, 2006). Thus, negative interactions will increase the extinction probabilities of a species and result in population size reduction (Hickman et al., 1993). A similar scenario can also occur in SMNP, where the original habitats of walia ibex, especially in the lowlands, have been occupied by livestock. Thus, the walia ibex population is confined to relatively inaccessible areas within gorges and escarpments towards the eastern part of the park (Hurni & Ludi, 2000).

The goal of our research was to determine areas of potential habitats for walia ibex in the SMNP to support sustainable conservation and management plans. Our specific objectives were to: (1) use presence data based on direct observations to describe habitat used by walia ibex; (2) use presence/absence data from pellet counts to assess and compare habitat useand preference of walia ibex and livestock; and (3) develop a descriptive, spatial model of habitat preference to highlight areas of SMNP that are critical for protection and management of walia ibex.

Material and methods

Study area

Simien Mountains National Park is located in the Amhara National Regional State within the North Gondar Administrative Zone (UTM 376047 E to 444522 E, and 1458552 N to 1467230 N; fig. 1). The SMNP is composed of a broad undulating plateau and the highest point is Ras Dejen (altitude: 4,543 m). The park is known for its impressive escarpments (Nievergelt, 1981). The SMNP borders were established in 1966 (Hurni & Ludi, 2000; ANRSPDPA [Amhara National Regional State Park Development and Protection Authority], 2009). UNESCO declared SMNP as a World Heritage Site in 1978 based on its importance as a refuge for rare and endemic animals and plants, as well as its exceptional natural beauty (Yalden & Largen, 1992; Hurni & Ludi, 2000; Puff & Nemomissa, 2001; Debonnet al., 2006). However, regulations adopted during the park's establishment allowed livestock grazing, agriculture, and human settlement in 80% of the park (Debonnet et al., 2006). The total area of the park is 412 km^2, and walia ibex habitats (mountainous regions and sub–afroalpine grasslands) are restricted to 94.1 km^2.

From 2000 to 2009, the mean annual rainfall at SMNP was 1,054 mm. The mean annual minimum and maximum temperatures were 8.7°C and 19.9°C, respectively (National Meteorological Agency, Addis Ababa, Ethiopia: http://www.ethiomet.gov.et/). Seasonal differences in temperature are minimal due to Ethiopia's proximity to the equator (Nievergelt, 1998).

The Simien Mountains form a contact zone between the Palearctic region in the north and the Ethiopian region in the south. This makes the flora and fauna of the area representative of both regions (Nievergelt, 1981). The mountain's geographical position and the presence of altitudinal belts as well as different topographic features in the SMNP results in a mosaic pattern of different habitats that can promote species diversity and richness (Puff & Nemomissa, 2001). The altitudinal variations in the SMNP can determine variations in the natural vegetation. The vegetation of the park consists mainly of *Erica arborea, Lobelia rhynchopetallum, Hypericumrevolutum, Helichrysum spp., Rosa abyssinica* and *Solanum* spp.

Endemic large mammals include walia ibex, Ethiopian wolf (*Caniss imensis*) and gelada baboon (*Theropithecus gelada*). The distribution of walia ibex in SMNP extends from Buyetras in the western parts of the park to the southeastern end of Sebatminch, which is 94.1 km^2 of the area. The density of walia ibex in its current range is 7.99 individuals/km^2, and counts of walia ibex during the last ten years (2002–2011) have indicated a gradual increase (ANRSPDPA, 2009).

Uncontrolled human use of natural resources in the park is the greatest threat to biodiversity. More than 75% of the SMNP is used by local human communities for grazing, agriculture and settlements, leaving only the highest peaks and inaccessible cliffs relatively undisturbed (Hurni & Ludi, 2000; Ludi, 2005). Barley is the main crop type in the area, and livestock species grazing in the park are cattle (7.49 individuals/km^2), sheep, goat and equine.

Data collection

We conducted our study of walia ibex at SMNP over 15 days every second month from October 2009 to November 2011. Our observations covered wet and dry seasons (wet: May–October; dry: November–April) and all hours of daylight. Data were collected by the primary investigator and two well–trained field assistants. Our study design was affected by the logistic hurdles presented by the rough terrain and remote locations of SMNP. First, we used a series of transects through portions of the park deemed most likely to contain walia ibex, as judged by anecdotal evidence and/or habitat characteristics. We followed transects to locate and observe walia ibex herds to document the sizes of groups and assess the habitats in which walia ibex were found. We used GPS to record the location and habitat classifications of individuals or herds, and morphological features —particularly horn shape of males, unique skin colour of some groups of individuals, and deformities such as swelling belly or broken horn— were used to identify the herd. We identified the topography (open plateau, bushy plateau, top of plateau, escarpment, or gorge/cave), and we visually scored the density of vegetation (sparse: < 25% cover, moderate: 25–50%, dense: 51–75%, very dense: > 75%) in the area of each individual or herd.

Second, we used randomly distributed, systematic transect surveys to describe habitat that was available for walia ibex. We established a total of 637 (319 during wet and 318 during dry seasons) quadrats to characterize vegetation along transects at 200 m intervals. We separated the parallel transects by 500 m in areas that allowed multiple transects (*e.g.,* plains and plateaus); we used single transects in gorges and escarpments and the direction of the transect was constrained by topography. Quadrats were square, and the boundaries of the 400 m^2 were marked by rope while we collected data. We visited each transect during wet and dry seasons, and each transect was visited every other month for two consecutive years. We determined the aspect and slope of each quadrat using a Clinometer (Gillen et al.,1984); we recorded the slope at each corner of the quadrat and used the mean of the four samples to represent the slope of the quadrat. The availability of water and food influences the distribution of animals (Kauffman & Krueger, 1984; Knight et al., 1988); thus, we visually estimated the distance from each quadrat to a cliff, water source, and nearest road and settlement (Rondinini et al., 2004). Ground cover (grass, herbs, shrubs, trees, rocks) was described as the percent cover of the quadrat.

We also used the vegetation transects and quadrats as sample locations to conduct observations of pellets of walia ibex and livestock. We identified pellets within each quadrat. Walia pellets were identified from livestock pellets (goat or sheep) based on their colour, shape and size. Sutherland (1996) suggested that pellet counts are the best indication of animal abundance when species are not easily observed. We used the presence of pellets rather than the number of pellets in our analysis, thus avoiding any problems (*e.g.,* unknown defecation rates or decay rates) that may arise from trying to determine relative abundance from pellet counts.

Statistical analyses

We used Chi–square (χ^2), independent sample t–tests and means of samples to compare distribution between wet and dry seasons. We used $\alpha = 0.05$ for statistical significance, and data were analysed using SPSS software version 16.0.

Habitat preference is the measure of a species' disproportional use, relative to availability (Krausman, 1999). To evaluate habitat preference, we used two methods to compare levels of habitat use (locations of sightings or pellets) and habitat availability (quadrats, as representative of SMNP) to assess habitat preference of walia ibex following Harrett (1982) and Steinein et al. (2005). First, we compared the vegetation cover of quadrats in which walia ibex pellets were found with the cover of quadrats without pellets. Second, we used data from pellet surveys to describe the potential of habitat characteristics to predict the probability (with the ranges of 0–1) of detecting walia pellets in a quadrat. We summarized the pellet counts at each sample point as presence/absence data (success: pellet found, failure: pellet not found) for use in logistic regression models that predicted the probability (Ψ) of detecting walia pellets in a quadrat:

$$\Psi_{(pellet)} = 1 / (1 + e^{-z})$$

where z is a linear model with intercept β_0 and factors $\beta_1 - \beta_n$:

$$Z = \beta_0 + \beta_1(x_1) + \beta_2(x_2) + \beta_3(x_3) + \dots + \beta_n(x_n)$$

We determined the best linear model to describe the probability of detecting walia pellets by *a priori*

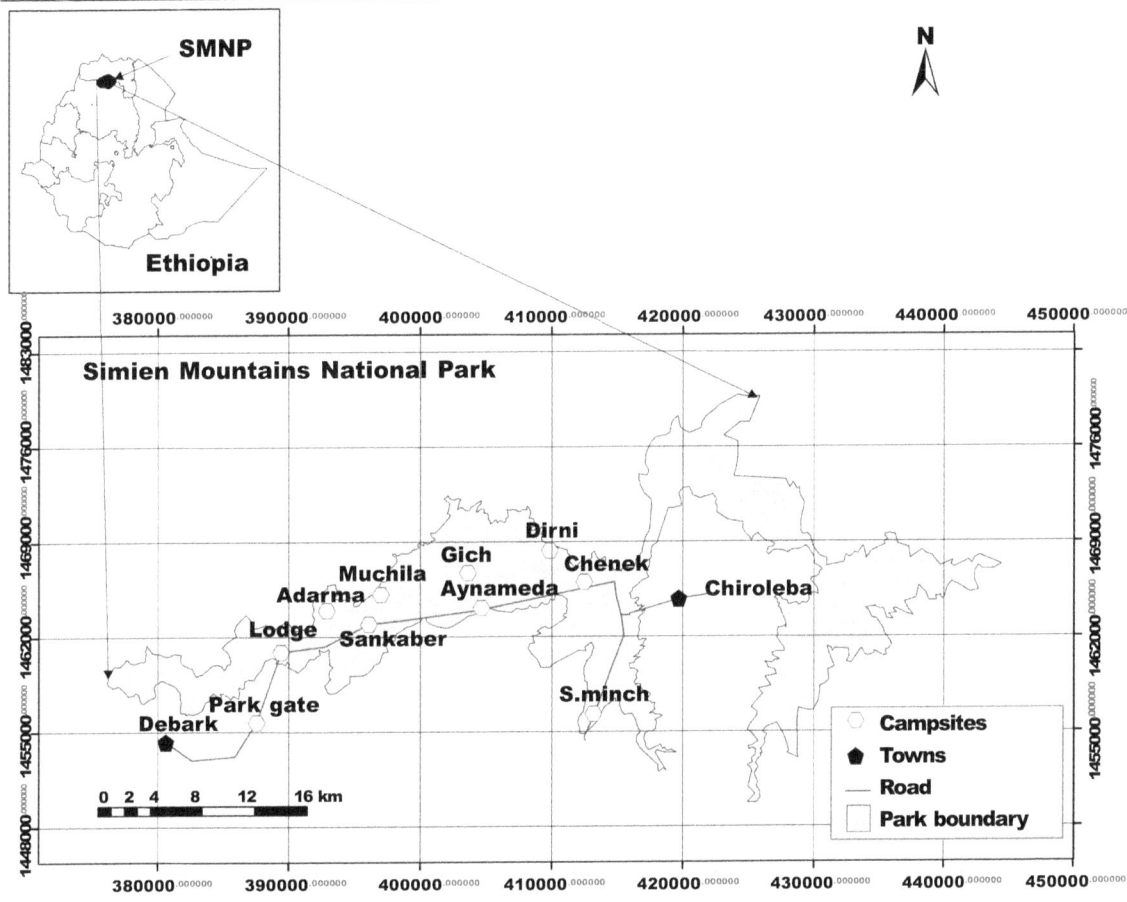

Fig. 1. The location of Simien Mountains National Park in Ethiopia, and Park boundary with associated towns, local campsites, and roads.

Fig. 1. La ubicación del Parque Nacional de las Montañas Simien en Etiopía y los límites del parque con las poblaciones, campamentos y carreteras cercanas.

describing 11 potential habitat variables: ground cover (grass cover, herb cover, shrub cover, tree cover and rock cover), altitude, slope, and distance from cliffs, water sources, roads and settlements. We used a backwards stepwise variable removal procedure to form the final model from the initial 11 variables. Prior to the analyses, we assessed colinearity ($r > 0.6$) among the variables under consideration.

We then applied the predictive model to our spatial data from the quadrats to visually determine areas of SMNP that had a high probability of walia presence. Quadrat–specific characteristics were then applied to the final regression model to calculate Ψ at each quadrat and characterize the probability of pellets of walia ibex across the study site. We assumed that areas of high probability of presence of pellets were also areas of high habitat preference; conversely, areas with a low probability of presence of pellets were areas of low habitat preference for walia ibex within their range at SMNP (Conroy, 1996). Finally, we used the ordinary kriging method with a spherical

semi–variogram model in ArcMap (ESRI, Redlands, CA, USA) to spatially extrapolate habitat preference of walia ibex across points not sampled at SMNP. We displayed the probability of presence of pellets in the park as a gradient from low to high probability of presence as adopted by Rondinini et al. (2004).

Results

Herd size was variable (range 1–32 ibex per observation) and increased by approximately seven ibex per observation as altitude increased from our lowest observation (3,543 m) to our highest (4,361 m; $F_{1,261} = 9.9$, slope = 0.007, $P < 0.01$, $r^2 = 0.04$, $n = 263$). Similarly, we observed more walia pellets in quadrats at higher altitudes ($F_{1,381} = 19.9$, slope = 0.003, $P < 0.0001$, $r^2 = 0.05$) than quadrats at lower altitudes.

Most of our 267 independent observations (wet season: 132, dry season: 135) of walia ibexes were in open plateaus (42%) and escarpments (32%; fig. 2).

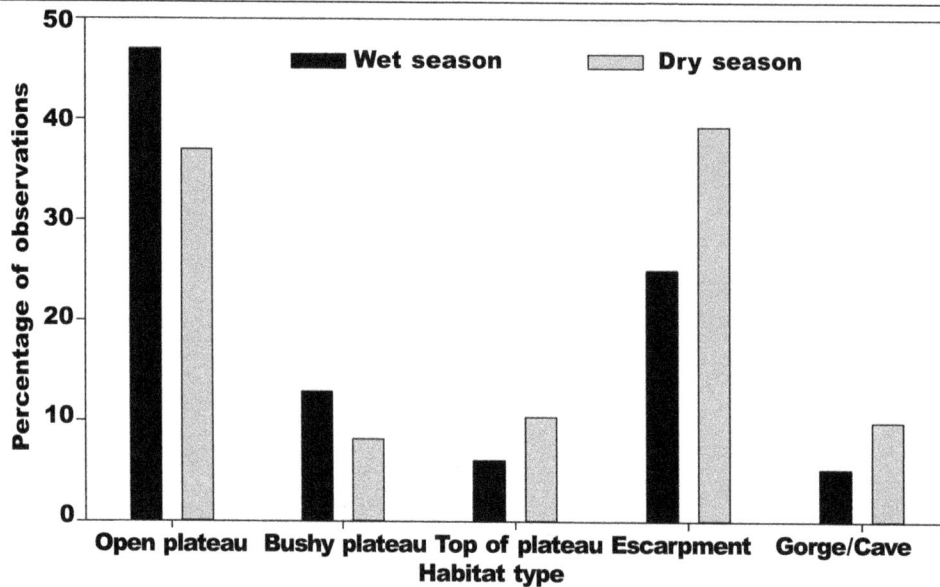

Fig. 2. Percentage of observations of walia ibex individuals or herds according to habitat type during wet and dry seasons at Simien Mountains National Park, Ethiopia, during 2009–2011.

Fig. 2. Porcentaje de observaciones de individuos o rebaños de íbice de Etiopía según el tipo de hábitat durante las estaciones seca y húmeda en el Parque Nacional de las Montañas Simien, en Etiopía, entre los años 2009 y 2011.

Use of open plateaus, bushy plataus, and gorges was higher during wet season, and the use of escarpments and tops of plateaus was higher in the dry season ($\chi^2 = 10.1$, df = 4, $P = 0.03$). More observations of walia ibex tended to be from areas with dense vegetation cover (34.2% of observations) than other vegetation densities (sparse: 26.7%, moderate: 21.8%, very dense: 17.3%), and these observations did not vary by season ($\chi^2 = 5.94$, df = 3, $P = 0.11$).

We collected samples from 319 and 318 quadrats during wet and dry seasons, respectively. Most of our quadrats (323, 50.7%) had west–facing aspects (north: 103, 16.2%; south: 94, 14.8%; east: 75, 11.8%; southwest: 16, 2.5%; northwest: 15, 2.4%; southeast: 7, 1.1%; northeast: 4, 0.6%). The mean slope at our quadrats was 23.3° (SE = 3.0) or 43.1% slope. Walia ibex and livestock pellets were not typically found in the same quadrat. In fact, 70% of quadrats with walia pellets had no livestock droppings, and 73% of quadrats with livestock droppings had no walia pellets.

Our quadrat samples were dominated by grass and rock cover. Quadrats with walia ibex pellets had less grass and trees but more rocks and shrubs than quadrats without pellets. In contrast, quadrats with livestock pellets had less grass and rocks and more trees than quadrats without livestock pellets (fig. 3). The only difference in habitat use between wet and dry seasons was that quadrats with pellets of walia ibex had 28.7% of rock cover in the wet season but only 14.4% in the dry season. Livestock habitat use varied

only in tree cover; quadrats with livestock pellets had 13.2% tree cover during the wet season and 9.1% during the dry season.

Prior to the regression analysis, we removed 'tree' from the variable set, as altitude and tree were correlated ($r = 0.64$, $P < 0.001$). No other variable pairs were correlated above the level $r = 0.6$. The logistic regression analysis suggested that the habitat characteristics of herb cover, shrub cover, altitude, and distance of the sampled habitat from cliffs and roads were the best factors to describe the probability of walia pellets ($\Psi_{(pellet)}$). The values of the associated intercept and regression coefficients were: $\beta_0 = -15.991$, $\beta_{\%herb} = 0.047$, $\beta_{\%shrub} = 0.030$, $\beta_{dist\ to\ cliff} = -0.001$, $\beta_{dist\ to\ road} = 0.001$, $\beta_{altitude} = 0.004$. The probability of presence of pellets ranged from 0.17 to 0.93 throughout the park. The spatial descriptive map indicated that the portion of the park with the highest probability of presence of walia ibex (as measured by pellets) occurred from Chenek–Buahit to Mesareri towards the eastern portion of the park (fig. 4).

Discussion

Walia ibex and livestock in SMNP

Ludi (2005) documented that walia habitats within SMNP had been affected by the increase of the human population within and around the park, which had

A

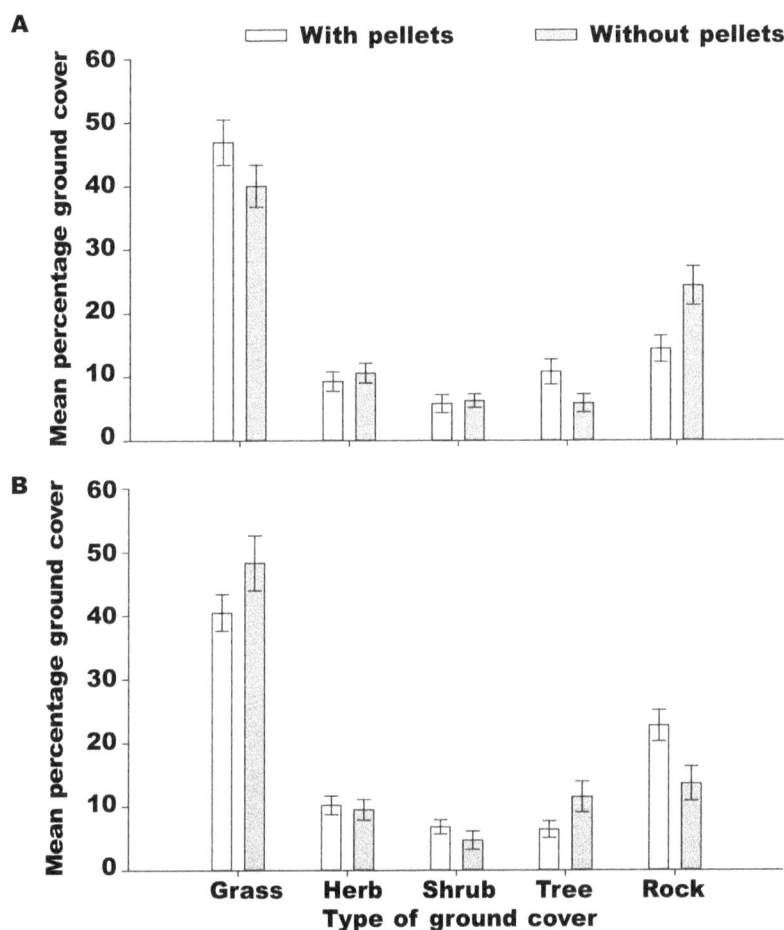

B

Fig. 3. Mean (95% confidence interval) percentage of types of ground cover in quadrats with and without pellets of livestock (A) and walia ibex (B) in Simien Mountains National Park, Ethiopia, during 2009–2011.

Fig. 3. Porcentaje medio (95 % intervalo de confianza) de los tipos de cubierta vegetal en los cuadrados con y sin excrementos de ganado (A) y de íbice de Etiopía (B) en el Parque Nacional de las Montañas Simien, en Etiopía, entre los años 2009 y 2011.

resulted in heavier pressure from grazing of livestock. We found that herd size of walia ibex increased with altitude. Our data suggest that walia ibex use shrubby areas of open plateaus extensively and use rocky escarpments preferentially. Nievergelt (1981) reported that walia ibex prefer rocky terrains with no human related disturbances.

We suggest that walia ibex are responding to the impacts documented by Ludi (2005) by moving farther from human populations and into habitats that are less likely to overlap with livestock grazing. Indeed, we found that pellets of walia ibex and livestock did not tend to be found in the same sampling quadrat, which may indicate ibex are avoiding livestock areas. As expected, livestock in the SMNP tend to use areas with more grass cover. Many species of wildlife are threatened by intensification of agriculture and overgrazing of livestock on their habitats (Jean Desbiez et al., 2009).

Hurni & Ludi (2000) also suggested that severe human–related disturbance at lowlands force the movements of walia ibex towards inaccessible habitats. Currently, the distribution of walia ibex is towards Sebatminch in the eastern portion of the park (fig. 1), which has more highlands available. Walia were previously found further west in the park, and more forage plants are available in the lowlands. But human disturbance and livestock grazing at lowlands, especially in the Gich area (fig. 1), may have contributed to displacement of walia ibex to the east and to the highest, steepest areas of the park (Hurni & Ludi, 2000). Our study was not designed to determine the relative quality of habitats for walia ibex, but we encourage biologists to consider the possibility that presence of livestock in former ibex range within the park has forced walia ibex to select habitats of lesser quality (Pelayo et al., 2007). We concur with

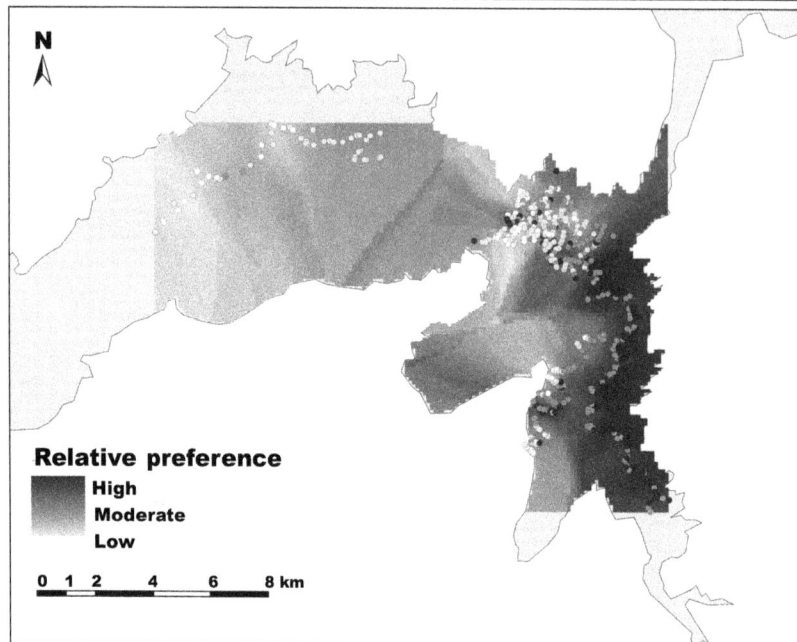

Fig. 4. Spatial representation of relative preference of walia ibex within Simien Mountains National Park (SMNP), Ethiopia, defined as the probability of finding a walia ibex pellet at a sample point. The predictive model was applied, *a posteriori*, to sample locations (dots) and then extrapolated from the sample point to the portion of SMNP in which samples were collected during 2009–2011. Dark areas indicate a higher preference for sample locations and landscape.

Fig. 4. Representación espacial de la preferencia relativa del íbice de Etiopía en el Parque Nacional de las Montañas Simien, en Etiopía, definida como la probabilidad de encontrar excrementos de íbice de Etiopía en un determinado punto muestral. El modelo predictivo se aplicó, a posteriori, a los lugares muestreados (puntos) y después se extrapoló a la porción del Parque en las que se recogieron muestras durante 2009 y 2011. Las áreas oscuras indican una preferencia alta para los lugares muestreados.

previous suggestions that assessment of habitat preference for species can provide critical information for problems in conservation and management of a species (Morris, 2003).

As most mammals are highly selective in their habitats, degradation of habitat by domestic animals is a common problem (Jean Desbiez et al., 2009). Human activities interfere with animal distribution and anticipate access to critical habitat change (Williamson et al., 1988), and the animal's behavioural response to the presence of humans has often been used as an indication to its susceptibility to disturbance (Beale & Monaghan, 2004). As described by Brown (1971), coexistence between pastoralists and wildlife requires the maintenance of a very low–density of livestock and human populations. However, it is not achievable in developing countries as the local people raise a large number of livestock and practise free grazing; such is the case in the SMNP.

Seasonal trends of habitat use and preference

Identification of potential suitable habitats for wildlife is an essential step to insure sustainable conservation of

species (Huettman & Calgary, 2003; Jean Desbiez et al., 2009). Human activity and wildlife food availability vary in space and quantity at the SMNP between wet and dry seasons. We observed only minor differences in the topography used by walia ibex during wet and dry seasons (fig. 2). However, these minor shifts may reveal important factors in the ecology of walia ibex. Regardless of season, ibex were found extensively in open plateaus. Nievergelt (1981) also reported that open areas with *L. rhyncopethalum* and *Festucamacrophylla* were used by ibex. We found that ibex were more often found on bushy plateaus during the wet season than the dry season. This may be because herbaceous plants such as *Thymus*, *Alchemillarothi* and *Simeniaacaulis,* the principal diet for walia ibex, become available in the bushy plateau during the wet season.

Similarly, we found that slopes and troughs had *Alchemilla rothi*, *Arabis alpine*, *Simenia acaulis* and *Festuca macrophylla*, all of which are used by walia ibex (Nievergelt, 1981). During the dry season, the abundance of herbs decreases. Thus, walia ibexes may have been forced to switch their diet to other plant species, causing them to move to escarpments and the top of the plateau. Indeed, herds of walia ibex were observed in less dense areas during the dry season.

Livestock pellets were found in areas with more tree cover during the wet season than during the dry season. We observed that livestock were restricted to tree–covered areas during the wet season because crops were grown in open areas. After harvesting, livestock were able to range into open fields to forage on leftover crops.

Predictive maps for conservation planning

We used pellet counts to survey habitat use of walia ibex, allowing us to assess vast spatial areas of the park at lower costs than radio–telemetry or other individual–based methods of habitat assessment. Walia ibex in the SMNP are a small population, and individuals are difficult to detect and capture. Furthermore, to determine whether they had used a specific sample location, pellet counts were more reliable than visual surveys of animals because individuals are constantly moving to new locations; detection of use at a given sample point would therefore be very low (Gu & Swihart, 2004). Pellet counts were more cost–effective and more logistically realistic than remote cameras. We designed our pellet survey to allow us to consider spatial processes, which would have been much more difficult logistically to conduct with direct observation of animals (Rhodes et al., 2005; Skarin, 2006). Consequently, we suggest that biologists consider pellet counts for similar assessments of habitat use and preference in remote areas that present logistical difficulties.

Walia ibex were previously described to be found in areas of the SMNP with altitudes of 2,800 to 3,400 m a.s.l. (Nievergelt, 1981). However, we found walia ibex to commonly occur at altitudes about 4,000 m a.s.l. and that their habitats have been shifted towards the eastern part of the park. Our predictive model shows the potential spatial extent of the park that could be defined as providing the remaining suitable habitat for walia ibex, and we believe this is a key piece of information in the process to establish a conservation plan for the species (Owen, 2009). We encourage biologists and managers of SMNP to consider further collection of demographic information, such as breeding success and survival. Such information can further refine and target management efforts, especially if demographic patterns can be related to habitat use (Peek, 1986).

The SMNP is a complex system that includes sensitive wildlife species and human activities, and management decisions to support the conservation of species of wildlife such as walia ibex are also complex. Our surveys of habitat use and preference serve to document recent changes in the range of walia ibex. Indeed, much of the process of habitat selection for a species may depend upon several limiting factors (Peek, 1986; Ottaviani et al, 2004).The disturbances within the park caused by human activities could be minimized by identifying zones to allow tourism, human use, and protection of critical biodiversity. Information provided by habitat use and selection data should enable more informed decisions about the conservation status of walia ibex and help to ensure its long–term survival.

Acknowledgements

We thank the field assistants and scouts in the Simien Mountains National Park for their help during data collection. D. Ejigu was a Visiting Scholar at the School of Natural Resources at the University of Nebraska–Lincoln, and L. Powell's contributions were supported by Hatch Act funds through the University of Nebraska Agricultural Research Division, Lincoln, Nebraska. The authors are greatly indebted to Addis Ababa University Postgraduate Program Office, CEPA, the Mohamed bin Zayed Species Conservation Fund, Chicago Zoological Society and Lleida University for funding. The authors are greatly indebted to the anonymous reviewers for their constructive comments and suggestions while reviewing the manuscript.

References

ANRSPDPA (Amhara National Regional State Parks Development and Protection Authority), 2009. *Simien Mountains National Park General Management Plan*. Bahir Dar.

Beale, C. & Monaghan, P., 2004. Behavioural responses to human disturbance: a matter of choice. *Animal Behaviour*, 68: 1065–1069.

Brown, L. H., 1971. The biology of pastoral man as a factor in conservation. *Biological Conservation*, 3: 93–100.

Conroy, M. J., 1996. Mapping of species for conservation of biological diversity: conceptual and methodological issues. *Journal of Ecological Applications*, 6: 763–773.

Debonnet, G., Melamari, L. & Bomhard, B., 2006. *Reactive Monitoring Mission to Simien Mountains National Park Ethiopia (10–17 May 2006)*.[online] Paris, France: World Heritage Centre, UNESCO. Available at: http://whc.unesco.org/archive/2006/mis9–2006.pdf. (Accessed on 5 September 2013).

Doswald, N., Zimmermann, F., & Breitenmoser, U., 2007. Testing expert groups for a habitat suitability model for the lynx (Lynx lynx) in the Swiss Alps. *Wildlife Biology*, 13: 430-446.

Feng, X., Ming, M. & Yi–Qun, W., 2007.Population density and habitat utilization of ibex (*Capra ibex*) in Tomur National Nature Reserve, Xinjiang, China. *Zoological Research*, 28: 53–55.

Fiorenza, P., 1983. *Encyclopedia of Big Game Animals in Africa with their Trophies*. Larousse and Co. Inc., New York.

Geist, V., 1971. *Mountain sheep*. Chicago Press, Chicago, Illinois.

Gebremedhin, B., Ficetola, G. F., Naderi, S., Rezaei, H. R., Maudet, C., Rioux, D., Luikart, G., Flagstad, Ø., Thuiller, W. & Taberlet, P., 2009. Combining genetic and ecological data to assess the conservation status of the endangered Ethiopian walia ibex. *Animal Conservation*, 12: 89–100.

Gillen, R. L., Krueger, W. C. & Miller, R. F.,1984. Cattle distribution on mountain rangeland in northeastern Oregon. *Journal of Range Management*, 37: 549–550.

Gu, W. & Swihart, R. K., 2004. Absent or undetected? Effects of non–detection of species occurrence on wildlife–habitat models. *Biological Conservation*, 116: 195–203.

Hardin, G., 1960. The competitive exclusion principle. *Science*, 131(3409): 1292-1297.

Harrett, R. H., 1982. Habitat preference of feral hogs, deer and cattle on a Sierra foothill range. *Journal of Range Management*, 35: 342–346.

Hickman, C. P., Roberts, L. S. & Larson, A., 1993. *Integrated Principles of Zoology, 9th edition*. Mosby, St. Louis.

Hirzel, A. H. & Le Lay, G., 2008. Habitat suitability modeling and niche theory. *Journal of Applied Ecology*, 45: 1372–138.

Hurni, H. & Ludi, E., 2000. *Reconciling conservation with sustainable development: a participatory study inside and around the Simen Mountains National Park, Ethiopia*. [online] Berne, Switzerland: Centre for Development and Environment (CDE), Institute of Geography, University of Berne. Available at: http://www.cde.unibe.ch/CDE/pdf/afr22_part1.pdf (Accessed on 10 January 2015).

Huettmann, F. & Calgary, J. L., 2003. An automated method to derive habitat preferences of wildlife in GIS and telemetry studies: A flexible software tool and examples of its application. *Zeitschriftfür Jagdwissenschaft*, 49: 219–232.

Jean Desbiez, A. L., Bodmer, R. E. & Santos, S. A., 2009. Wildlife habitat selection and sustainable resources management in a Neotropical wetland. *International Journal of Biodiversity and Conservation*, 1: 11–20.

Kauffman, J. B. & Krueger W. C., 1984. Livestock impacts on riparian ecosystems and streamside management implications. *Journal of Range Management*, 37: 430–432.

Knight, M. H., Knight, A. K. & Bornman, J. J., 1988. The importance of borehole water and lick sites to Kalahari ungulates. *Journal of Arid Environments*, 15: 269–281.

Krausman, P. R., 1999. Some basic principles of habitat use. In: *Grazing behavior of livestock and wildlife*: 85–90 (K. Launchbaugh, K. Sanders, & J. Mosley, Eds.). University of Idaho, Moscow, USA.

Krausman, P. R. & Morrison, M. L., 2003. *Wildlife Ecology and Management, Santa Rita Experimental Range (1903 to 2002)*.USDA Forest Service Proceedings RMRS–P–30, Tucson, Arizona.

Last, J., 1982. *Endemic mammals of Ethiopia*. Ethiopian Tourism Commission. Addis Ababa.

Ludi, E., 2005. Simien Mountains study 2004. *Intermediate report on the 2004 field expedition to the Simien Mountains in northern Ethiopia. Dialogue Series*. NCCR North–South, Berne, Switzerland.

Morris, D. K., 2003. How can we apply theories of habitat selection to wildlife conservation and management? *Wildlife Research*, 30: 303–319.

Namgail, T., 2006. Winter habitat partitioning between Asiatic ibex and blue sheep in Ladakh, northern India. *Journal of Mountain Ecology*, 8: 7–13.

Nievergelt, B., 1981. Ibexes in an African Environment: Ecology and Social Systems of Walia Ibex in the Simien Mountains National Park, Ethiopia. *Ecological Studies*, 40.

– 1998.Observations on the walia ibex, the klipspringer and the Ethiopian wolf.*Walia*1998 (Special Issue): 44–51.

Ottaviani, D., Lasinio, G. J. & Boitani, L., 2004. Two statistical methods to validate habitat suitability models using presence–only data. *Ecological Modeling*, 179: 417–443.

Owen, M., 2009. Habitat Suitability Modeling for Eld's deer (Rucervus eldiisiamensis) northwest Cambodia. M. Sc. Thesis, Imperial College, London.

Peek, J. M., 1986. *A review of wildlife management*. Prentice Hall, New Jersey.

Pelayo, A., Jorge, C. & Christian, G., 2007. The Iberian ibex is under an expansion trend but displaced to suboptimal habitats by the presence of extensive goat livestock in Central Spain. *Biodiversity and Conservation*, 16: 3361–3376.

Puff, C. & Nemomissa, S., 2001. The Simien Mountains (Ethiopia): Comments on plant biodiversity, endemism, phytogeographical affinities and historical aspects. *Systematics and Geography of Plants*, 71: 975–991.

Rhodes, J. R., McAlpine, C. A., Lunney, D. & Possingham, H. P., 2005. A spatially explicit habitat selection model incorporating home range behavior. *Ecology*, 86: 1199–1205.

Rondinini, C., Stuart, S. & Boitani, L., 2004. Habitat suitability models and the shortfall in conservation planning for African vertebrates. *Conservation Biology*, 193: 1488–1497.

Skarin, A., 2006. Reindeer use of Alpine summer habitats. Ph. D. Thesis, Swedish University of Agricultural Sciences, Upsala.

Steinein, G., Wegge, P., Fjellstad, J., Jnawali, S. R. & Weladji, R. B., 2005. Dry season diets and habitat use of sympatric Asian elephants (Elephans maximus) and greater one–horned rhinoceros (Rhinocerus unicornis) in Nepal. *Journal of Zoology*, 265: 377–385.

Sutherland, W. J.,1996. *Ecological census techniques, a handbook*. Cambridge University Press, UK.

Williamson, D., Williamson, J. & Ngwamotsoko, K. T., 1988. Wildebeest migration in the Kalahari. *African Journal of Ecology*, 26:269–280.

Yalden, D. W. & Largen, M. J., 1992. The endemic mammals of Ethiopia. *Mammal Review*, 22: 115–150.

22

The effect of habitat degradation, season and gender on morphological parameters of lesser jerboas (*Jaculus jaculus* L.) in Kuwait

M. Al–Mutairi, F. Mata & R. Bhuller

Al–Mutairi, M., Mata, F. & Bhuller, R., 2012. The effect of habitat degradation, season and gender on morphological parameters of lesser jerboas (*Jaculus jaculus* L.) in Kuwait. *Animal Biodiversity and Conservation*, 35.1: 119–124.

Abstract

*The effect of habitat degradation, season and gender on morphological parameters of lesser jerboas (*Jaculus jaculus L.*) in Kuwait.*— Arid environments suffer anthropogenic interference causing habitat degradation. This degradation can influence animal populations. We randomly captured a total of 198 lesser jerboas *Jaculus jaculus* in three seasons (autumn, spring and summer) in two relatively close areas (intact and degraded). All animals were sexed, and weight, body and tail length, and thigh thickness were taken. We found significant differences in weight ($p < 0.001$), which was lower in summer ($p < 0.05$) when fewer food resources were available. Thigh thickness was greater in the intact habitat ($p < 0.01$), explained by the greater amount of food resources and also by the higher numbers of predators in this area, prompting escape behaviour. Females in the intact area were heavier and had longer bodies and tails. This was related to greater availability of time for mothers to search for food in this area.

Key words: Habitat, Degradation, Kuwait, Lesser jerboa, *Jaculus jaculus* L.

Resumen

*El efecto de la degradación del hábitat, la estación del año y el sexo en algunos parámetros morfológicos del jerbo egipcio (*Jaculus jaculus L.*) en Kuwait.*— En los climas áridos las interferencias antropógenas causan la degradación del hábitat, que puede afectar a las poblaciones animales. Se capturaron 198 jerbos egipcios al azar en tres estaciones del año (otoño, primavera y verano) y en áreas intactas y degradadas cercanas. Los individuos capturados se sexaron y se les registró el peso, la longitud del cuerpo, la longitud de la cola y el grosor del muslo. Se encontraron diferencias significativas en el peso ($p < 0,001$), que fue inferior en verano ($p < 0,05$), cuando hay menos recursos alimentarios. El grosor del muslo fue mayor en el hábitat intacto ($p < 0,01$), lo que se atribuye a una mayor disponibilidad de alimento y también a una mayor presencia de depredadores en la zona, que provocaba conductas de huida. Las hembras fueron más pesadas y tuvieron mayor longitud del cuerpo y de la cola en las zonas intactas, como consecuencia de una mayor disponibilidad de tiempo para buscar alimento por parte de las madres en esta zona.

Palabras clave: Hábitat, Degradación, Kuwait, Jerbo egipcio, *Jaculus jaculus* L.

Matrah Al–Mutairi & Ravneet Bhuller, School of Biological Sciences, Univ. of Bristol, Woodland Road, Bristol BS8 1UG, UK.– Matrah Al–Mutairi, Kuwait Inst. for Scientific Research, P. O. Box 24885, Safat 13109, Kuwait .– Fernando Mata & Ravneet Bhuller, Hartpury College, Univ. of the West of England, Hartpury, GL19 3BE, UK.

Introduction

Desert mammal species have developed complex combinations of behavioral and physiological adaptations to ameliorate the impact of extreme temperatures and limited free water. These adaptations typically consist of being nocturnal, semi–fossorial, and having the ability to extract all their water from their food (Whitford, 2002).

Human activities have modified natural habitats in many ways. The most dramatic changes involve widespread degradation of entire areas, such as shift from forest to agricultural use or urbanization. However, a more pervasive influence is the construction of linear open areas —truck paths and roads — through previously continuous habitat. These open areas may provide ecological situations that differ profoundly from those of the surrounding habitat and cause habitat fragmentation and deterioration of food resources for desert mammals.

Limited food resources would result in smaller body size. Small desert mammals are often limited in their food choices in that they are forced to select more digestible, energy–rich diets such as those based on seeds, as compared to large animals (Demment & Van Soest, 1985). Heavier individuals have higher overwinter survival rates for the two genders: males and females. This is primarily because larger body size should increase energy conservation and probability of survival (Schorr et al., 2009). Larger individuals typically have more fat reserves and maybe less dependent upon food availability (French, 1988). Larger rodents use their fat storage while fasting at a slower rate than smaller individuals and as a result, they can spend longer times at higher body temperatures during hibernation, enabling enables them to survive (Geiser, 2004).

Several ecological and biological changes have been attributed to climate change. Visser (2010), referring to small mammals, stated that climate change is affecting the seasonal timing of reproduction and hibernation, as well as body size and species'' distribution ranges. Brown et al. (1993) assumed that each animal taxon has an optimal body weight; if reached for mammals smaller than 100 g, their foraging range will decrease and their energy will be diverted to reproduction which increases the population. In addition, population growth is related to the carrying capacity of the ecosystem. With fixed resources, the carrying capacity of any ecosystem should decrease with increasing body size. Therefore, if the resources are limited, body size would be reduced to maintain the population (Savage et al., 2004).

The United Nations Environmental Program (UNEP) defined desertification as land degradation in arid, semi–arid and dry–sub humid areas caused by adverse human impact. The causes of desertification are a combination of several factors that include: scarcity and decrease of mean rainfall as a result of global changes, overgrazing, the increase of human population that inhabits the dry lands, industrial misuse, land exploitation, domestic use of fragile vegetation especially for fuel, and many more factors that vary from one region into another (Thomas, 1993). In Kuwait, nearly all plants and animals outside protected areas are lost. The country has lost most of its valuable and vulnerable plants in the second half of the last century due to urbanization, human impact and climate change, factors that also reduced wildlife. Unfortunately, plant and animal communities are disappearing quickly from other areas with open access. Habitat loss and land deterioration are not the only factors contributing to the diminished wildlife populations in Kuwait. Other factors include unjustified hunting and intentional destruction of their shelters. Mammals are the animals that suffer most from these natural and anthropogenic factors.

The aim of this study was to analyse differences in morphological parameters of lesser jerboa (*Jaculus jaculus* L.) populations living in intact habitat with abundant food resources, and populations living in degraded habitats with scarce food resources. The effect of season and gender on the morphological parameters of lesser jerboas was also analysed.

Material and methods

The data were collected in autumn 2010, spring 2011 and summer 2011 in the semi–arid lands of the State of Kuwait. Two study sites were designated for the study purposes: intact and degraded habitat. The intact study area was the Kabd Research Station, which is located 35 km southwest of Kuwait City. Ground elevation in this area ranges between 70 and 130 m above sea–level (Misak et al., 2002). Dense vegetation covers the area and is dominated by the *Arfaj* shrubs (*Rhanterium epapposum* Oliv.) and other plants. It covers an area of 40 km². The area is fenced and protected, and human disturbance is minimal because free access is prohibited. This allows free foraging behavior for the lesser jerboas governed only by natural predation risk.

The degraded study area is an unprotected area with free access and many camps for sheep and camel herders. The area has poor vegetation cover of the same type found in protected areas, but found in scattered patches only. It is located almost 45 km to the west of Kuwait city. It is separated from the intact study area by almost 10 km and has a ground elevation ranging between 100 and 140 m.

Jerboas were captured over 24 nights for the three seasons (eight nights per season) for each of the sites, totaling 48 nights. The aim was to capture four jerboas per night to sample around 100 in each area. To achieve these figures, the capture effort almost doubled in the degraded areas where around five hours on average were spent per night, in contrast with around three hours on average in the protected areas. Animals were captured using a slowly–driven vehicle with powerful spotlights immediately after sunset when the animals start foraging for food. The lesser jerboas were blinded by the powerful light and they stopped moving provided no sounds were produced. To capture jerboas, a soft net with a long handle was used (a similar approach was followed by Happold, 1967).

Body weight, body length, tail length and thigh thickness of captured jerboas were measured. All jerboas were released at the point of capture immediately after

Table 1. Lesser jerboa mean weights (g), with 95% confidence intervals, in different seasons. The letters in superscript indicate a significant difference ($p < 0.05$).

Tabla 1. Pesos medios (g) de jerbo egipcio con intervalos de confianza del 95%, en diferentes estaciones del año. Las letras en forma de exponente indican una diferencia significativa (p < 0,05).

Season	Weight (g)
Autumn 2010	60.09[a] ± 1.98
Spring 2011	61.55[a] ± 1.67
Summer 2011	55.94[b] ± 2.00

taking the measurements that took only a few minutes. Juveniles, pregnant and lactating females were discarded from the study. The captures took place more than 100 m apart on each night and more than 500 m on different nights, providing a low probability of recapturing the same individual. Observations of gender and body weight measurements were done in the autumn. Body length, tail length and thigh thickness were additionally measured in the spring and summer.

An ANOVA followed by the Fisher's least significant differences post–hoc test was used to analyse differences in the body weights of jerboas in three different seasons (autumn 2010, spring 2010 and summer 2011). T–tests for independent samples were also used to analyse differences between gender and between habitats in the variable weight, and body length, tail length and thigh thickness (data from autumn 2010 only). Gender was used as the dependent variable in a logistic regression via generalized linear models, where the independent variables used were categorical habitat (intact and degraded), and covariates were all the measurements (data from autumn 2010 only). A backwards stepwise procedure for selection of significant variables and covariates was implemented, and those found significant ($p < 0.05$) were left in the model. The statistical package used was the SPSS© Statistics 19.

Results

A total of 198 lesser jerboas were caught for this study. In autumn 2010, n = 29 lesser jerboas were caught from the intact habitat and n = 26 from the deteriorated habitat. In spring 2011, n = 39 lesser jerboas were caught in the intact study site while n = 35 were captured in the deteriorated site, and in summer 2011, the captures were n = 36 and n = 33, in the intact and deteriorated habitats, respectively.

With regards to season, a significant difference ($p < 0.05$) was found in body weight between summer 2011 and the other seasons (autumn 2010, spring 2011), as summarised in table 1. With regards to gender, significant differences were found between male and female jerboas for body weight ($p < 0.001$), body length ($p < 0.001$) and tail length ($p < 0.001$). No significant differences ($p > 0.05$) were found for thigh thickness (as summarised in table 2).

With regards to habitat, significant differences were found between intact and degraded habitats for body weight ($p < 0.001$) and thigh thickness ($p < 0.01$), and no significant differences ($p > 0.05$) were found for body length and tail length, as summarised in table 3.

Table 2. Means with 95% confidence interval for the parameters body weight (g), body length (cm), tail length (cm) and thigh thickness (cm) in male and female lesser jerboas: [a] $p < 0.001$; [b] $p > 0.05$.

Tabla 2. Medias con intervalos de confianza del 95% para los parámetros peso (g), longitud del cuerpo (cm), longitud de la cola (cm) y grosor del muslo en los machos y las hembras de jerbo egipcio: [a] p < 0,001; [b] p > 0,05.

	Gender	
	Male	Female
Weight[a]	55.52 ± 1.87	63.49 ± 1.71
Body[a]	28.80 ± 0.60	30.61 ± 0.54
Tail[a]	17.94 ± 0.43	19.10 ± 0.39
Thigh[b]	9.45 ± 0.48	9.91 ± 0.44

Table 3. Means with 95% confidence interval for the parameters body weight (g), body length (cm), tail length (cm) and thigh thickness (cm) in lesser jerboas in the intact and degraded habitat: [a] $p < 0.001$); [b] $p < 0.01$; [c] $p > 0.05$.

Tabla 3. Medias con intervalo de confianza del 95% para los parámetros peso (g), longitud del cuerpo (cm), longitud de la cola (cm) y grosor del muslo en el jerbo egipcio, en hábitat intacto y degradado: [a] p < 0,001; [b] p < 0,01; [c] p > 0,05.

	Habitat	
	Intact	Degraded
Weight[a]	63.58 ± 1.74	55.43 ± 1.83
Body[c]	29.96 ± 0.56	29.44 ± 0.59
Tail[c]	18.72 ± 0. 40	18.32 ± 0.42
Thigh[b]	10.27 ± 0.45	9.08 ± 0.47

Table 4. Logistic regression parameters to model gender in lesser jerboas, as a function of the factor habitat (intact or degraded) and the covariate weight (g).

Tabla 4. Parámetros de la regresión logística para la modelar el sexo de los jerbos egipcios en función del factor hábitat (intacto vs. degradado) y de la covariable peso (g).

	β	SE (β)	p value	95% CI (β)	OR (e^β)	95% CI OR (e^β)	
				Variables in the equation			
Intercept	−27.49	−7.45	< 0.001	−42.09 −12.89	$1.16 \cdot 10^{-12}$	$5.28 \cdot 10^{-19}$	$2.53 \cdot 10^{-6}$
Habitat							
Intact	−4.27	1.43	< 0.01	−7.07 −1.47	0.014	0.001	0.230
Degraded	0						
Weight	0.50	0.14	< 0.001	0.24 0.77	1.652	1.269	2.153

The body weights and thigh thickness of jerboas were found to be greater in the intact habitat.

In the logistic regression, the factor 'habitat' ($p < 0.01$) and the covariate 'body weight' ($p < 0.001$) were found to be significant to model gender. Table 4 shows the full description of the model parameters and the 95% confidence intervals for the respective values are also stated, together with the odd ratios. The graphs in figure 1 show the probability of being females in dependency of the variable and covariate considered in the model.

Discussion

The study shows that lesser jerboas in the intact, protected area with thicker vegetation cover have significantly greater body weight than those in the degraded habitat. The change of morphological measurements of mammals in different habitats is well documented in several studies (*e.g.* Heaney, 1978; Jennings et al., 2010). The abundance of biodiversity and food resources, with minimal disturbances, favours morphological changes towards increased body size in mammals. The extremely disrupted habitats in the State of Kuwait have been well documented (*e.g.* Omar, 2000; Omar et al., 2001; Misak et al., 2002). The effect of disrupted habitat on wildlife has also been monitored, mostly on the abundance of animals in different habitats rather than on morphological parameters (*e.g.* Al–Sdirawi, 1985; Taha et al., 2000; Delima et al., 2002; Zaman et al., 2005).

This pioneer study in Kuwait focused on determining the effects of habitat degradation on the morphology of lesser jerboas. The magnitude of damage on habitats in both study areas was fully described by Misak et al. (2002) and Omar et al. (2001). Within the enclosed intact habitat, *Rhanterium epapposum* shrubs and associated plant communities thrive, and they have a good ground cover compared to the surrounding non–protected and degraded areas.

There is a direct relationship between the range of available and accessible food, and the body mass of animals, particularly mammals (Schoener, 1968; Harestad & Bunnell, 1979; Peters & Wassenberg, 1983). An increased body mass usually means an increase in the availability of large food quantities (positive correlation).

Another reason for differences in body weight between intact habitat and degraded habitat populations can be the dispersion and abundance of seeds in desert soils, which comprise the main source of energy in desert rodents' diets (Brown et al., 1994; Baker & Patterson, 2010). Seeds beneath shrubs are concentrated in densities that 5 to 10 times greater than those in open areas between shrubs (Nelson & Chew, 1977; Thompson, 1980). Therefore, when annual plants reach the end of their cycle, there are plenty of seeds available for consumption by desert rodents.

Body mass shows a highly significant positive correlation with the amount of annual rainfall (Abramsky et al., 1985). It may seem that the differences in body mass and other morphological parameters are related to possible differences in rainfall rates in the two areas, but the two study areas were located close to each other; they are separated by only 10 km and have similar elevation, averaging 100 m in the protected and 120 m in the degraded areas. The possibility of weight differences being attributed to rainfall differences can thus be ruled out.

Other body dimensions, such as body length and tail length, showed no difference in this study when comparing the degraded habitat to the intact habitat. The difference was seen in the thickness of the thigh (hind limb), with greater thigh thickness in jerboas in the intact habitat. This could be explained by the fact that lesser jerboas use bipedal hopping as an escape mechanism from predators. They can jump up to 2 m high and jump while running in different directions. The increased thickness of the muscle tissues in the hind limb improves the animal's jump performance (Alexander et al., 1981; Perry et al.,

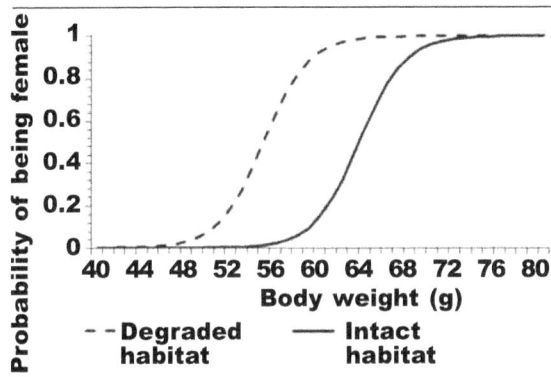

Fig. 1. Logistic regression modelling the weight (g) dynamics of lesser jerboa in dependency of its gender and habitat

Fig. 1. Regresión logística modelando la dinámica del peso (g) del jerbo egipcio en función del sexo y tipo de hábitat.

1988; Taraborrelli et al., 2003; James et al., 2007). Intact areas contain a higher number of predators (Zaman et al., 2005) than the degraded habitats. Therefore, lesser jerboas in the protected areas might be under a higher predation risk than animals living in unprotected, degraded habitats.

Sexual dimorphism in body size has been well defined as the difference in mean body size of males and females. This occurs in rodents and other animals (Ralls, 1976, 1977) and is traditionally explained by sexual selection acting on males (Lammers et al., 2001). Ralls (1976) explained that larger body sizes in females was the result of selection, favouring smaller males in mating systems where male reproductive success is related to encounter rate with females and where smaller males may be favoured when food is limited (a trade–off between foraging and mate acquisition favouring smaller males with regards to mate acquisition). Dimorphism in monogamous species is less pronounced than in species with more promiscuous habits (Dewsbury et al., 1980). Sachser et al. (1999) determined that the rodent *Galea musteloide* has a promiscuous mating system in smaller males, as reproduction is decided by sperm competition rather than male–male contest, which appears to result in greater offspring survival. The lesser jerboa is a solitary species; males and females meet during the mating season and the female is left alone to breed and nurse the offspring. Some studies (Krasnov et al., 2005), nevertheless, suggest that sexual differences in body size are related to foraging behaviour and the home range of each sex. Therefore, the more active the foraging animal, the larger it is. This indicates that female lesser jerboas have a wider home range and forage for food more frequently than males. This is a reasonable interpretation of the results, since females have to carry and nurse the offspring. Female lesser

jerboas have to store as much fat as possible to feed their young. Body length and tail length are related to the animal's body size. Many others species of desert rodents are social; the males are subject to sexual selection, and they also play a role in nurturing the young.

The results also revealed that weight of the lesser jerboa changes with the seasons. Greater body weights were seen in spring and autumn than in summer. This could be because of the availability of feed resources, which depends on the annual rain cycles. The annual rain cycle in Kuwait is concentrated in autumn and winter, with year accumulations lower than 100 mm (Zaman, 1997). The annual plants flourish with the onset of rainfall in arid lands in the autumn and complete their life cycles in the spring. Seeds of annual plants germinate and produce seeds in a very short life span. Annual plants form most of the vegetation cover in Kuwait (Zaman, 1997; Brown, 2002), and the protein levels of annual grasses decrease immediately after blooming, with seed formation. Lesser jerboas can also extract moisture from these annual plants that are abundant in the autumn and spring. Consequently, body condition is better in autumn and spring in preparation for the winter hibernation and summer shortage.

The findings in this study clearly show that a degraded habitat has a negative effect on the body weight and dimensions of lesser jerboas. Lower body weight may decrease the winter survival rate and negatively impact on the lesser jerboa population size and distribution in the degraded habitat.

References

Abramsky, Z., Brand, S. & Rosenzweig, M., 1985. Geographical ecology of gerbilline rodents in sand dune habitats of Israel. *Journal of Biogeography*, 12: 363–372.

Alexander, R., Jayes, A., Maloiy, G. & Wathuta, E., 1981. Allometry of the leg muscles of mammals. *Journal of Zoology*, 194: 539–552.

Al–Sidrawi, F., 1985. *Conservation and protection of the wildlife in Kuwait (phase I): review and assessment of information on desert fauna. Final Report.* Kuwait Inst. for Scientific Research, Kuwait.

Baker, M. & Patterson, B., 2010. Patterns in the local assembly of Egyptian rodent faunas: Areography and species combinations. *Mammalian Biology*, 75: 510–522.

Brown, G., 2002. Species richness, diversity and biomass production of desert annuals in an ungrazed Rhanterium epapposum community over three growth seasons in Kuwait. *Plant Ecology*, 165: 53–68.

Brown, J., Kotler, B. & Mitchell, W., 1994. The foraging theory, patch use and the structure of a Negev desert granivore community. *Ecology*, 75: 2286–2300.

Brown, J., Marquet, P. & Taper, M., 1993. Evolution of body size: consequences of an energetic definition of fitness. *American Naturalist*, 142: 573–584.

Delima, E., Peacock, J. & Khalil, E., 2002. *Bibliographic list of the wildlife fauna of Kuwait. Technical Report.* Kuwait Inst. for Scientific Research, Kuwait.

Demment, M. & Van Soest, P., 1985. A nutritional explanation for body–size patterns of ruminant and nonruminant herbivores. *American Naturalist*, 125: 641–672

Dewsbury, D., Baumgardner, D., Evans, R. & Webster, D., 1980. Sexual dimorphism for body mass in 13 taxa of Muroid rodents under laboratory conditions. *Journal of Mammalogy*, 61: 146–149.

French, A., 1988. The patterns of hibernation. *American Scientist*, 76: 569–575.

Geiser, F., 2004. Metabolic rate and body temperature reduction during hibernation and daily torpor. *Annual Review of Physiology*, 66: 239–274.

Happold, D., 1967. Biology of the jerboa, *Jaculus jaculus butleri* (*Rodentia, Dipodidae*), in the Sudan. *Journal of Zoology*, 151: 257–275.

Harestad, A. & Bunnell, F., 1979. Home range and body weight a re–evaluation. *Ecology*, 60: 389–402.

Heaney, L., 1978. Island area and body size of insular mammals: evidence from the tri–colored squirrel (*Callosciurus prevosti*) of Southeast Asia. *Evolution*, 32: 29–44.

Isaac, J., 2005. Potential causes and life–history consequences of sexual size dimorphism in mammals. *Mammal Review*, 35: 101–115.

James, R. & Navas, C. & Herrel, A., 2007. How important are skeletal muscle mechanics in setting limits on jumping performance? *The Journal of Experimental Biology*, 210: 923–933.

Jennings, A., Zubaid, A. & Veron, G., 2010. Range behavior, activity, habitat use, and morphology of the Malay civet (*Viverra tangolunga*) on peninunsular Malaysia and comparison with studies on Borneo and Sulawesi. *Mammalian Biology*, 75: 437–446.

Krasnov, B., Morand, S., Hawlena, H., Khokhlova, I. & Shenbrot, G., 2005. Sex–biased parasitism, seasonality and sexual size dimorphism in desert rodents. *Oecologia*, 146: 209–217.

Lammers, A., Dziech, H. & German, R., 2001. Ontogeny of sexual dimorphism in *Chinchilla lanigera* (Rodentia: Chinchillidae). *Journal of Mammalogy*, 82: 179–189.

Misak, R., Al–Awadhi, J., Omar, S. & Shahid, S., 2002. Soil degradation in Kabd area, southwestern Kuwait City. *Land degradation & development*, 13: 403–415.

Nelson, J. & Chew, R., 1977. Factors affecting seed reserves in the soil of a Mojave Desert ecosystem, Rock Valley, Nye County, Nevada. *American Midland Naturalist*, 97: 300–320.

Omar, S., 2000. *Vegetation of Kuwait: a comprehensive illustration guide to the flora and ecology of the desert of Kuwait*. Kuwait Inst. for Scientific Research, Kuwait.

Omar, S., Misak, R., King, P., Shahid, A., Abo–Rizq, H., Grealish, G. & Roy, W., 2001. Mapping the vegetation of Kuwait through reconnaissance soil survey. *Journal of Arid Environments*, 48: 341–355.

Perry, A., Blickhan, R., Biewener, A., Heglund, N. & Taylor, C., 1988. Preferred speeds in terrestrial vertebrates: are they equivalent? *Journal of Experimental Biology*, 137: 207–219.

Peters, R. & Wassenberg, K., 1983. The effect of body size on animal abundance. *Oecologia (Berl.)* 60: 89–96.

Ralls, K., 1976. Mammals in which females are larger than males. *Quarterly Review of Biology*, 51: 245–276.

– 1977. Sexual dimorphism in mammals: avian models and unanswered questions. *American Naturalist*, 111: 917–938.

Sachser, N., Schwarz–Weig, E., Keil, A. & Epplen, J., 1999. Behavioural strategies, testis size, and reproductive success in two caviomorph rodents with different mating systems. *Behaviour*, 136: 1203–1217.

Savage, V., Gillooly, J., Brown, J., West, G. & Charnov, E., 2004. Effects of body size and temperature on population growth. *American Naturalist*, 163: 429–441.

Schoener, T., 1968. Sizes of feeding territories among birds. *Ecology*, 49: 123–141.

Schorr, R., Lukacsb, P. & Florant, G., 2009. Body mass and winter severity as predictors of overwinter survival in Preble's Meadow jumping mouse. *Journal of Mammalogy*, 90: 17–24.

Taha, F., White, P., Cowan, P., Delima, E., Al–Hadad, A., Al–Ragam, O. & Al–Mutawa, S., 2000. *Collection and holding of living desert fauna specimens for the scientific center. Final Report, 5909*. Kuwait Inst. for Scientific Research, Kuwait.

Taraborelli, P., Corbalán, V. & Giannoni, S., 2003. Locomotion and Escape Modes in Rodents of the Monte Desert (Argentina). *Ethology*, 109: 475–485.

Thomas, D., 1993. Sandstorm in a teacup? Understanding Desertification. *The Geographical Journal*, 159: 318–311.

Thompson, S., 1980. *Microhabitat use, foraging behavior, energetics and community structure of heteromyid rodents. Dissertation*. Univ. of California, Irvine, California, USA.

Visser, M., 2010. Climate change: Fatter marmots on the rise. *Nature* 466: 445–447.

Whitford, W., 2002. *Ecology of desert systems*. Elsevier Science, London.

Zaman, S., 1997. Effects of rainfall and grazing on vegetation yield and cover of two arid rangelands in Kuwait. *Environmental Conservation*, 24: 344–350.

Zaman, S., Peacock, J., Al–Mutairi, M., Delima, E., Al–Othman, A., Loughland, R., Tawfiq, H., Siddiqui, K., Al–Dossery, S. & Dashti, J., 2005. *Wildlife and vegetation survey (Terrestrial Cluster 3) Addendum 2. Final Report, 7885*. Kuwait Ins. for Scientific Research, Kuwait.

Permissions

All chapters in this book were first published in ABC, by Natural Science Museum of Barcelona; hereby published with permission under the Creative Commons Attribution License or equivalent. Every chapter published in this book has been scrutinized by our experts. Their significance has been extensively debated. The topics covered herein carry significant findings which will fuel the growth of the discipline. They may even be implemented as practical applications or may be referred to as a beginning point for another development.

The contributors of this book come from diverse backgrounds, making this book a truly international effort. This book will bring forth new frontiers with its revolutionizing research information and detailed analysis of the nascent developments around the world.

We would like to thank all the contributing authors for lending their expertise to make the book truly unique. They have played a crucial role in the development of this book. Without their invaluable contributions this book wouldn't have been possible. They have made vital efforts to compile up to date information on the varied aspects of this subject to make this book a valuable addition to the collection of many professionals and students.

This book was conceptualized with the vision of imparting up-to-date information and advanced data in this field. To ensure the same, a matchless editorial board was set up. Every individual on the board went through rigorous rounds of assessment to prove their worth. After which they invested a large part of their time researching and compiling the most relevant data for our readers.

The editorial board has been involved in producing this book since its inception. They have spent rigorous hours researching and exploring the diverse topics which have resulted in the successful publishing of this book. They have passed on their knowledge of decades through this book. To expedite this challenging task, the publisher supported the team at every step. A small team of assistant editors was also appointed to further simplify the editing procedure and attain best results for the readers.

Apart from the editorial board, the designing team has also invested a significant amount of their time in understanding the subject and creating the most relevant covers. They scrutinized every image to scout for the most suitable representation of the subject and create an appropriate cover for the book.

The publishing team has been an ardent support to the editorial, designing and production team. Their endless efforts to recruit the best for this project, has resulted in the accomplishment of this book. They are a veteran in the field of academics and their pool of knowledge is as vast as their experience in printing. Their expertise and guidance has proved useful at every step. Their uncompromising quality standards have made this book an exceptional effort. Their encouragement from time to time has been an inspiration for everyone.

The publisher and the editorial board hope that this book will prove to be a valuable piece of knowledge for researchers, students, practitioners and scholars across the globe.

List of Contributors

A. Martínez-Ortí
Museu Valencià d'Història Natural, Passeig de la Petxina 15, 46008 Valencia, España (Spain)
Dept. de Zoologia, Fac. de Biologia, Univ. de València

L. Cádiz
Dept. de Ciencias Experimentales y Matemáticas. Fac. de Ciencias Experimentales, Univ. Católica de Valencia, c/. Guillem de Castro 94, 46001 Valencia, España (Spain)

Sergi Herrando
Catalan Ornithological Inst., Natural History Museum of Barcelona, Psg. Picasso s/n., E–08003 Barcelona (Spain)

Anne Weiserbs & Jean-Yves Paquet
Aves–Natagora, Rue Nanon 98, 5000 Namur (Belgium)

Javier Quesada
Natural History Museum of Barcelona. Psg. Picasso s/n., E–08003 Barcelona (Spain)

Xavier Ferrer
Dept. of Animal Biology, Univ. of Barcelona, Av. Diagonal 645, E–08028 Barcelona (Spain)

Jennifer Rivera
Inst. 'Clodomiro Picado', Univ. de Costa Rica, San José, Costa Rica

Mahmood Sasa
Inst. 'Clodomiro Picado', Univ. de Costa Rica, San José, Costa Rica

Emilio Barba
Inst. 'Cavanilles' of Biodiversidad y Biología Evolutiva, Univ. de Valencia, AC 22085, E–46071 Valencia, España (Spain)

Pablo Vera
Inst. 'Cavanilles' of Biodiversidad y Biología Evolutiva, Univ. de Valencia, AC 22085, E–46071 Valencia, España (Spain)

Juan S. Monrós
Inst. 'Cavanilles' of Biodiversidad y Biología Evolutiva, Univ. de Valencia, AC 22085, E–46071 Valencia, España (Spain)

Alexandre Mestre
Dept. de Microbiología y Ecología, Univ. de Valencia, c/ Dr. Moliner 50, E–46100 Burjassot, España (Spain)

Juan Rueda
Dept. de Microbiología y Ecología, Univ. de Valencia, c/ Dr. Moliner 50, E–46100 Burjassot, España (Spain)

Ruth Rodríguez-Pastor
Associate Research Unit, CSIC, Museu de Ciències Naturals de Barcelona, Psg. Picasso s/n., E–08003 Barcelona, Espanya (Spain)

Juan Carlos Senar
Associate Research Unit, CSIC, Museu de Ciències Naturals de Barcelona, Psg. Picasso s/n., E–08003 Barcelona, Espanya (Spain)

Alba Ortega
Associate Research Unit, CSIC, Museu de Ciències Naturals de Barcelona, Psg. Picasso s/n., E–08003 Barcelona, Espanya (Spain)

Jordi Faus
Associate Research Unit, CSIC, Museu de Ciències Naturals de Barcelona, Psg. Picasso s/n., E–08003 Barcelona, Espanya (Spain)

Francesc Uribe
Associate Research Unit, CSIC, Museu de Ciències Naturals de Barcelona, Psg. Picasso s/n., E–08003 Barcelona, Espanya (Spain)

Tomas Montalvo
Servei de Vigilància i Control de Plagues Urbanes, Agència de Salut Pública de Barcelona, Av. Príncep d'Astúries 63, 3r 2a, E–08012 Barcelona, Espanya (Spain)

Alejandro Martínez-Abraín
Depto. de Bioloxia Animal, Bioloxia Vexetal e Ecoloxia, Univ. da Coruña, Campus da Zapateira s/n., 15071 A Coruña, España (Spain)
Population Ecology Group, IMEDEA (CSIC-UIB), c/ Miquel Marquès 21, 07190 Esporles, Mallorca, Espanya (Spain)

David Conesa
Grup d'Estadística Espacial i Temporal en Epidemilogia i Medi Ambient, Dept. d'Estadística I Investigació Operativa, Univ. de València, c/ Dr, Moliner 50, 46100 Burjassot, Valencia, España (Spain)

Anabel Forte
Grup d'Estadística Espacial i Temporal en Epidemilogia i Medi Ambient, Dept. d'Estadística I Investigació Operativa, Univ. de València, c/ Dr, Moliner 50, 46100 Burjassot, Valencia, España (Spain)

C. Román–Valencia
Lab. de Ictiología, .Univ. del Quindío, A. A. 2639, Armenia, Colombia

R. I. Ruiz–C
Lab. de Ictiología, .Univ. del Quindío, A. A. 2639, Armenia, Colombia

D. C. Taphorn B
Lab. de Ictiología, .Univ. del Quindío, A. A. 2639, Armenia, Colombia

C. García–Alzate
Lab. de Ictiología, .Univ. del Quindío, A. A. 2639, Armenia, Colombia
Programa de Biología, Univ. del Atlántico, km 7 antigua vía a Puerto Colombia, Barranquilla, Colombia

D. C. Taphorn B
1822 N. Charles St., Belleville, 62221 Illinois, USA

Alejandro Martínez–Abraín
Depto. de Bioloxía Animal, Bioloxía Vexetal e Ecoloxía, Univ. da Coruña, Fac. De Ciencias, Campus da Zapateira s/n., 15071 A Coruña, España (Spain)
Population Ecology Group, IMEDEA (CSIC–UIB), Miquel Marquès 21, 07190 Esporles, Mallorca, España (Spain)

Miguel Delibes–Mateos
Instituto de Investigación en Recursos Cinegéticos (IREC, CSIC–UCLM–JCCM), Ronda de Toledo s/n., 13071 Ciudad Real, España (Spain)

Adolfo Delibes
Junta de Castilla y León, c/ Rigoberto Cortejoso 14, 41014, Valladolid, España (Spain)

Olga Jordi
Dept. of Ornithology, Aranzadi Sciences Society, Zorroagagaina 11, E20014 Donostia (San Sebastián), España (Spain)

Alfredo Herrero
Dept. of Ornithology, Aranzadi Sciences Society, Zorroagagaina 11, E20014 Donostia (San Sebastián), España (Spain)

Asier Aldalur
Dept. of Ornithology, Aranzadi Sciences Society, Zorroagagaina 11, E20014 Donostia (San Sebastián), España (Spain)

Juan F. Cuadrado
Dept. of Ornithology, Aranzadi Sciences Society, Zorroagagaina 11, E20014 Donostia (San Sebastián), España (Spain)

Juan Arizaga
Dept. of Ornithology, Aranzadi Sciences Society, Zorroagagaina 11, E20014 Donostia (San Sebastián), España (Spain)

Antonio García–Quintas
Centro de Investigaciones de Ecosistemas Costeros (CIEC), Cayo Coco, Ciego de Ávila, 69400 Cuba

Alain Parada Isada
Centro de Investigaciones de Ecosistemas Costeros (CIEC), Cayo Coco, Ciego de Ávila, 69400 Cuba

Alberto Martínez–Ortí
Museu Valencià d'Història Natural i iVBiotaxa, L'Hort de Feliu–Alginet, P. O. Box 8460, 46018 València, Espanya (Spain)
Dept. de Zoologia, Fac. de Ciències Biològiques, Univ. de València, c/ Dr. Moliner 50, 46100 Burjassot, València, Espanya (Spain)

Vicent Borredà
Museu Valencià d'Història Natural i iVBiotaxa, L'Hort de Feliu–Alginet, P. O. Box 8460, 46018 València, Espanya (Spain)
Dept. de Zoologia, Fac. de Ciències Biològiques, Univ. de València, c/ Dr. Moliner 50, 46100 Burjassot, València, Espanya (Spain)

J. M. Lucas
Depto. de Zoología y Antropología Física, Fac. de Veterinaria, Univ. de Murcia, Campus de Espinardo, 30100 Murcia, España (Spain)

J. Galián
Depto. de Zoología y Antropología Física, Fac. de Veterinaria, Univ. de Murcia, Campus de Espinardo, 30100 Murcia, España (Spain)

P. Prieto
Parque Natural de Cazorla, Segura y las Villas, c/ Martinez Falero 11, 23470 Cazorla, Jaén, España (Spain)

Çağla Kılıç
Biology Section, Dept. of Biology, Fac. of Arts and Science, Ordu Univ., 52200 Cumhuriyet Campuss, Ordu (Turkey)

Onur Candan
Biology Section, Dept. of Biology, Fac. of Arts and Science, Ordu Univ., 52200 Cumhuriyet Campuss, Ordu (Turkey)

M. Sarasa
Grupo Biología de las Especies Cinegéticas y Plagas (RNM–118), Sevilla, España (Spain)

A. H. Plasencia–Vázquez
El Colegio de la Frontera Sur (ECOSUR), Libramiento Carretero Campeche km 1.5, Av. Rancho, Polígono 2–A, Parque Industrial de Lerma, C.P. 24500, San Francisco de Campeche, Campeche, México

G. Escalona–Segura
El Colegio de la Frontera Sur (ECOSUR), Libramiento Carretero Campeche km 1.5, Av. Rancho, Polígono 2–A, Parque Industrial de Lerma, C.P. 24500, San Francisco de Campeche, Campeche, México

L. G. Esparza–Olguín
El Colegio de la Frontera Sur (ECOSUR), Libramiento Carretero Campeche km 1.5, Av. Rancho, Polígono 2–A, Parque Industrial de Lerma, C.P. 24500, San Francisco de Campeche, Campeche, México

Nicolas Bech
Univ. of Perpignan, Via Domitia, CNRS, UMR 5244, Evolutionary and Ecology of Interactions (2EI), Perpignan, F-66860, France
Present address: Equipe Ecologie, Evolution Symbiose, Lab. EBI Ecologie & Biologie des Interactions, Univ. de Poitiers–UFR Sciences Fondamentales et Appliquées, UMR CNRS 7267, Bât. B8, 40 avenue du Recteur Pineau, F-86022 Poitiers Cedex, France

Sophie Beltran
Univ. of Perpignan, Via Domitia, CNRS, UMR 5244, Evolutionary and Ecology of Interactions (2EI), Perpignan, F-66860, France
Present address: Equipe Ecologie, Evolution Symbiose, Lab. EBI Ecologie & Biologie des Interactions, Univ. de Poitiers–UFR Sciences Fondamentales et Appliquées, UMR CNRS 7267, Bât. B8, 40 avenue du Recteur Pineau, F-86022 Poitiers Cedex, France

Jérôme Boissier
Univ. of Perpignan, Via Domitia, CNRS, UMR 5244, Evolutionary and Ecology of Interactions (2EI), Perpignan, F-66860, France

Jean François Allienne
Univ. of Perpignan, Via Domitia, CNRS, UMR 5244, Evolutionary and Ecology of Interactions (2EI), Perpignan, F-66860, France

Jean Resseguier
ONCFS, Dept of Studies and Research, F-66500, Prades, France

Claude Novoa
ONCFS, Dept of Studies and Research, F-66500, Prades, France

Gary D. Grossman
Warnell School of Forestry & Natural Resources, Univ. of Georgia, Athens GA 30602, USA

Antonio J. Carpio
Dept of Zoology, Univ. of Cordoba, Campus de Rabanales Ed. Darwin, 14071 Córdoba, Spain

Jesús Castro–López
Dept of Zoology, Univ. of Cordoba, Campus de Rabanales Ed. Darwin, 14071 Córdoba, Spain

José Guerrero–Casado
Dept of Zoology, Univ. of Cordoba, Campus de Rabanales Ed. Darwin, 14071 Córdoba, Spain

Leire Ruiz–Aizpurua
Dept of Zoology, Univ. of Cordoba, Campus de Rabanales Ed. Darwin, 14071 Córdoba, Spain

Francisco S. Tortosa
Dept of Zoology, Univ. of Cordoba, Campus de Rabanales Ed. Darwin, 14071 Córdoba, Spain

Joaquín Vicente
Inst. en Investigación de Recursos Cinegéticos, IREC (CSIC, UCLM, JCCM), Ronda de Toledo s/n., 13071, Ciudad Real, Spain

Rui Sun
College of Food and Biology Engineering, Qilu Univ. of Technology, Jinan, 250353, China

Yaomin Zheng
Inst. of Remote Sensing Applications, Chinese Academy of Sciences, Beijing 100101, China

TingLei
College of Nature Conservation, Beijing Forestry Univ., Beijing 100083, China

Guofa Cu
College of Nature Conservation, Beijing Forestry Univ., Beijing 100083, China

A. Martínez–Ortí
Museu Valencià d'Història Natural. Passeig de la Petxina 15, 46008 València
Dept. de Zoologia, Fac. de Biologia, Univ. de València, Av. Dr. Moliner 50, 46100 Burjassot, València (Spain)

F. Robles
Dept. de Geologia, Fac. de Biologia. Univ. de València, Av. Dr. Moliner 50, Burjassot, València (Spain)

Dessalegn Ejigu
Dept. of Biology, College of Sciences, Bahir Dar Univ., P. O. Box 79, Bahir Dar, Ethiopia

Afework Bekele
Dept. of Zoology, College of Natural Sciences, Addis Ababa Univ., Ethiopia

Larkin Powell
School of Natural Resources, Univ. of Nebraska–Lincoln, Lincoln, Nebraska, USA

Jean–Marc Lernould
Conservation des Especesetdes Populations Animales, Schlierbach, France

Matrah Al–Mutairi
School of Biological Sciences, Univ. of Bristol, Woodland Road, Bristol BS8 1UG, UK.– Matrah Al–Mutairi, Kuwait Inst. for Scientific Research, P. O. Box 24885, Safat 13109, Kuwait

Ravneet Bhuller
School of Biological Sciences, Univ. of Bristol, Woodland Road, Bristol BS8 1UG, UK.– Matrah Al–Mutairi, Kuwait Inst. for Scientific Research, P. O. Box 24885, Safat 13109, Kuwait
Hartpury College, Univ. of the West of England, Hartpury, GL19 3BE, UK

Fernando Mata
Hartpury College, Univ. of the West of England, Hartpury, GL19 3BE, UK

www.ingramcontent.com/pod-product-compliance
Lightning Source LLC
Chambersburg PA
CBHW050437200326
41458CB00014B/4975